Elementary Statistics

Elementary Statistics

John A. Ingram

Florida Technological University

Cummings Publishing Company

Menlo Park, California   Reading, Massachusetts   London   Amsterdam   Don Mills, Ontario   Sydney

To Jim, Beth, Jay, and Kathy

Title page photo: How does a stockbroker know just how much to pay for a certain stock or bond? Statistical data is fed constantly into computers where it is coded and printed on tape for easy reference.

Acknowledgments

Book Design: Design Office/Bruce Kortebein

Photographs
Cover, title page, chapter openers: Marshall Berman
Page 3:   A. C. Nielsen Company
Page 32:   Roger Malloch, Magnum
Page 121:   NASA
Page 175:   Marshall Berman
Page 197:   Union Oil
Page 249:   Marshall Berman
Page 314:   Marshall Berman

Copyright © 1977 by Cummings Publishing Company, Inc.
Philippines Copyright 1977.

All rights reserved. No part of this publication may be reproduced, stored in a retrieval system, or transmitted, in any form or by any means, electronic, mechanical, photocopying, recording, or otherwise, without the prior written permission of the publisher.
Printed in the United States of America.
Published simultaneously in Canada.
Library of Congress Catalog Card Number 76-24506

ISBN 0-8465-2660-3
abcdefghij-HA-7987

Cummings Publishing Company, Inc.
2727 Sand Hill Road
Menlo Park, California 94025

Preface

This book is written for beginners. The first purpose of it is to provide a clear introduction to the fundamental statistics concepts for students who have a limited mathematics background. Some mathematics is used, but only to the extent of arithmetic and the basics of elementary algebra. The theme of this work is using good judgment and exercising a sense of reasonableness. Details are important; the discussion is intuitive and graphic. The book includes many diagrams, displays, examples, and pictures. Seven essay examples include practical applications from television polls, auto gas mileage tests, and more.

The content includes many of the first principles of classical statistics. Yet the text is concise; it includes only as much as can be mastered easily in one term (of three semester hours or four to five quarter hours) by students who have little experience with numbers. The work covers the basic concepts necessary for elementary courses in psychology, education, sociology, business, and the sciences.

The text contains thirteen chapters. The first three chapters are about descriptive statistics. Chapters 4 through 7 concern probability concepts and common probability experiments. Then Chapters 8 through 13 are basic (classical) statistical inference. The appendix section includes a math appendix, statistical tables, and the answers to odd-numbered problems.

Topics discussed in this text include averages, measures of variation, frequency distributions, graphs and charts, counting and probabilities, binomial distribution, the normal curve, random sampling, normal inference, t-tests, linear regression and prediction, Pearson correlation, simple analysis of variance, and the chi-square statistic. One basis for the inclusion of topics was a nationwide survey of instructors on the "content of first statistics courses in two-year institutions of higher education."

A minimum course should include Chapters 1 through 9. Depending on the class makeup, the remaining topics may be more, or less, important. For business majors, I suggest stressing Section 3.4 and Chapter 12; for social science students, Chapters 6 and 13; for education majors, Chapters 3 and 6 and Sections 10.4 and 11.1 are important. Appendix A has ample material for a review of the mathematics used in this book.

A variety of learning activities is offered. There are numerous procedural guides, examples, and definitions, and review sections appear at the end of most chapters. The student is encouraged to follow through by working related problems, including regular section problems and also review problems at the end of some chapters. The review problems are a mixture of problems from several chapters.

The supplemental workbook begins with the display of a substantial amount of real data. Exercises that use this data are given for major concepts in the text. Also, procedures are outlined for a number of experiments within your class. The workbook has a section of formulas that include all "rules" given in the text. This can be useful for working problems and/or for closed book tests. Chapter overviews identify key ideas. True-false, multiple choice, and completion-type exercises afford immediate feedback on important concepts. An instructor's guide is also available.

My thanks go to others whose results are used in this work. For permission to reprint tables, thanks to F. Mosteller and R. Rourke; and to D. B. Owen for works published by Addison-Wesley; to the executors of the Biometrika Trust; and to the Chemical Rubber Company. A special thanks to my co-worker Bernie Lantz, who generated Table V by computer. And more thanks to the several agencies including A. C. Nielsen Company, Gibbs-Cook Caterpillar Corporation, and the Union Oil Company of California for information used in some of the essay examples. I've a warm spot in my heart for the publishing team at Cummings, who know just how to put it all together. Last but not least a special thank you to my family for their encouragement; to my wife, Lois, who has typed the manuscript (several times), and to my children Jim, Beth, Jay, and Kathy who have been very patient.

John A. Ingram

Contents

Chapter One — Introduction — 1

1/ A Very Brief History — 1
2/ Basic Terms — 2
3/ How to Use This Book — 6

Chapter Two — Description and Graphical Displays — 9

1/ Averages — 9
2/ Variation — 15
3/ Graphical Displays — 22
4/ Essay Example — 31

Chapter Three — Description from Grouped Data — 37

1/ Grouping Data — 37
2/ Numerical Description — 44
3/ Percentiles — 49
4/ Weighted Means and Index Numbers — 55
5/ Summary for Chapters 2 and 3 — 61
   Review Problems — 62

Chapter Four

Probability Concepts — 69

1/ Describing Experiments — 69
2/ Probability Rules — 80
3/ Probability Problems — 88

Chapter Five

Binomial and Related Experiments — 97

1/ Identifying Binomial Experiments — 98
2/ Related Experiments — 108
3/ Describing Distributions — 114
4/ Essay Example — 119

Chapter Six

The Normal Distribution — 123

1/ Continuous Experiments—The Uniform Distribution — 125
2/ The Standard Normal or Z-Distribution — 130
3/ The Normal Distribution with Applications — 136
4/ Normal Approximations — 145
5/ Summary for Chapters 4, 5, and 6 — 151
Review Problems — 152

Chapter Seven

Sampling and the Central Limit Theorem — 157

1/ Sampling in Experiments — 158
2/ Sampling Distributions — 162

3/	The Central Limit Theorem with Applications	168
4/	Essay Example	174

Estimation and Sample Size — 179 — Chapter Eight

1/	Estimation of the Mean—Large Samples	181
2/	Sample Size	185
3/	Estimation and Sample Size in Binomial Experiments	191
4/	Essay Example	196

Testing Hypotheses (Statistical Decisions) — 201 — Chapter Nine

1/	Introduction to Testing	202
2/	Tests on Means—Large Samples	208
3/	Binomial Experiments—Tests on Proportions	213
4/	Summary for Chapters 7, 8, and 9	218
	Review Problems	220

Inference Using the t-Statistic — 225 — Chapter Ten

1/	The t-form and Degrees of Freedom	225
2/	Estimation and Sample Size Considerations	231
3/	Tests on Means for Small Samples	236
4/	Paired Differences Tests	240
5/	Essay Example	247

Chapter Eleven

Tests on Means—Analysis of Variance — 251

1/ Comparing Means for Two Samples — 251
2/ Relation Between the t- and F-Statistics — 258
3/ The Analysis of Variance Procedure — 262
4/ Tests on Means—Unequal Numbers — 270
5/ Summary for Chapters 10 and 11 — 275
Review Problems — 276

Chapter Twelve

Regression, Paired Observations, and Linear Models — 283

1/ Simple Linear Regression — 283
2/ The Regression Equation by Least Squares — 287
3/ Pearson's Coefficient of Correlation — 293
4/ Other Linear Models—Time Series — 299
5/ Inference in Regression — 305
6/ Essay Example — 312
7/ Summary for Chapter 12 — 316
Review Problems — 317

Chapter Thirteen

Inference on Categorical Data and Frequencies — 323

1/ The Chi-Square Statistic — 324
2/ Inference in Binomial Experiments — 327
3/ Contingency Tables and Independence — 334
4/ Summary for Chapter 13 — 340
Review Problems — 341

Math Essentials	347	Appendix A
1/ Math Skills	348	
2/ Algebraic Manipulations	361	
3/ Statistical Concepts	369	
Problems	372	
Statistical Tables	376	Appendix B

Selected Answers to Problems	413
Index	441

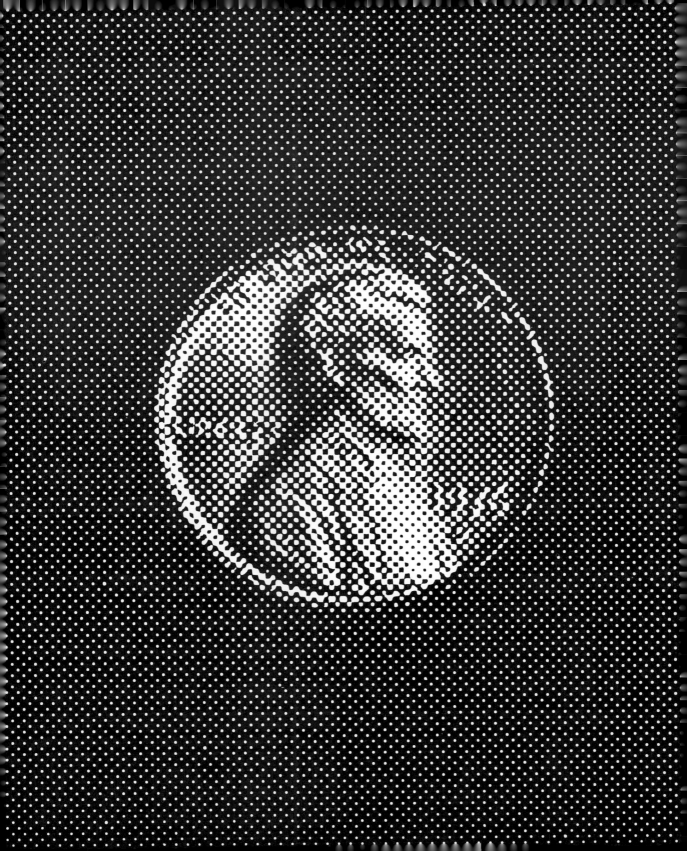

Introduction

CHAPTER 1

1/ A Very Brief History

The science of statistics has grown from ancient times in a rather natural way. Early historical-biblical records indicate a systematic use of data in counting animals, measuring land, casting lots, etc. With the formation of governments came taxation and population counts. By the late 1600's a few insurance companies were using facts on births and deaths as a means for projecting average life-spans. At about this same time a small group of mathematicians were discussing probabilities and chance relating to some gambling problems. These men, including Blaise Pascal, Pierre de Fermat, and James Bernoulli, developed some of the first concepts leading toward a modern theory of probabilities.

In the early 1700's, Abraham DeMoivre developed the principle of least squares and an equation describing the normal (bell-shaped) curve. During this same period Carl Friedrich Gauss derived a normal curve from his studies of measurement errors. Later in the same century, the influence of Adolph Quetelet (pronounced Kĕt′lā′) brought about a fairly wide use of statistical methods. Thomas Bayes proposed a philosophy of subjective probabilities, the basis for a "decision science" which has become useful, especially in business and economics work.

Several Englishmen were prominent in statistics in the latter 1800's. Among these were Francis Galton and Karl Pearson. Galton, a cousin of Charles Darwin, used statistics in his studies of heredity. Pearson made several contributions to statistics including the idea of a standard deviation and frontier work in correlation analysis.

In the 20th century, R. A. Fisher contributed substantially to the application of statistical methods, especially in agriculture. One important contribution was made by William Gossett, who introduced the Student or t-statistic for inference on small samples from normal distributions with unknown variance.

In America a number of events have led to a wide interest in statistics. Our Constitution provides for a recurring population census; accounts of the early census appear in [3]. The enormous task of compiling census data, which took many months for the 1920 and 1930 censuses, led to some statistical sampling in the 1940 census. In 1970 most of the information came from samples. Another widespread impetus for sound statistical methodology came from voter polling; samples that were not probability

When we want to know something about a population but can't observe every member in it, we can take a random sampling to try to obtain a reasonable description of that population.

oriented led some pollsters to predict the wrong person as President in both the 1936 and the 1948 elections. Since then carefully chosen samples, based on probability, have been very common in voter polls.

Many Americans have contributed significantly to statistics. Among these, George Snedecor initiated the idea of a "statistical laboratory" as a professional center for service to the university and the community. Snedecor's stat lab idea, first used at Iowa State College (now Iowa State University), is common in many countries today. In 1933 Jersey Neyman and E. A. Pearson published some results basic to much of classical hypothesis testing. Around 1940 W. A. Shewhart was instrumental in developing scientific sampling processes for use in quality control of industrial products.

Increased use of data in the 1940's and 1950's must relate to the development of the electronic computer. Conceived by Charles Babbage, an Englishman, about 1812, it was not until 1940 to 1960 that the computer came of age. Scientists from numerous American universities and business machine corporations have developed the present-day electronic wizards. For example, 1970 census data was computer summarized within a few short weeks. Before, this had taken months, even years! Also the speed with which election data is summarized by national television networks is due to computerization. Many future developments in statistics will undoubtedly be linked to advances in the computer sciences.

The wide applicability of statistics is indicated by the names of subsections within the American Statistical Association—business and economic statistics, physical and engineering sciences, social statistics, statistical computing, statistical education, and the biometrics section. *Statistics concerns problems in which there is data, quantification, and the occurrence of chance events.*

2/ Basic Terms

Many words with particular meanings are used in this book. These include *statistics, sample,* and *population.* We will discuss these terms and others in formal definitions. For now several ideas are described in the following example.

This example is of television program ratings, specifically the Nielsen Television Index (NTI). The NTI provides continuing estimates of television viewing and nationally sponsored network program audiences, including national ratings 52 weeks per year. Data regularly reported includes audiences by regions, daily ratings, national and multi-network

Basic Terms

area ratings, and more. Here are some descriptions as given by the A. C. Nielsen Company [2].

> The Nielsen Rating you may see reported in the newspaper is simply a statistical estimate of the number of homes tuned to a program. We repeat: it has nothing to do with program quality. For example, a rating of 20 for a network TV program means that 20% of U.S. homes are estimated to be tuned in to that program.
>
> Since over 70 million U.S. households (97% of the total) now have TV sets, a rating of 20 means that an estimated 14.0 million TV households tuned in.
>
> Note that when we described the rating, we used the words, "statistical estimate." That's because a rating is subject to a margin of statistical error. It is based not on a count of all TV households, but on the count within a sample of TV households selected from all TV households (the population). The findings within the sample are then "projected" to national totals.
>
> *"Why use a sample?"*
>
> Simply because a complete count—program-by-program—of those 70 million TV homes would cost countless millions of dollars. Furthermore, any count —complete or from a sample—has to be taken regularly so that broadcasters and sponsors can stay in tune with people's likes and dislikes, which often change over time.
>
> It is far more efficient to draw a sample, and then project the results.
>
> *"How does sampling work?"*
>
> Most expert statisticians could give you some very comprehensive answers to that question. Probably too comprehensive, in fact, for anyone but another expert statistician. So let's explain sampling by using a photograph of a [woman].
>
> Picture No. 1 is composed of several hundred thousand dots. Let's consider these dots as our total population and draw several samples.
>
> The three pictures represent samples of 250, 1,000, and 2,000 dots. These samples represent a specific kind of sample design called "area probability sampling" because the black and white dots in the samples are distributed in proportion to their distribution in the original picture. (More black dots in the hair, more white dots in the face, etc.) Think of homes (which add up to

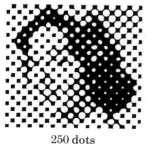
250 dots

1,000 dots

2,000 dots

Picture No. 1

Introduction

our population) and you have the sampling method used by Nielsen for arriving at national TV ratings.

Now . . . if you put the book down and step back a few feet, you'll notice a very interesting thing as you look at these small pictures. Your eye will adjust to the overall image and stop trying to "read" the dots. See how the 250-dot sample provides a recognizable picture? Recognizable, yes, but obviously not much detail. So, let's take a look at the 1,000 dot sample . . . again from a few feet away. Now we find that the woman is very recognizable; in fact if all we wanted was a reliable idea of what she looked like, this sample would be quite adequate.

Another interesting thing about sampling, the 1,000-dot photograph is about twice as sharp as the 250-dot photograph because it has four times as many dots. And so it is with sampling: to double the accuracy, one must quadruple sample size.

These are some of the basic laws followed in constructing Nielsen's 1,200 home television sample.

'W-H-A-A-T . . . do you mean to say that only 1,200 homes . . . ?' . . . The real question should be: 'Does a 1,200-household sample provide a sufficiently reliable estimate of the national TV audiences?' . . . If the Nielsen sample, constructed as it is, produces a rating of 20% for a number of programs, the TRUE rating lies somewhere between 18.7 and 21.3 for two out of three programs. . . .

"Well then . . . How is the national Nielsen sample drawn?"

Using the U. S. Census Bureau's counts of all housing units in the nation, we randomly select about 1,500 housing units, using scientific sampling procedures. Housing units that are occupied and have a TV set are asked to become a part of our sample. The whole process takes thousands of manhours of work, and costs literally hundreds of thousands of dollars.

Remember the 1,000-dot photograph? Just as a random selection of black and white dots turned out to be representative of the whole photograph, the Nielsen sample now contains all types of households: city, town, farm, rich, poor, etc., each selected at random according to population density across the U. S.

In short, the Nielsen sample now provides what in effect is a scale model of all U. S. TV households. . . .

"How does Nielsen measure local audiences . . . Say, in Pittsburgh or Little Rock?"

Since there are over 200 local TV markets, which require a total of nearly 100,000 sample households, obviously the cost of a metered system is prohibitive—except in the three largest markets, Chicago, New York and Los Angeles. In the others, we ask cooperating households to keep a television viewing diary for one week. After processing, the information is released in reports for each market. Aside from this difference, the same statistical laws and sampling principles apply in making these periodic local measurements.

"In closing . . ."

Nielsen ratings provide a reliable estimate of TV audience size and characteristics. In no way are they intended to measure program quality. The rating techniques are based on sampling laws which are statistically valid. . . . Ratings benefit the television audience because they provide a barometer of people's likes and dislikes.

The following is a postnote to the example for those who regularly watch TV [1].

Top 15 programs ranked by Households and Total persons.

Program rankings vary somewhat according to whether they are based on household audiences or people audiences. However, for the most part the same programs appear on both lists. Readers should note that the size of a program's audience is but one of a number of ways to assess its worth or success.

TOP 18 PROGRAMS

	Two weeks ending January 11, 1976			Two weeks ending January 11, 1976		
Rank	Program name	Household Avg. Aud. %	Rank	Program name	Total persons %	0,000
1	All in the Family	36.2	1	All in the Family	26.4	52,87
2	NFL Championship Game (s)	33.3	2	Rose Bowl Game (s)	22.6	45,24
3	Maude	31.9	3	Maude	22.4	44,76
4	Rose Bowl Game (s)	30.6	4	Phyllis	21.9	43,81
5	Phyllis	30.4	5	Six Million Dollar Man	21.9	43,74
6	Rhoda	30.3	6	Happy Days	21.6	43,19
7	Kojak	28.1	7	Kojak	21.3	42,66
8	Orange Bowl Game (s)	27.9	8	Rhoda	21.3	42,65
9	M*A*S*H	27.7	9	NFL Championship Game (s)	21.2	42,49
10	Sanford and Son	27.2	10	Orange Bowl Game (s)	20.9	41,75
11	Mary Tyler Moore Show	27.0	11	Happy Aniversary, C. Brown (s)	20.9	41,74
12	CBS Reports: Inquiry Pt. 4 (s)	26.8	12	Welcome Back, Kotter	20.5	41,03
13	One Day at a Time	26.5	13	M*A*S*H	20.5	41,02
14	The Jeffersons	26.3	14	Mary Tyler Moore Show	19.7	39,44
15	Little House on the Prairie	26.1	15	The Jeffersons	19.6	39,31
16	Chico and the Man	25.9	16	Tom Sawyer (s)	19.5	39,08
17	Six Million Dollar Man	25.4	17	One Day at a Time	19.5	38,94
18	Carol Burnett Show	25.3	18	The Waltons	18.7	37,42

(s)—special or preempting program

This ends one thumbnail sketch of statistics in action; there will be others later—called essay examples.

Did you notice those key words—statistics, sampling, and population? The population included all 70+ million U. S. households that now have TV sets. Television viewing interests are collectively described by viewer ratings. For economic reasons, opinions are obtained from only a part of the population. Here is where sampling is used. The sample should be good, meaning it should reflect the viewing preferences of the population. And this is where statistics comes in. The Nielsen Agency, for example, scientifically obtains numbers like 20% from their samples. Then other statistical procedures are used to project these statistics to a population characteristic.

Statistics, numbers obtained through sampling, are used in two ways: (1) for description and (2) for inference (projections). Our study begins in Chapters 2 and 3 with descriptive techniques. Remember the pictures of the woman? Description concerns visual and numerical evaluations on sampled data. You have seen many such expressions published in periodical literature: sports averages, vital statistics, weather readings, etc. Most are presented as visuals, as graphics, or simply as numbers. Chapters 4 through 7 concern probability concepts, that is, introduce chance quantities into sampling experiments. Finally, remember that the Nielsen people aimed for reasonable projections from the sample to the population, their "statistical estimates." We will get a basis for understanding much of the logic behind such projections beginning in Chapter 8, with estimation.

3/ How to Use This Book

The material is divided so that one, or at most two, class periods can be devoted to a section. Regular sessions for working and discussion of problems can help fix concepts in your mind. The data in the workbook can be used for class or individual experiments. (*Note:* My experience is that beginners are often overly concerned with getting exactly the answers shown in the answer section. But many problems can be worked in more than one way so answers can differ slightly. Rules for rounding off are given in Appendix A, p. 356. As a general rule I have used two decimal places of accuracy in calculations, including the use of tabled values. A number of suggestions for simultaneously simplifying calculations and improving accuracy appear in the text.)

Readers may find this approach more meaningful by periodically viewing actual applications. The essay examples in Chapters 2, 5, 7, 8, 10, and 12 describe a few. Others appear in *Statistics: A Guide to the Unknown* [3], and *Statistics by Example* [4]. These could be considered for case studies or for special-interest projects. Both are quite good and are consistent with the level of this work.

The appendices contain aids for problem solutions. I have found Appendix A—Math Essentials—meaningful near the end of Chapter 3. Learning how to use statistical tables is a must. Directions for use and examples appear at the first mention of each table. The Rules section (see the Workbook) lists procedures in the order of their first presentation. If used for working study problems, it becomes a potential tool for "closed book" exams.

This work is centered on problem solving and applications and primarily concerns using statistical concepts. The approach is intuitive and offers a variety of experiences for a first study of statistics.

REFERENCES

[1] *A Look at Television,* Nielsen Newscast (Northbrook, Ill: A. C. Nielsen Company), 1974.

[2] *Everything You've Always Wanted to Know about T.V. Ratings* (Northbrook, Ill: A. C. Nielsen Company), 1974.

[3] Morris H. Hansen, "How to Count Better: Using Statistics to Improve the Census" in Judith Tanur, Fred Mosteller and others, editors, *Statistics: A Guide to the Unknown* (San Francisco: Holden-Day Publishers), 1972.

[4] Frederick Mosteller, W'm H. Kruskal and others, editors, *Statistics by Example* (Reading, Mass: Addison-Wesley Publishing Company), 1973.

[5] *Nielsen Television Index—1974–75 NTI/NAC Reference Supplement* (Northbrook, Ill: A. C. Nielsen Company), 1974.

Description and Graphical Displays

CHAPTER 2

Two classes of measures are commonly used for describing data. The first class, *averages,* describes location values, and although as many as a dozen averages are common, only three have wide use: the arithmetic mean, the median, and the mode. The second general class of measures describes *variation* in observations. These include the range and the standard deviation. Often graphical displays are used to give visual description about data. Collectively these are called *descriptive statistics.*

1/ Averages

When I ask students the average of the numbers 0, 1, 2, and 5, invariably the first answer given is 2. As you may know, this is the arithmetic *mean,* a perfectly fine average. Yet another equally good description would be 1.5, the *median* value. This points out two things. First, to many people, the word *average* brings to mind the arithmetic mean. Secondly, the mean is not the only average, and others, including the median, may be equally or more appropriate in some cases.

Suppose you were asked to find the "mean" price charged for a half-gallon of milk given the price in five stores: 69, 80, 76, 80, and 70 cents. Most of you already know the process requires summing the values, then dividing by five, the number of stores. So the value of the mean for this data is

$$\frac{(69 + 80 + 76 + 80 + 70)}{5} = 75 \text{ cents.}$$

Did you say, "No problem, that's easy enough"? I hope so. But what if we had prices from 50 stores or maybe even more? Actually we don't have to show each number and all the details for computing this value because we can describe the process in general.

Statistical description helps us to identify measurable characteristics. Although these items are very different, they can be grouped into a single class—footwear.

Description and Graphical Displays

Rule

The value of the arithmetic mean for a collection of numerical observations is

$$\overline{X} = \frac{\sum\limits^{n} X}{n} \quad \text{where}$$

$\sum$ = the summation symbol. Read this as "sum whatever follows."
X = any of the numerical observations
n = the number of observations in the sum, the sample size
$\overline{X}$ = a statistical symbol for the arithmetic mean.
(See Appendix A, p. 361 for more information about summations.)

For the milk prices, X = prices at the five individual stores. We've already found $\overline{X} = 75$ cents.

One might rightly ask, "Why introduce all these extra symbols when there are only five observations?" I agree—for this problem. But in experience we usually have nearer 50 or even 500 observations! Suppose the prices for a consumer food index are obtained for milk from 50 stores, X: 69, 80, 78, 80, 70, 75, 80, 80, 80, 76, 78, 78, 80, 80, 80, 75, 80, 72, 78, 75, 80, 80, 82, 80, 80, 78, 80, 76, 70, 70, 80, 75, 80, 80, 78, 75, 70, 75, 80, 70, 80, 80, 80, 78, 80, 75, 78, 80, 75, and 80 cents. The Rule indicates a process, sum the prices and divide by the number of prices, 50, which is a shorter description than

$$\overline{X} = \frac{(69 + 80 + 78 + 80 + 70 + \cdots + 80)}{50}.$$

And the Rule is general; it applies to other samples of observations as well as to this one. It can serve to guide the calculations of the mean value for any sample.

Now we can formalize some of the basic terms described in Chapter 1.

Definitions

A POPULATION is the collection of all things that have one or more specified characteristics. A SAMPLE is any part (but less than all) of a population.

For the milk prices, we have described two samples from a population of the prices at many stores; many more samples are possible. We will treat all data sets in this chapter as samples. Subsequent calculations on the numbers give sample characteristics called *statistics*. The mean, $\overline{X}$, is a common one.

Besides the fact that many people regularly use the mean as an average, what good is it? Well, it's important in statistics because it does something that no other average does—it provides a balance value. That is, the mean is to a set of numbers like the point of a fulcrum is to a teeter-totter. The balance attained by the mean shows that the total for its algebraic distances

Figure 1 / The mean as a balance value for the prices 69, 80, 76, 80, and 70 cents

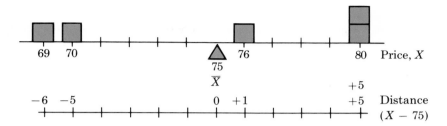

from each of the observations is zero. See Figure 1, which illustrates this property for five milk prices. The balance in Figure 1 is expressed by

$$\sum_{}^{5}(X - 75) = (-6) + (-5) + (1) + (5) + (5) = 0,$$

again a summation, but now of differences from the mean. This property leads to an additional use for the mean in the most popular procedure for describing variation. We'll see how this works in the next section. Now let's look at another average, the *median*.

> The MEDIAN is a value located at the middle position when the observations are arranged in order by size.

Definition

The first step in the calculation is to be sure the observations have been ordered. Thus 69, 80, 76, 80, and 70 cents are not ordered by size, but 69, 70, 76, 80, and 80 are. There is an odd number of prices, $n = 5$. This is important because the calculation procedure differs depending whether n is odd, or even. The following two examples will show you the difference.

> The prices for milk, in order, are 69, 70, 76, 80, and 80 cents. Since $n = 5$, $(n + 1)/2 = 6/2 = 3$. The value in the third position (counting from either extreme) is 76 cents, the median value.

Example

> Suppose we have an even number of observations, 69, 70, 76, 80, 80, and 80 cents. Now $n = 6$ and $(n + 1)/2 = 7/2 = 3.5$. So we seek a value between the two in the third and fourth positions in the ordered set as the median. The value chosen is $(76 + 80)/2 = 78$ cents. (Actually any value between 76 and 80 could be taken as a median, because 50% of the observations would be smaller and 50% bigger, but we'll use the midpoint.)

Example

To generalize, first order the observations by size (smallest to largest or vice versa). If n is odd, the value in the middle position, counting from either extreme, is the median. If n is even, the median is the sum of the two middle values divided by two.

Since they are determined by different procedures, the values for the mean and median are usually different. Yet for many samples these averages will be nearly equal. For example, with five observations on milk prices, we found $\overline{X} = 75$ cents while the median = 76 cents. With our six observations, $\overline{X} = 75.8$ (check this for yourself) and the median = 78.

A third average, the *mode,* is easily determined.

Definition

> The MODE is that value which appears most frequently among a set of numerical observations.

Computation is simplified by ordering the observations according to size.

Example

> For the example with the prices 69, 70, 76, 80, and 80 cents, the mode is 80 cents.

Often all observed values occur an equal number of times. Then the mode does not exist. Thus the set 69, 70, 76, and 80 does not have a mode, nor does 69, 69, 70, 70, 76, 76, 80, and 80. A shortcoming of this average is that for many samples there is no mode; this average does not exist.

A practice to be encouraged is the making of independent "quick checks" on important calculations. The *midrange* can serve this purpose for the mean and median. This calculation requires recognition of only the largest and the smallest observations.

Rule

> The midrange of a collection of numerical observations is
> $$\text{midrange} = \frac{\text{largest value} + \text{smallest value}}{2}.$$

The advantage over calculations for other averages is speed. This is especially true when the sample size is even moderately large.

Examples

1. For the sample $-3, -1, 6, 12, 0, -6, 8, 14, -10, 8, 7, 4, 16, -2, 11, -6, 3, 0, -3, 9, 14, 15, -10, 0, -4,$ and 7, the largest (most positive) value is 16, the smallest (most negative) value is -10. These give

   $$\text{midrange} = \frac{16 + (-10)}{2} = \frac{6}{2} = 3.$$

   For comparison, the mean is 3.4 (rounded) and the median is 3.5 (you should check these for yourself).

2. As a quick "ball-park" estimate for reasonable values for the mean and the median for the milk price data—69, 70, 76, 80, and 80 cents—

the midrange value is $(69 + 80)/2 = 74.5$ cents. Recall that $\overline{X} = 75$ and the median = 76 cents.

The midrange is most useful as a coarse check on values for these common averages. That is, the midrange should be close in value to both the median and the mean.

We close this section with comparisons of the three averages. No one is always best. For example since extreme values distort the mean more than the median or mode, the latter are preferred when most observations have similar values, but a few are much larger or smaller. Examples 1 and 2 below show this. Otherwise the mean is a more reliable measure than the others because, for numerous samples taken from the same population, the values obtained for the mean are generally closer to one another than those for either the median or the mode. The mode is the least reliable, and for some sets there is no mode. Its frequent use is required in opinion polling where the most common opinion is being sought. Most important the mean, because of certain unique and desirable properties, is used extensively for more than just statistical description.

The following examples display a preferred average. Based on what you've learned so far, do you agree with my choices?

Examples

1. The life of a set of radial tires (in thousands of miles): 38.6, 41.4, 37.4, 64.8; median = 40 (i.e., 40,000 miles).

2. Years of experience for corporate executives: 24, 22, 21, 28, 31, 41, 24, 25, 18, 27, 26, 24; mode = 24, or median = 24.5. The mean ($\overline{X} = 25.9$) is distorted toward the single large value, 41, and so is not a good choice.

3. Class standing for college students (1 = freshman, 2 = sophomore, etc.) 1, 3, 2, 2, 4, 5, 2, 3, 1, 2, 1, 1, 3, 1; mode = 1.

4. Wages for employees paid by the hour at Disney World: $2.41, $2.78, $3.12, $2.56, $2.18, $2.64, $2.44, $2.18, $2.34, $2.96; mean = $2.56, or median = $2.50.

5. Batting averages in baseball: .300, .289, .265, .274, .216, .188, .210, .213; mean = .244, or median = .240.

Problems

1. Given the following observations compute $\overline{X}$, the median, and the mode: 6, 5, 3, −3, −2, 3.

Description and Graphical Displays

2. Find mean, median, and mode values for the scores on these five rounds of golf: 68, 75, 71, 62, 69. Check the reasonableness of your answers by using the midrange.

3. Indicate the most appropriate average(s), mean, median, or mode, for each of the following:
   a. life of light bulbs (in hours) 426, 437, 398, 412, 416, 431
   b. number of hits per game 0, 0, 1, 2, 0, 0, 1, 0, 4, 0
   c. opinion of reasonable term (in years) for legislators 1, 1, 3, 6, 2, 2, 3, 2, 2, 4, 1, 2, 4
   d. age of college students 19, 19, 20, 21, 46, 20, 23, 18
   e. miles per gallon of gas 18.6, 18.1, 18.3, 18.5, 18.0
   f. drugstore sales $3.27, $.57, $2.76, $1.19, $4.23, $2.54

4. a. Find mean, median, and modal values for 302, 305, 299, 306, and 303.
   b. Now find the mean by coding, i.e., first subtract 299 (the least value) from each observation, compute the mean for the coded values, then decode by adding 299. Check the result with your previous answer.

5. For X = the number of service calls per month on color TV's, the values for one dealer for a year (12 months) were

   X: 26, 34, 29, 41, 33, 24, 27, 36, 34, 40, 40, 32.

   a. What was the total number of calls for the year; that is, find $\sum X$.
   b. Find $\overline{X}$, then show $\sum (X - \overline{X}) = 0$.

6. The term *grade* in one course is determined from the average for the scores on four one-hour exams. Which average—mean, median, or mode—would be *preferred* (a) by Beth, (b) by Ann, (c) by Paul? Explain.

Student	Exam grades			
Beth	87	63	72	72
Ann	83	86	84	86
Paul	81	57	76	74

7. The quality control inspector for the Never-Too-Short Ruler Company found that five of their meter sticks have lengths of 1.01, 1.07, 1.03, 1.02, and 1.02 meters. Find the mean, median, and modal values. As a practical matter why might the company be concerned if the average length were say 0.99 meters? (Note: 1 meter is 39.37 inches, but the math is easier using meters rather than inches.)

2/ Variation

A second class of measures describes differences among observations. These measures of variation include the range and the standard deviation.

Although averages give useful information, the bulk of our work will concern the differences between numbers. For example, you recognize the importance of variability in scores every time a group test is returned. Suppose your score is 86. Knowing that the average, i.e., the mean, is 78 is informative. But also one feels much more secure knowing that an 86 was an achievement above $\overline{X}$, good enough for a high letter grade. Some quantitative appraisal of differences from the mean is commonly the basis for letter grades. There is little need to motivate the existence of differences in data; our interest is to make quantitative sense from these differences.

The *range* gives us a rough gauge of variability among observations.

> The RANGE is the difference between the values of the largest and the smallest observations.

Definition

For the exam scores in Table 1, the range is $100 - 62 = 38$.

Table 1 / Exam Grades for 20 Students

62	72	76	82	84	87	95
67	73	78	83	85	90	100
71	74	79	83	86	93	

The range is attractive because of its computational simplicity; it needs very little information from the sample. Yet, because it is derived from only the extreme values, it tells little about intermediate ones. For example, the sets (1) 62, 100, 100, 100 and (2) 62, 71, 85, 89, 100 have identical ranges ($100 - 62 = 38$), yet the intermediate values are quite different. The range, like its counterpart among averages, the midrange, can serve a useful role in a first approximation of more-exacting measures.

A second measure of variation, the *standard deviation,* is used to describe differences among all the observations. As the name implies this is a standard unit of deviation. It is similar to the units on a ruler in that a standard deviation unit can perform a role similar to that of an inch. In both cases a unit describes a certain distance. With a ruler the measurement

Description and Graphical Displays

is the same for each unit of distance. But the measure described by each unit of standard deviation depends on the distribution (relative position) of the individual values. See Figure 2. One essential difference between the inch and the standard deviation as units is the item to which each unit applies. For example, in Figure 2a the inch is the unit for identifying an individual's height. Here each person is considered individually—separately from all others. In contrast the standard deviation is a unit for describing differences in height among numerous individuals. Here the height for any individual is described as some multiple (positive or negative) of one standard deviation unit. For example, the individual pointed out in Figure 2b has a height that is +2 standard deviation units. So his height is above the average, which is at zero (0) standard deviations. The standard deviation describes his height relative to others in this picture.

Figure 2/ The inch and the standard deviation as "units"

a. The inch as a unit describing individual heights

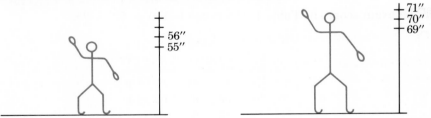

b. The standard deviation as a unit describing differences in heights for numerous individuals

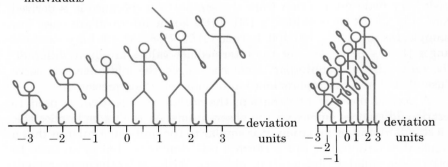

For widespread values   For narrowly spaced values

This new unit contains several characteristics. First, the units are *difference* units, meaning comparison is made by the subtraction of some number. That number, as you might guess from our work in the previous section, is the mean. So the standard deviation as a unit has meaning only if we are describing individuals, or more generally, observations, within the context of a defined group. By comparing the illustrations in Figure 2b, we might suspect that different groups of observations can have standard deviation units of different sizes. This is generally the case. The nature of the dispersion of individual values about the central value is described as their *distribution*.

Obviously the standard deviation unit is more complicated than the inch. It is, however, quite important, even central, to our study. Further explanations of its use and relevant concepts will take place in this chapter and in several others. The rest of this section concerns the techniques for computing a value for the standard deviation unit for any given group of observations. The next section contains graphical displays of distributions. Chapter 3 includes a discussion of how to construct a distribution for any specific group of values. After that our main objective is to show how this new unit, the standard deviation, can be used to give meaning to individual values within a group.

First, we describe how to get a value for this unit. The following steps are used to compute the value of the standard deviation for a set of sample observations.

1. Compute a value for the arithmetic mean, $\overline{X}$.

2. Determine differences by subtracting the mean value individually from each observation. These individual differences are symbolized by $(X - \overline{X})$.

3. Square each difference, then sum to find a total.

$$\sum_{}^{n} [(X - \overline{X})^2]$$

4. Divide this total by the number of observations less one, $n - 1$. The result is the *sample variance, s^2*.

5. Convert back to the original scale by taking the square root of the sample variance, which gives us the standard deviation, $s = \sqrt{s^2}$.

Stated as a rule, we have

The sample variance is   Rule

$$s^2 = \frac{\sum_{}^{n} [(X - \overline{X})^2]}{n - 1}$$

Description and Graphical Displays

and the standard deviation is $s = \sqrt{s^2}$
where X = the individual values
$\overline{X}$ = the sample mean
n = the total number of observations in the sample.

Observe that in a sense the sample variance, s^2, is a type of average. That is, s^2 is the sum of squared differences divided by $n - 1$. The divisor, $n - 1$, indicates the *degrees of freedom* (for a discussion of degrees of freedom, see Chapter 10, Section 1). Also this is a first use of the sample mean beyond its use as an average. The example below demonstrates the calculation procedures.

Example

These values represent the times required for an automatic bottling machine to fill a one-gallon container: 7.3, 7.0, 7.0, 7.3, 6.8, and 7.2 seconds. The sample variance, s^2, and the standard deviation, s, are computed as follows:

First, $\overline{X} = \dfrac{(7.3 + 7.0 + 7.0 + 7.3 + 6.8 + 7.2)}{6} = \dfrac{42.6}{6} = 7.1$ seconds.

Then,

Time, X		$(X - \overline{X})$	$(X - \overline{X})^2$
6.8	6.8 − 7.1 =	−0.3	.09
7.0	=	−0.1	.01
7.0	=	−0.1	.01
7.2	=	0.1	.01
7.3	=	0.2	.04
7.3	=	0.2	.04
Totals 42.6		0.0	.20

giving $s^2 = \dfrac{\sum_{}^{6}[(X - 7.1)^2]}{6 - 1} = \dfrac{.20}{5} = .04$, $s = \sqrt{.04} = 0.2$ seconds.

In this example, the total for the unsquared differences is zero. Remember our earlier discussion of this balance. It can serve as a check so long as the sample mean is well rounded. (For example, if $\overline{X} = 7.09$ were rounded-off to 7.1, this column would total nearly, but not exactly, zero.) The process of taking a square root may appear simple here, but actually it warrants some practice. (See the discussion of extracting square roots from Table VIII given in Appendix A, p. 358).

Variation

The grades for the 20 students given in Table 1 serve as a second illustration of computing values for the mean, variance, and standard deviation.

The data from Table 1 is presented below. Totals are provided for calculations on $\overline{X}$ and s^2.

Example

Grade, X	$(X - \overline{X})$	$(X - \overline{X})^2$
62	−19	361
67	−14	196
71	−10	100
72	− 9	81
73	− 8	64
74	− 7	49
76	− 5	25
78	− 3	9
79	− 2	4
82	1	1
83	2	4
83	2	4
84	3	9
85	4	16
86	5	25
87	6	36
90	9	81
93	12	144
95	14	196
100	19	361
Totals 1620	0	1766

A mean value is first computed,

$$\overline{X} = \frac{\sum\limits_{}^{20} X}{20} = \frac{1620}{20} = 81.$$ Then

$$s^2 = \frac{\sum\limits_{}^{20} [(X - 81)^2]}{20 - 1} = \frac{1766}{19} = 92.9, \text{ and } s = \sqrt{92.9} = 9.6^*$$

---

*See Table VIII to find the square root. Again, Appendix A, p. 358 describes how to use Table VIII.

The steps used in the calculation of s will be used many times and therefore are worth review.

1. Compute a value for the arithmetic mean, $\overline{X}$. Using the example, $\overline{X} = 81$ points.
2. Subtract the mean from each individual observation. Here $X - 81$. The total for these differences should be essentially zero.
3. Square the individual differences and sum. This gives

$$\sum_{}^{20} [(X - 81)^2] = 1766.$$

4. Divide by the sample size less one $(n - 1)$, the degrees of freedom. The resulting sample variance is $s^2 = 92.9$.
5. Take the square root to obtain the sample standard deviation unit. Here $s = 9.6$.

In our example the mean is $\overline{X} = 81$ points and the standard deviation is $s = 9.6$ points.

These calculations may seem superfluous if you have an electronic calculator that automatically extracts standard deviations and square roots, but these procedures will be extended to include two very similar versions later in our work. For those with calculators, machine versions appear in the workbook and also in reference [4].

Earlier I stated that the range can be used for a quick approximation of the standard deviation. The following rule describes how to find a rough, but quick estimate of the correct value [7].

Rule

For any set of n observations:

if n is near the number	then the standard deviation is roughly approximated by dividing the range by
5	2
10	3
25	4
100 (or more)	5

Again, this rule gives only approximate values, but is a quick "ball-park" check on the more sophisticated measure, s. For example, in the preceding example, we calculated $s = 9.6$ points for $n = 20$ grades. Using the rule, a check value is range/4 = $(100 - 62)/4 = 9.5$ points. For the bottling times data where $s = 0.2$ seconds and $n = 6$, the rule gives range/2 = $(7.3 - 6.8)/2 = 0.25$ seconds. Using this procedure to obtain rough estimates for a standard deviation value will give an independent check on the exact calculation. The rule is *not* a substitute for the five-step procedure, but it can help identify gross mathematical or procedural errors.

The next step in describing variation relates the standard deviation unit back to the individual observations. For example, a test score of say 86 can be scaled as a number of standard deviation units above or below the mean. This or any other observation can then be placed in a position relative to all of the others. However, before using s, several related concepts must be described.

Other concepts relate to our earlier comparison of s to the inch as a unit. For the bottling machine data we got $s = 0.2$ seconds, then for the data on grades, $s = 9.6$ points. Like the unit on a ruler, the basic unit of difference, s, has a fixed size, but only as long as we are evaluating a single sample. However, beyond this point the two contrast because the unit *inch* not only defines a standard unit or scale, but it also defines the measure the scale describes. The standard deviation can describe measure only through one's knowledge of the dispersion, or *distribution*, of scores. Thus our next concern is to give meaning to a distribution of scores. This involves both graphical (visual) and numerical description for numerous sets of data.

Problems

1. The following are sales in thousands of dollars for a sample of four used car salesmen: 17.5, 22.5, 27.0, 23.0. Find (a) the range, (b) the mean, (c) the sample variance, and (d) the standard deviation.

2. For the pressure readings of 29.7, 29.8, 29.9, 29.9, 30.0, 30.0, 30.2, and 30.3 inches of mercury, show that $\overline{X} = 30.0$ and $s = 0.2$ inches.

3. Find the mean and standard deviation for the sample observations 7, -5, 0, 2, -2, 1, 4, 6, 2, and 0. Check the reasonableness of your values using the midrange and the range.

4. The answer to each of the following requires a short calculation.
    a. For the observations 3, 3, 3, and 3, find a value for s^2, then s.
    b. Which is bigger, s_1 for 100, 101, and 102 or s_2 for 0, 1, and 2?
    c. For a range of 12, and an n equal to 20, roughly estimate s.

d. Given $\sum_{}^{13}(X - \overline{X})^2 = 24$, obtain a value for s^2.

e. For the sample values 1, 2, 2, and 3, is $s^2 = 2/4 = .5$ correct?

5. For the observations 301, 306, 306, and 303, compute $\overline{X}$ and s. Now subtract 300 from each observation and recompute $\overline{X}$ and s. Explain how coding might be used to simplify calculations for $\overline{X}$ and s in general.

6. Use the same procedure as in Problem 5 to find $\overline{X}$ and s for the amounts of dinner checks: $17.50, $19.50, $15.00, and $16.00. In this case code by subtracting $17.50 from each amount. Use the midrange and range to make independent checks on your values.

7. The heights of the starting five players on a collegiate basketball team were 5'8", 6'0", 6'3", 6'4", and 6'8". Convert the heights to decimal feet, round to the nearest tenth of a foot, then find $\overline{X}$ and s.

8. For the observations $-8, -5, -5, -4$, and -3,
   a. compute $\overline{X}$, then show that $\sum(X - \overline{X}) = 0$.
   b. compute $\sum[(X - \overline{X})^2]$, then find s^2. Can s^2 be negative? Under what conditions does $s^2 = 0$?

9. Use the lengths 1.01, 1.07, 1.03, 1.02, and 1.02 meters to compute the standard deviation in lengths for these meter sticks. Round your answer to the nearest centimeter ($\frac{1}{100}$ meter).

3/ Graphical Displays

An old adage states that a picture is worth a thousand words. This may be true if the picture is clear and is well defined. The purpose of this section is to introduce some of the more common displays used to illustrate data. Our goal will be to discern characteristics that make each type of display a good statistical picture.

Although displays could be made to illustrate features of five to ten observations, this is seldom done. For so few observations, the individual values should be viewed. A bar graph, shown in Figure 3, is often used to illustrate more numerous observations. Several features of Figure 3 are characteristic of many graphical displays. Proper labels are the first requirement for a good statistical picture. Second, the title should indicate the type of display and the characteristic being described. The horizontal scale label renames the variable characteristic and shows quantitative or

Figure 3/ A bar graph showing prices for milk

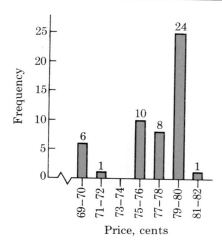

qualitative levels. The vertical scale often displays a count, or frequency. Sometimes for clarity, values for the vertical scale are placed on the figure. Both vertical readings appear in Figure 3. Some balance in total horizontal to total vertical distances is required so that the picture is neither too thin nor too wide. A common ratio is about 1 to 1; that is, the vertical distance covered by the figure is about equal to the horizontal distance. Again, the main objective is clarity. We will use the 1 to 1 ratio here.

What information is contained in Figure 3? Here are some descriptions; no doubt you can give more. All prices were 69 cents or more. The jagged lines cutting the horizontal axis indicate there is no interest in lesser values. Breaks in the vertical scale are undesirable because they can lead to distorted comparisons. The groupings 69–70, 71–72, and so forth, called *classes*, are formed to group together similar values. These classes have a common span, or *class width*, of 2 units. This number, the class width, is computed by subtracting the lower values for any two adjacent classes. For example, $71 - 69 = 2$. All classes in Figure 3 have the same class width. Equal class width is not only visually appealing, but also allows quick comparisons of classes by observing the relative heights of the bars. The class 73–74 is different in that no (0) prices were in this class, so it was left blank.

A *histogram* is much like a bar graph except that the bars touch. The histogram in Figure 4 depicts the racing speeds given in Table 2. In this figure classes are separated by break points called *class boundaries*. These values are halfway between the class limits. For example, 172.95 equals $(172.9 + 173.0)/2$. See Table 2. Again, the principal rule to follow is clarity.

Description and Graphical Displays

Example

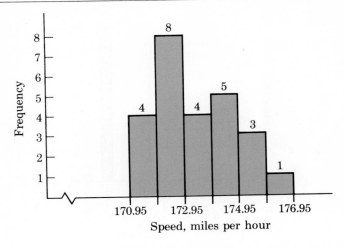

Figure 4/ A histogram of racing speeds

Definition

CLASS LIMITS are the largest and the smallest values that can be observed in a class.

Had we so chosen the classes could have been labeled by their limits—171.0–171.9, 172.0–172.9, 173.0–173.9, etc.

The name *frequency distribution* in Table 2 indicates a list that shows the number, or frequency, of observations given to the various classes.

Table 2/ Frequency Distribution of Racing Speeds

Speed Class	Frequency
171.0–171.9	4
172.0–172.9	8
173.0–173.9	4
174.0–174.9	5
175.0–175.9	3
176.0–176.9	1
Total	25

Although the bar graph and the histogram have similar forms and use, the histogram is most appropriate when the data have a continuous scale meaning that it can take any value on the real number line. When the data have values that can be ordered as bigger or smaller (an ordinal scale), or name values only (a nominal scale), the bar graph is used.

Another display that is especially useful for illustrating business or other time data is the *frequency polygon,* also called a *line graph.* Figure 5 shows oil production in Venezuela [10].

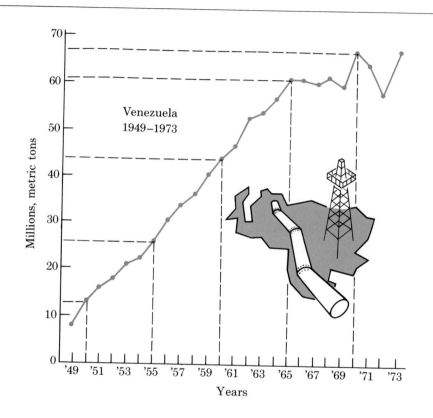

Figure 5/ A line graph showing production of petroleum products in Venezuela, 1949–1973

The horizontal lines on the figure help to read the quantity, and the vertical lines to read the dates. The map, pipeline, and oil rig are used to convey the subject of the graph and simultaneously to give some appeal to an otherwise drab picture. Frequency polygons are used to display counts or measurements over time, such as population figures, sales, sports records, weather (temperatures, rainfall), stockmarket levels, and so forth. The frequency polygon can be developed from a bar graph, a histogram, or from the associated frequency distribution. Using information from either Figure 4 or Table 2 yields the frequency polygon on racing speeds shown in Figure 6. The points are plotted above the class marks.

Figure 6/ A frequency polygon of racing speeds

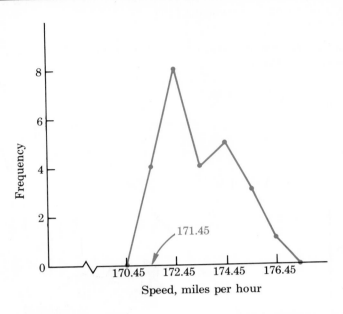

Definition
    The CLASS MARK, or MIDPOINT, is the arithmetic mean of the class limits.

    For example, in the class defined by the limits 171.0–171.9, the class mark is $(171.0 + 171.9)/2 = 171.45$. Points are scaled to the correct frequency, then connected by line segments. For completeness the graph is projected at either extreme on the horizontal axis to what would be the next midpoint if the same class width were used.

    The frequency polygon is something of a "working picture" since only the bare essentials are displayed, yet numerous facts can be quickly observed. For example, in Figure 6 the 172 mph class has the greatest frequency; the 171 and 173 mph classes each have half as many counts as the 172 mph class; and zero cars traveled 177.0 mph or more.

    A *frequency curve* has a form very similar to the frequency polygon. The frequency curve differs only in that the points are connected to form a smooth curve. Both serve essentially the same purpose.

    Pictograms use figures rather than bar or line diagrams to tell their story. Figure 7 displays a common use of pictograms—to illustrate population numbers. Here the world's human population at various periods in time is shown.

    Other pictograms show defense strength (e.g. pictures of ships, or infantry, etc.), the labor force, housing numbers, unemployment figures, car sales, energy figures, and so forth. Numerous examples of this and

Figure 7/ World population [2]

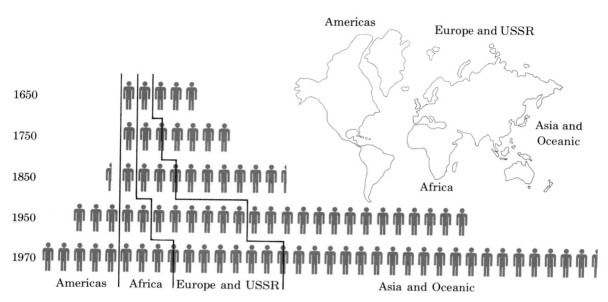

Legend: Each figure represents 100 million population. All part-figures are 50 million.

other displays can be found in current literature, especially business periodicals. A display can often serve the same purpose as a bar graph, but adds visual appeal. Its disadvantage is that scales are often inexact or are simply deleted.

Another pictogram illustrates increases in the number of air passengers in the rapidly expanding airline industry. See Figure 8. Although a pictorial display can be visually appealing, pictograms generally require a legend or other means for making relative comparisons. The picture should afford some quantitative exactness. Without the legend and/or numerical values, Figure 8 would have little meaning. Here the relative lengths of the lines of people give the comparison. Did you find Figure 8, which has a scale, easier to interpret than Figure 7?

Displays of financial allocations, including expenditures by government agencies, business firms, etc., often use the pie chart. See Figure 9. Again, the major goal is to present a clear picture while using few words. The distribution of pieces is restricted only in that the total, as percentages, must be 100% (in Figure 9 it's 100¢). This type of figure is rather difficult to use because of our inability to visually compare the sectors with any degree of accuracy.

Description and Graphical Displays

Figure 8/ Revenue passengers carried [9]

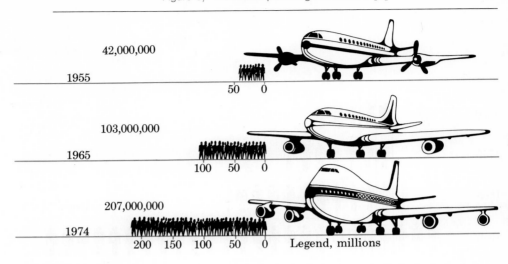

The graphic displays we've just seen are among the most common. It should be noted, however, that many statistical illustrations in popular literature (current periodicals, newspapers, etc.) are purposely distorted. For a discussion of common distortions and visual fallacies see [1] and [3], and the essay examples in the next section.

Figure 9/ A pie chart on the family food dollar [8]

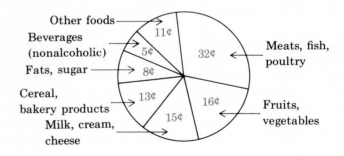

Problems

1. For the histogram which follows:

Problems

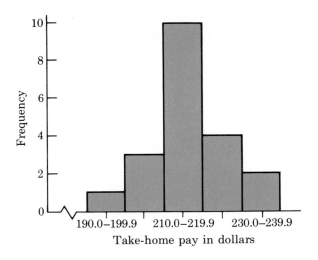

a. make an appropriate title.
b. indicate lower and upper boundary values for the class 190.0–199.9.
c. state the limits for the class with a frequency of 3.
d. what is the class mark for the class 210.0–219.9?
e. construct a frequency curve for this same data.

2. Given the following frequency polygon, construct a histogram, including the proper labels.

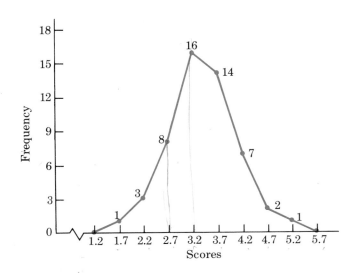

29

Description and Graphical Displays

3. In a certain community there are 104 families with no children, 123 with one child, 287 with two children, 98 with three children and 88 with four or more children. Use this information to complete the following figures:

a. A frequency distribution

number of children	frequency
0	
1	
2	
3	
4 or more	

b. A spike graph

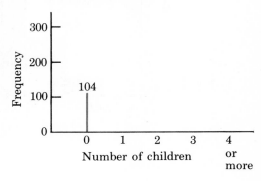

4. Sometimes a graphical display is intended to give a general picture rather than to show exact or specific values. For example the line graph below shows recurring lows in unemployment during June for both years. Describe other patterns that you observe in this figure.

5. The following pie charts relate the amount of work (⊟), leisure (⌇), and sleep (⊟), for various time periods in history. How has the amount of leisure changed for weekdays? For Saturdays? For Sundays?

Increasing leisure

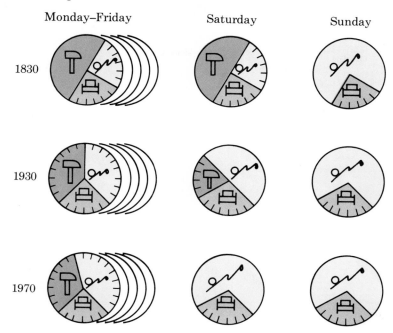

4/ Essay Example

We have discussed a number of descriptive forms, including averages, variation, and displays such as frequency distributions, histograms, and line charts. The examples in this section describe some common pitfalls in the use of descriptive statistics. These are just a few examples; statistical fallacies are all too common in published materials.

Example

The first job of the new business manager for the Indians, a pro football team, is to prepare a budget of player salaries for the next season. The former business manager's report indicated that the average salary for the team members was $40,000. Since the owners indicated a 10% across-the-board raise, he figures $40,000 + .10 · $40,000 = $44,000. So, for 40 players, $1,760,000 ($1.76 million) should be about right for player salaries.

Description and Graphical Displays

The business manager is assuming that "average" is synonymous with arithmetic mean, because he is using,

Total dollars = "average" salary · number of players.

Yet, the former business manager could have intended either the mean, the median, the mode, or some other "average." Unfortunately, only the first, the mean, would give a meaningful total. If in fact the average given was the median, then $1.76 million could be far off. Suppose the true mean was $45,000. Then the total salary requirement this year would actually be $1.98 million—roughly a $220,000 error!

This example shows that one must be very careful to identify *which* average is being used. The average should be appropriate for its intended use. If a descriptive number is used, it should be clearly defined. Otherwise get your own facts. The assurance of correct results is worth the extra effort.

The next example concerns the use of facts that show a distorted picture. Many displays can be made from a single set of data and, in the wrong hands, the facts can become misleading. Scales, labels, omissions, and even the type of display can grossly change the appearance of a data picture.

Cheating Charts. For our discussion the data base is population figures for a rapidly growing metropolitan area [6]. The actual statistics are:

Metro Area Population

Year	Population
1955	318,487
1960	388,940
1965	425,213
1970	500,000
1975	625,000 (est.)

Numerous distortions can be related to manipulations of the scales. We will look at three cases. The first distortion is produced by a broken scale. Figure 10a shows actual values while Figure 10b displays a scale that begins well above true zero. This is a common distortion that is used to "save space." Watch for it because it makes actual differences appear greater and trends to appear steeper.

The distortions shown in Figure 10c and 10d are not uncommon in reports or in advertisements. The irregular spacing, that is unequal time intervals, in Figure 10c might be taken to mean that the scale-maker does not want to show what happened during the omitted periods. Even the dates placed at right angles to the axis make this irregularity harder to detect.

The distortion shown in Figure 10d also heightens or accents a trend. The fact that the graph looks either quite narrow or extremely long may be an indicator of this distortion. A visual appraisal should show whether the relationship of the overall dimensions (horizontal to vertical) is approximately one to one. If it is not, the display may contain a shrunken scale.

It is not safe to assume that users will work at reading a diagram or at checking the reasonableness of statistical calculations. We should get the facts and give the facts as clearly as possible.

Figure 10 / Population growth in a metropolitan area

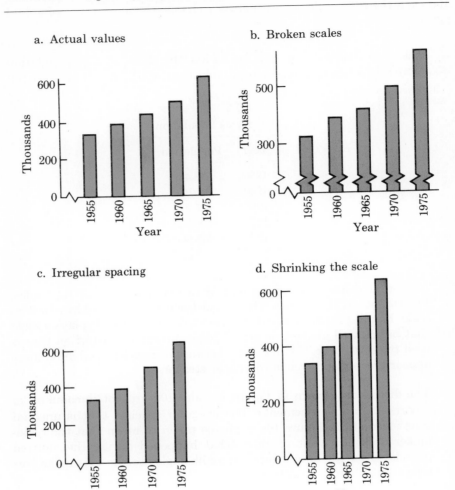

REFERENCES

[1] Stephen K. Campbell, *Flaws and Fallacies in Statistical Thinking* (Englewood Cliffs, N.J.: Prentice-Hall, 1974).
[2] *Encyclopedia Americana,* vol. 22 (New York: Americana Corporation, 1974).
[3] Darryl Huff, *How to Lie with Statistics* (New York: Norton, 1954).
[4] John Ingram, *Introductory Statistics* (Menlo Park, Ca: Cummings, 1974).
[5] Arthur Lockwood, *Diagrams* (New York: Watson-Guptill, 1969).

References

[6] Research Department, Orlando Area Chamber of Commerce, *Statistical Data* (Orlando, Florida: Orlando Chamber, 1972).

[7] George W. Snedecor and William G. Cochran, *Statistical Methods,* sixth edition (Ames, Iowa: Iowa State University Press: 1967).

[8] Mary E. Spear, *Practical Charting Techniques* (New York: McGraw-Hill, 1969).

[9] *Statistical Abstract of the United States, 1975* (Washington, D.C.: U.S. Department of Commerce, Bureau of the Census, 1975).

[10] Statistical Office, Department of Economic and Social Affairs, *Statistical Yearbook,* 8th, 17th, and 26th Issues (New York: The United Nations, 1956, 1965, and 1974).

Description from Grouped Data

CHAPTER 3

The numerical calculations in Chapter 2 used small sets of around ten observations. The computations of mean, standard deviation, and other descriptive measures used the raw scores. For data sets even slightly larger, calculations using these individual values become long and tedious so description for larger data sets often begins with grouping or classing the data. Subsequent calculations by "grouped data" procedures use less arithmetic to give the same descriptive statistics as the earlier "raw score" forms.

1/ Grouping Data

The frequency distribution is our principal vehicle for grouping. The essential characteristics are displayed in Table 1. The first column gives

Table 1 / A Frequency Distribution of Children for 40 Families

Number of children	Frequency (number of families)	Relative frequency
0	5	5
1	10	10
2	16	16 / 40
3	5	5
4 or more	4	4
	$n = 40$	$40/40 = 1$

the name of the variable, or its symbolic representation, and a list of possible values. The second column indicates the number of individuals or things in the class or grouping. The relative frequencies, if included, show what fraction the group represents from the total.

Which apples make the best cider? We can be sure of getting consistent results only if we sort and arrange by certain characteristics. Sorting and arranging by specific characteristics makes both apples and statistical data easier to use.

Description from Grouped Data

Grouping condenses the data. Ideally this allows visual appraisal of measures of central tendency and dispersion. Also it speeds hand calculation of the mean and standard deviation values. On the other hand, grouping into classes results in a loss of individual identity. For example in the class "4 or more" in Table 1, it is uncertain whether the class has four families each having four children, or if some have five children or more. Description resulting from this class is at best approximate. In the end the convenience of using grouped procedures should be weighed against any potential inaccuracies from its use.

In Table 1 the variable includes only a small number of values; 0, 1, 2, 3, and 4 or more. In some experiments the possibilities are so numerous that classes must be described by intervals of values. This may be the case if our sample includes even 20 or so observations. A good analogy for grouping is the preparation of apples for use. Initially the individual apples are picked, then they are sorted, say by size; after that they are readied for processing, bulk distribution, or some other use. Grouping scores follows a similar procedure and, after grouping, both apples and scores are manageable—more easily used. Use the following steps to group integer-valued (whole number) observations.

1. Determine the smallest and the largest values in the set. Compute their difference, the *range,* which then defines the span that includes all of the observations. To insure that the classes will include both extremes, it is necessary to add one unit of measure to the range.

2. Generally about six to fifteen classes are used. The number of classes can be determined by arbitrarily choosing a class width. This tells the span of values for any class. Division by this number into the range + 1 "unit" gives the approximate number of classes. The quotient should be rounded *up* to the nearest integer so that all scores are included. Numbers frequently used for class width are .1, .2, .5, 1, 2, 5, or 10.

3. After classes have been formed make a count (tally) of the number of scores in each class and display in either graphical or tabular form. This is the frequency distribution.

Table 2 / Number of Times Individuals Were Observed Littering

0	1	6	2	9	0	0	7	2	3
4	7	3	10	0	7	1	5	3	6
4	17	0	8	1	3	3	13	0	0

The data in Table 2 lacks order so the scores are first ordered by size, as in our analogy to the apples, and then placed into a frequency distribution. The range equals the highest value minus the least value, or $17 - 0 = 17$ (see the boxed numbers in Table 2). Arbitrarily choosing a class width of 3, we need to divide that into the range + 1 "unit", or $3\overline{)17 + 1} = 6$ classes. (*Note:* A process of trial-and-error led me to a workable class width of three with six classes for $n =$ thirty observations.) The classes are shown in the first column of Table 3. For example, zero (0) is tallied in the first class, 1 in the first class, 6 in the third class, and so on. The tally column

Table 3/ Frequency Distribution for Number of Times Individuals Were Observed Littering

Class	Tally	Frequency	Relative frequency
0–2	⊪⊪ ⊪⊪ 11	12	12/30 or .40
3–5	⊪⊪ 111	8	8/30 or .27*
6–8	⊪⊪ 1	6	6/30 or .20
9–11	11	2	2/30 or .07
12–14	1	1	1/30 or .03*
15–17	1	1	1/30 or .03
		$n = 30$	1.00

*.2666, a repeating decimal, is rounded to .27; similarly .0333 becomes .03.

is used for locating each value in its appropriate class. Note that for a class width of three, six is the minimum number of classes required in order to include all values. This is shown in that the extreme values of the sample 0 and 17, are the extreme class limits. The first class has limits of 0 and 2, and the limits for the remaining classes follow in logical order.

The relative frequencies in column four of Table 3 were obtained by dividing the correct class frequency by "n." This column should total one (i.e., 100%). In some cases rounding off numbers may produce a total that is slightly more or slightly less.

Many statistical questions require a logical count of the values included within some interval, e.g., three *or more,* five *or less, at least* twelve, *over* three, etc. The frequency distribution can be used to answer numerous "How many?" or "What percentage?" types of questions about counts. Several are posed for the littering data. You should check the answers using the information given in Table 3.

The frequency distribution in Table 3 can be used to answer all but one of the following:

Description from Grouped Data

Question   How many persons littered more than twice?
Answer   "More than twice" includes possibilities 3, 4, ..., 17, so using Table 3, $8 + 6 + 2 + 1 + 1 = 18$ persons littered more than two times.

Question   How many littered two or fewer times?
Answer   "Two or fewer" includes the possibilities 0, 1, and 2, hence the answer includes the twelve persons in the class 0–2.

Question   What percentage of these thirty persons littered between zero and five times, inclusive?
Answer   Using relative frequencies for the classes 0–2 and 3–5 gives us $.40 + .27 = .67$ or 67%.

Question   What percentage did not litter (i.e., littered zero times)?
Answer   This question cannot be answered by the frequency distribution alone. However, going to the original observations in Table 2, there are seven readings of zero. The answer is $7/30 = .23$, or 23%.

The last question illustrates how grouping hinders our ability to accurately answer some questions. Yet the original observations, if available, will provide the necessary information. In deciding whether or not to group observations one should weigh the importance of accuracy against ease in computation. In any case, a sample of ten or fewer observations should *not* be grouped.

A slight variation in procedure is required for grouping numbers that include decimals. These differences are illustrated by setting a frequency distribution on the rate of profit (in percents) on invested capital for twenty firms. See Table 4. Profit rates have been rounded off to one decimal place

Table 4 / Profit Data (Percentage) on Invested Capital for $n = 20$ Firms

16.9	12.1	13.0	10.4	17.3
17.5	17.6	15.6	17.7	14.5
17.4	17.3	15.0	12.6	16.6
16.3	18.1	22.2	15.0	18.3

so the basic "unit" of change is 0.1 (percent). Grouping follows the steps outlined on page 38. We begin with an arbitrary choice for class width. This is worth some thought, though, because a judicious choice here can simplify later calculations that will use the class width. Trial-and-error selection of a class width should lead to a reasonable value, one that gives six to fifteen classes. Choosing a class width of 2.0 gives us $2\overline{)(22.2 - 10.4) + .1}$ = $2\overline{)11.9}$ = 5.95, or six classes.* One should always round up (here to six classes) since rounding down would not provide enough distance to include all of the observations. The frequency distribution and related histogram appear in Figure 1. Notice that the choice of ten for the first lower limit leads to a requirement of seven classes. See Figure 1a. This is a choice of convenience for if, the classes had been set 10.4–12.3, . . . , 20.4–22.3, then only six classes would be used. Either will give reasonable description.

Class boundaries (see Section 2.3) are used to display the profit rate classes in Figure 1b. The class limits, 10.0–11.9, 12.0–13.9, . . . , 22.0–23.9, would serve equally well. The main criterion for a choice is visual clarity.

Figure 1/ Displays of profit rate on invested capital for $n=20$ firms

a. As a frequency distribution

Profit rate	Frequency
10.0–11.9	1
12.0–13.9	3
14.0–15.9	4
16.0–17.9	9
18.0–19.9	2
20.0–21.9	0
22.0–23.9	1
	$n = 20$

b. As a histogram

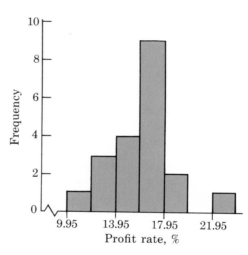

---

*It is noteworthy that six classes with a class width of 2.0 is not the only acceptable format. Clarity of display and convenience for further calculations should be the major concerns in directing your choice.

Problems

1. For the sample distribution that follows:
   a. what are the class limits for the class 3.00–5.99; what are its class boundaries?
   b. what is the class width?
   c. if the two most extreme values were 3.08 and 14.65, what is the range of values?

Class interval	Frequency
3.00–5.99	1
6.00–8.99	4
9.00–11.99	10
12.00–14.99	6

2. A sample of 120 spools of fishing line produced by a single machine yields a smallest breaking strength of 3.1 pounds and a biggest breaking strength of 18.6 pounds. Using a class width of 2.0, calculate the appropriate number of classes and display all class intervals.

3. Set a frequency distribution for the scores given below. (Suggestion: Use 3 as the lower limit for the first class.)

   Count scores

8	24	19	17	21
19	15	20	17	14
7	3	27	26	16

4. Among a group of thirty women at a party, fifteen are married, eight are single, four are divorced, and three are widowed.
   a. Make a frequency distribution based on the marital status of these women.
   b. Convert your frequency distribution into a bar graph.

5. This table below describes the weekly take-home pay for twenty employees of a single firm. Using the classes $200.00–209.99, $210.00–219.99 ... $240.00–249.99:

a. set up a frequency distribution table and plot an appropriate histogram.
b. find how many employees are taking home $220.00 or more?
c. find how many are taking home $220.00–$239.99, inclusive?
d. find what percentage of employees are taking home at least $240.00 per week?

$219.12	$226.81	$224.79	$213.48
231.50	248.39	237.24	226.67
220.76	221.67	231.47	226.67
225.42	215.38	204.12	221.19
233.98	240.89	227.02	227.11

6. The following table shows twenty-five dime store sales (in cents) to children ten years of age or younger. Treating them as whole numbers, make a frequency distribution.

56	76	63	35	53
47	62	68	42	49
82	65	45	56	56
45	46	53	57	59
63	66	50	58	67

7. For the frequency distribution given:
a. what are the class boundaries for the class 60–69?
b. how many of the scores are between 40–49, inclusive?
c. if the actual values of the scores in the class 50–59 are 51, 53, 54, 58, what percentage of the scores are less than 55?
d. what percent of the scores are over 50?

Class	Frequency
30–39	2
40–49	6
50–59	4
60–69	12
70–79	21
80–89	7
90–99	8
Total	60

Description from Grouped Data

8. Develop a frequency distribution table for these fifty milk prices: 69, 80, 78, 80, 70, 75, 80, 80, 80, 76, 78, 78, 80, 80, 80, 75, 80, 72, 78, 75, 80, 80, 82, 80, 80, 78, 80, 76, 70, 70, 80, 75, 80, 80, 78, 75, 70, 75, 80, 70, 80, 80, 80, 78, 80, 75, 78, 80, 75, and 80 cents per half gallon. What percent of the stores were charging 75 cents or less for each half gallon of milk? How many were charging between 70 and 75 cents, inclusive?

2/ Numerical Description

When observations are grouped the exact values for the mean and standard deviation cannot be obtained, but we can find approximations using a slight modification of earlier procedures. The major change is that the class mark is used to approximate the value for all observations in a class; the actual values are *not* used.

Rule

Approximation of the mean from grouped observations uses

$$\overline{X} = \frac{\sum_{}^{k}(Xf)}{n} \quad \text{where} \quad n = \sum_{}^{k} f \quad \text{with}$$

$f =$ class frequencies
$X =$ class marks
$k =$ the number of classes.

This procedure gives an approximation to the value of the mean. The first example uses the littering data introduced earlier.

Example

Using Table 5 and the rule,

$$\overline{X} = \frac{\sum_{}^{6}(Xf)}{30} = 135/30 = 4.5.$$

This value seems reasonable since the concentration of frequencies is in the first three classes.

Values for the other common averages can be approximated from the frequency distribution. The mode is rather easily approximated.

Table 5 / Approximation of the Mean Number of Times Individuals Were Observed Littering

Class	X	Frequency, f	Class values, (Xf)
0–2	1	12	12
3–5	4	8	32
6–8	7	6	42
9–11	10	2	20
12–14	13	1	13
15–17	16	1	16
		$n = 30$	135

Rule

The mode is approximated from a frequency distribution by using the class mark of the class having the highest frequency.

For the data in Table 5 the class 0–2 has the highest frequency so the mode is approximated by using the class mark "1." Finding the approximation for the median is presented in the next section.

The approximation of values for the sample variance and standard deviation extend the procedures used on ungrouped (raw) scores. Since the identity of the individual observations is lost in grouping, the class mark is used to approximate the value for everything in a class. The procedure is outlined in a rule, then illustrated.

Rule

For grouped data the sample variance, s^2, and the standard deviation, s, are approximated from the frequency distribution using

$$s^2 = \frac{\sum_{}^{k}[(X - \bar{X})^2 \cdot f]}{n - 1} \quad \text{and} \quad s = \sqrt{s^2} \text{ where}$$

X = class marks
f = class frequencies
k = the number of classes
n = the total number of scores.

This rule is very much like that in Section 2.2 which describes s^2 for calculation from raw scores. The present rule requires one additional step to accommodate the frequencies resulting from grouping.

Example

Variance and standard deviation are approximated for the littering data. Recall that $\bar{X} = 4.5$.

Description from Grouped Data

Table 6 / Approximating Variance and Standard Deviation for the Number of Times Individuals Were Observed Littering

Class	Marks, X	Frequencies, f	$(X - 4.5)$	$(X - 4.5)^2$	$[(X - 4.5)^2 \cdot f]$
0–2	1	12	-3.5	12.25	147.00
3–5	4	8	-0.5	.25	2.00
6–8	7	6	2.5	6.25	37.50
9–11	10	2	5.5	30.25	60.50
12–14	13	1	$13 - 4.5 = 8.5$	72.25	72.25
15–17	16	1	11.5	132.25	132.25
		$n = 30$			$\sum = 451.50$

Using the information from Table 6 gives

$$s^2 \doteq \frac{\sum_{}^{6}[(X - 4.5)^2 \cdot f]}{30 - 1} = \frac{451.50}{29} \doteq 15.57, \text{ and } s = \sqrt{15.57} \doteq 3.9$$

(For a discussion about using Table VIII to approximate square roots, see Appendix A, beginning on page 358.) A coarse check on s using the rule in Section 2.2 gives $s \approx \text{range}/4 = (17 - 0)/4 = 4.25$. The accurate value, 3.9, is reasonable.

Example

A second example illustrates the computational procedures. It also demonstrates the correct round-off procedure for decimal numbers.

Suppose we want the descriptive values $\bar{X}$, the mode, s^2, and s for a distribution of bicycling times in a marathon race, but the individual times (actual observations) are not available. See Table 7.

The numbers in Table 7 give

$$\bar{X} \doteq 1018/50 = 20.36, \quad \text{mode} \doteq 19.00$$

$$s^2 \doteq \frac{\sum_{}^{7}[(X - 20.36)^2 \cdot f]}{50 - 1} = \frac{531.5200}{49}, \text{ and } s = \sqrt{10.8473} = 3.29.$$

If nothing else this example illustrates the need for calculators or computers for data treatment. "Coding" can greatly simplify these calculations. See [5].

Table 7 / Cycling Times, in Hours, for Contestants in a Marathon Race with Calculations for Numerical Description

Class (hours)	Marks X*	Frequency f	(Xf)	$(X - 20.36)$	$(X - 20.36)^2$	$[(X - 20.36)^2 \cdot f]$
14.00–15.99	15.00	4	60.00	−5.36	28.7296	114.9184
16.00–17.99	17.00	9	153.00	−3.36	11.2896	101.6064
18.00–19.99	19.00	12	228.00	−1.36	1.8496	22.1952
20.00–21.99	21.00	10	210.00	.64	.4096	4.0960
22.00–23.99	23.00	7	161.00	2.64	6.9696	48.7872
24.00–25.99	25.00	5	125.00	4.64	21.5296	107.6480
26.00–27.99	27.00	3	81.00	6.64	44.0896	132.2688
		$n = 50$	1018.00			531.5200

*Marks are rounded off, e.g., $(14.00 + 15.99)/2 = 14.995$ is rounded off to 15.00 for ease in calculation.

A NOTE TO THE INSTRUCTOR: For those who prefer to give meaning to the concept of standard deviation early, you will have little loss of continuity by giving Sections 6.2 and 6.3 next, and then returning to Section 3.3 (Percentiles). However, in Sections 6.2 and 6.3 students will need to be alerted to the population symbols for the mean μ, and the standard deviation, σ.

Problems

1. Given the following frequency distribution, compute approximate values for $\overline{X}$, s^2, s, and the mode.

Class	Frequency
0–2	9
3–5	4
6–8	11
9–11	20
12–14	34
15–17	14
18–20	8

Description from Grouped Data

2. The hourly wages for a sample of electricians were collected and organized into a frequency distribution.
   a. Compute values for $\bar{X}$ and s. Approximate the mode.
   b. Determine what percentage of these electricians earn at least $8 per hour? Less than $8 per hour? Between $6 and $10 per hour inclusive of $6, but not of $10?

Hourly rate*	Frequency
$2.00–$3.99	3
4.00–5.99	11
6.00–7.99	20
8.00–9.99	10
10.00–11.99	6
	$n = 50$

   *Use class marks $3, $5, etc.

3. The weekly accident rates from selected firms gave the number of injuries per 1000 man-hours of work. Compute approximations for s^2 and s using $\bar{X} = 3.4$. Also approximate the mode.

Number of injuries	Frequency
1.5–1.9	1
2.0–2.4	3
2.5–2.9	8
3.0–3.4	13
3.5–3.9	15
4.0–4.4	7
4.5–4.9	2
5.0–5.4	1
Total	50

4. The annual travel expenses for 200 executives are distributed as shown in the table below. Plot a histogram illustrating this distribution and guess the value of the mean. Then compute the approximate mean value to see how close you were. Does the mode reasonably approximate your value for $\bar{X}$?

Percentiles

	Annual expenses	Number
	$1,000.0 up to $1,200.0	16
	1,200.0 up to 1,400.0	43
	1,400.0 up to 1,600.0	98
	1,600.0 up to 1,800.0	31
	1,800.0 up to 2,000.0	12
		200

Handwritten annotations:
10–12, 12–14, 14–16, 16–18, 18–20

10 – 11.5   1100
12 – 13.5   1300
14 – 15.5   1500
16 – 17.5   1700
18 – 19.5   1900

5. Use the frequency distribution in Problem 8, p. 44, to approximate the mean price and the unit deviation (s) for the 50 prices for half gallons of milk. The values determined by summing the ungrouped prices are $\overline{X} = 77.4$ and $s = 3.5$.

3/ Percentiles

Percentiles and percentile ranks are essential statistics in education, psychology, and in related disciplines that try to interpret the results of tests. No doubt most of us have been subjected one or more times to a college entrance exam, or similar test, and subsequently were presented not only with the raw scores, but with the equivalent percentile ranks. Let's explore how these numbers are computed and how to convert between raw scores and ranks.

The percentile is an extension of the median, or middle divider. Earlier we described the median as a value that divides the observations so that at least half are less than or equal in value to the median, and at least half are equal to or greater in value than the median. Other names for the median are second quartile point (Q_2), fifth decile point (D_5), and fiftieth percentile point (P_{50}). All of these terms denote the same positional value.

Just as the median splits a set of observations into two parts, the three quartile points divide a set of observations into four quarters. Thus a value higher than the third quartile point is in the fourth, or upper quarter. Similar division by nine decile points splits a group into parts of ten; percentile points make division into parts of one hundred, called percents.

Computation of the median and other percentiles from grouped data uses an extension of the histogram. The data in Table 8 is used to illustrate (1) visual approximation of percentiles and then (2) their quantitative determination.

Description from Grouped Data

Table 8/ SCAT (aptitude) Test Scores for a Class of 125 Students

Scores	Frequency	Cumulative "less than" Frequencies	
40–49	4	Less than 40	0
50–59	11	Less than 50	4
60–69	23	Less than 60	15
70–79	47	Less than 70	38
80–89	31	Less than 80	85
90–99	9	Less than 90	116
	$n = 125$	Less than 100	125

The cumulative "less than" frequencies in Table 8 indicate that no one scored less than 40, only four persons achieved less than 50, etc. The cumulative distribution is plotted in Figure 2 with points plotted above the breaks, the boundary values. Notice that the total frequency, 125 scores, is accumulated through 99.5 (i.e., the class 90–99); all persons achieved less than 100 on this test. Percentile ranks are visually estimated (possibly with the aid of a ruler) from Figure 2. Consider the median, or P_{50}. Since there are 125 scores, we go up the vertical scale to $125/2 = 62.5$ for the 50th percentile. Using dashed lines we then construct a rectangle, two sides of which are

Figure 2/ Cumulative "less than" frequency polygon (ogive) for SCAT test scores

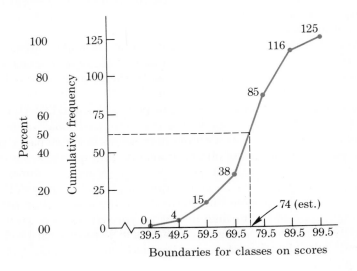

Percentiles

the positive axes. By observation the projection touches the horizontal or $X-$ axis at about 74. This approximates the median; $P_{50} \doteq 74$. Reversing this process leads from the value 74 to an approximate percentile rank of 50. That is, project from 74 on the horizontal scale (perpendicular to the ogive) and then to the vertical scale, getting 50 (%). Accurately scaled graph paper should be used if you want any precision in such approximations.

Calculation of the median and other percentiles uses an approximating procedure based on the class intervals. See Table 8. To illustrate, we will find the median. This is the fiftieth percentile, a value where the number "less than" the frequency is 62.5. Realizing that the class 70–79 contains this "less than" value gives a first approximation, 69.5. The question remains, "How far into the class must one go to achieve the median?" Here we make the assumption that within any class the values are dispersed evenly. Then, upon entering the class 70–79 using lower values, we have surpassed 38 scores. We must still go beyond 69.5 by $62.5 - 38 = 24.5$ scores of the 47 in the interval. But this is 24.5/47 parts of an interval with a width of $80 - 70 = 10$ units. Therefore the median value adds $(24.5/47) \cdot 10 = 5.2$ to 69.5, giving $P_{50} = 74.7$. These procedures are summarized by a rule.

Rule

Determination of percentiles (from the class that contains the nth percentile) is

$$P_n = L + c.w. \cdot \left(\frac{d}{f}\right)$$

where
$L =$ the lower boundary
$c.w. =$ the class width
$d =$ the "distance" in frequency count from entry to the percentile value
$f =$ the frequency of that class.

Follow through the next example. Use the rule or the first example as a guide.

Example

Suppose we seek the value at the 90th percentile for the SCAT scores. Using Figure 2 we can make a first approximation of about 86. (You should check this). Using $125 \cdot .90 = 112.5$, Table 8 shows P_{90} is contained in the class 80–89. Eighty-five scores are less than 80. Then

$$P_{90} = 79.5 + 10\frac{(112.5 - 85)}{31} = 79.5 + 8.9 = 88.4.$$

You will be asked to find other percentiles using the SCAT data in Problem 1 of this section.

Description from Grouped Data

Percentile ranks, rather than the raw scores, are usually reported for standardized test results. The process of determining percentile ranks is just the reverse of finding percentiles. For example, in Figure 2 we are now reading a known raw score from the horizontal scale and converting this to a comparable percentile rank on the vertical (percent) scale. The percentile rank determination is a conversion from a raw score to a percentage value. To illustrate, we have just seen that $P_{90} = 88.4$ is the raw score at the 90th percentile on the SCAT distribution. Conversely, 88.4 should have percentile rank 90 (rounded to the nearest percent). The following rule outlines the procedure. (*Note:* This is very similar to the preceding rule but with the roles of scores and frequencies reversed).

Rule

The percentile rank for a raw score taken from a frequency distribution is determined as

$$\text{Percentile rank} \doteq \frac{\text{(Approximate) cumulative frequency}}{n} \cdot 100$$

with the

$$\text{(Approximate) cumulative frequency} = \left[F_c + \left(\frac{D}{c.w.}\right) \cdot F_s\right]$$

where
 $F_c =$ the cumulative frequency to entry of the class which contains this score
 $D =$ the "distance" in raw score units from entry to this score
 $F_s =$ the frequency for the class that contains this score
 $c.w. =$ class width.

Unfortunately, with grouped data an *exact* number of scores below any raw score cannot be determined, so we must approximate the cumulative frequency. In Table 8, the entry nearest to, (but below) 88.4 for which an exact cumulative frequency is known is the "less than 80" entry. This class begins at 79.5, so we associate this value with the cumulative frequency $F_c = 85$. Next we determine the distance from the entry, 79.5 to the known score, $D = 88.4 - 79.5 = 8.9$ units. Then divide this by a class-width value of 10 units. If we assume that the frequencies are evenly dispersed through this class,

$$\text{cumulative frequency} \doteq 85 + \left[\frac{(88.4 - 79.5)}{10} \cdot 31\right] = 85 + \frac{8.9 \cdot 31}{10}$$

$$= 112.59.$$

(Here 31 is the frequency count for the class 80–89.) Then the percentile rank for 88.4 is, using the rule and with $n = 125$, $(112.59/125) \cdot 100 = 90.07$.

Again $P_{90} \doteq 88.4$. Using Figure 2, the percentile rank can be traced, approximately, by reading the value projected from the horizontal scale through the ogive to the vertical scale.

As a check to make sure you understand, follow the percentile rank determination for a SCAT score of 75. (Recall that $P_{50} = 74.7$ so the answer should be just over 50%.) Using Table 8, the cumulative frequency $\doteq 38 + (75 - 69.5/10)47 = 63.86$, so percentile rank $\doteq (63.85/125) \cdot 100 = 51(\%)$.

Percentile ranks from a test battery are commonly preferred over high school records as a measure of potential; the former allows comparison with a norm group as well as with others, classmates, who have taken the same test. High school grade averages generally reflect numerous undefined variables, including the relative difficulty of courses in that high school, the marking standards of individual teachers, etc. You might find it interesting to visit with a counselor to check the demonstration of percentile ranks for some of your own entrance scores.

Problems

1. Using the distribution of SCAT scores in Table 8, determine values for P_{31}, P_{55}, Q_2, D_9 and Q_3. Check your answers by reading the approximate values from Figure 2.

2. Use Table 8 to help you find the percentile ranks for the raw scores 55, 64, 73, and 92 on the SCAT test. Use Figure 2 to check (approximate) the values you compute.

3. For the following distribution of scores:

Interval	Frequency
61–80	2
81–100	23
101–120	36
121–140	66
141–160	43
161–180	26
181–200	4
	$n = 200$

Description from Grouped Data

   a. construct a cumulative "less than" ogive similar to that in Figure 2.
   b. approximate the median, i.e., P_{50}, using the ogive you just constructed, then compute P_{50} and compare the values you obtained.
   c. find the values of P_{25} and P_{75}.
4. Using the distribution in Problem 3, determine percentile ranks for the raw scores 95, 121, 125, and 190. The figure you constructed for Problem 3a can provide a quick check on your values.
5. For the following ogive:

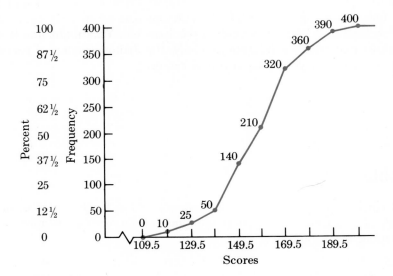

   a. construct a frequency distribution, including the cumulative "less than" frequency table, similar to that in Table 8.
   b. estimate values for the median and for the percentile rank for a score of 175.
   c. evaluate your answers to part b above by using the frequency distribution you constructed for part a and the rules.

4/ Weighted Means and Index Numbers

As the number of observations increases, so does the work involved in computing their mean. Consider the computation of a college student's term grade-point average. Under the four-point grading system, each hour of A work gets four quality points, B three, C two, D one, and an F zero quality points.

To determine the grade-point average, we total the quality points, then divide by the total hours. Using the data in Table 9,

Example

Table 9/ A Record of Academic Achievement

Courses	Grade	Grade points	· Hours =	Quality points
Economics	B	3	3	9
Humanities	B	3	2	6
Statistics	B	3	4	12
Biology	C	2	4	8
Sociology	A	4	3	12
			Totals: 16	47

$$\text{Grade average} = \frac{\sum(\text{Quality points})}{\text{Total hours}} = \frac{47}{16} = 2.94.$$

This is a *weighted mean*, $\bar{X}_W$.

The weighted mean is determined by

Rule

$$\bar{X}_W = \frac{\sum_{}^{k}(WX)}{\sum_{}^{k} W}$$

where
 $X =$ values for the variable
 $W =$ weights
 $k =$ the number of classes.

Description from Grouped Data

For the example, X is grade points while the weight, W, is hours. In this case the weighted mean, $\overline{X}_W = 2.94$, gives exactly the same value as would ungrouped procedures. The quality points assigned to the grade for Economics using ungrouped, or raw scores, is $3 + 3 + 3 = 9$, which is precisely the same as that obtained using three 3s, $3 \times 3 = 9$. That is, rather than adding the individual grades for each hour of course work, the total quality points for that class is obtained as the product of the class grade times the credit hours. This is equivalent to finding the mean for grouped data, but with the frequencies replaced by weights.

It may be hard to decide which variable is X and which is W. This potential dilemma can be resolved by answering the question "What is being averaged?" The answer defines X. Through the process of elimination we then find the weights, W.

Example

An individual has three small credit loans outstanding in the amounts of $100 at 12%, $50 at 18%, and another of $50 at 21%. If we seek the average rate of interest then X becomes percents (%) and weights, W, are dollar amounts. The average of the interest rates, is

$$\overline{X}_W = \frac{\$100 \cdot .12 + \$50 \cdot .18 + \$50 \cdot .21}{\$100 + \$50 + \$50} = .1575, \text{ or } 15.75\%.$$

Although most of us have used weighted means in the form of grade averages, other forms are common, especially in business work.

A special class of weighted averages, called *index numbers,* has wide use in business and economics. These include common lists such as the Consumer Price Index, Gross National Product, New York Stock Exchange readings, and the prime rate of interest. In general index numbers are devices for comparing sizes of groups of two or more related quantities. In its simplest form, index numbers are comparisons of two things such that a relative change is expressed by giving one number as a percentage of the other; or stated most simply, as the ratio of two numbers.

Example

The U. S. population in 1970 was 203,000,000 people; in 1975 it was 215,000,000. That is, the population in 1975 was $(215/203) \cdot 100 = 105.9\%$ of what it was in 1970. During this five year period our population had increased by 5.9% over the 1970 value.

An index number is obtained by combining two or more variables (e.g., prices for several food items) in total or by averages. Several related index numbers are then set in an *index* with each number expressed as a percentage of a convenient base. The most familiar example is probably the Consumer Price Index.

Weighted Means and Index Numbers

Example

The Consumer Price Index (CPI) is a monthly index prepared by the U. S. Bureau of Labor Statistics. It is based on prices for about 400 items covering foodstuffs, clothing, shelter, medical expenditures, utility rates, etc. The current base year is 1967, i.e., assumes an index value of 100(%). A partial index covering the six years, 1970–1975, appears in Table 10 [3]. For example, relative to the base period,

Table 10 / A Partial Consumer Price Index Covering 1970–1975, Base = 1967

Year	CPI value
1970	116.3
1971	121.3
1972	125.3
1973	129.8
1974	147.7
1975	159.3

1967 = 100.0*, the price of consumer goods, overall, had increased in 1975 by 59.3% in 8 years!

The CPI is one of the most widely used of indexes. Many union contracts contain an "escalator clause," in essence an automatic adjustment of wages based on changes in the cost of living as measured by the CPI. For example, using Table 10, under such a clause a worker earning $2.00 per hour in 1970 would, without requesting it, receive $(121.3/116.3) \cdot \$2.00 \doteq \$2.09$ per hour in 1971.

Most indexes relate to either (1) price, (2) quantity, or (3) value. The type is determined by the purpose for the comparison, e.g., for dollar sales (price), for production rates (quantity), or for value of goods (value).

Although there are many index computing forms, most stem from two basic types—aggregate or relatives. We will do calculations for each type of form for a common basket of goods. See Table 11.

---

*"base 1967 = 100.0(%)" is not an equality in the sense of a mathematical equality. It is a fairly common notation used in business statistics and economics to indicate the standard for an index is the year (or period) for which the index number is 100.0.

Description from Grouped Data

Table 11 / A Basket of Foods with Periodic Prices, 1975

Item	Quantity	March 21* P_0	April 21 P_1	May 23 P_2
Bread	Kentucky Farm (20 oz.)	$ .57	$ .57	$ .57
Soft drinks	16 oz. Pepsi (8-pack)	1.57	1.67	1.57
Cereal	Kellogg's Frosted Flakes (10 oz.)	.61	.61	.57
Milk	one-half gallon	.90	.90	.80
Hamburger	1 lb. lean beef USDA Choice (3 lb. container)	1.18	1.18	1.09
	Totals	$4.83	$4.93	$4.60

*The zero subscript, in P_0, indicates that the base is March 1975 prices. All prices were obtained at the same supermarket.

Rule

A simple aggregate price index is determined by

$$I_n = \frac{\sum P_n}{\sum P_0} \cdot 100.$$

Then for the "basket" in Table 11,

$$I_0 = \frac{\$4.83}{\$4.83} \cdot 100 = 100(\%) \text{ for March (base)},$$

$$I_1 = \frac{\$4.93}{\$4.83} \cdot 100 = 102.1\% \text{ for April, and}$$

$$I_2 = \frac{\$4.60}{\$4.83} \cdot 100 = 95.2\% \text{ for May.}$$

For this basket of food in the given quantities, the price rose 2.1% from March to April, then fell in May to 4.8% below the March base.

Users claim two major defects in the aggregate index form. The first is that little consideration is given to the relative importance of the items, and the second is that the units (here of prices) for individual items will affect the index. As a matter of convenience we assumed the quantities shown in Table 11, but, for example, is a box of cereal consumed as often as a half gallon of milk? If not, other more realistic weights should be used. The result is a "weighted" index form.

The effect of the second defect, units, can be that one or a few items with higher prices, if the prices are unstable, can dominate the entire index. For example, an index involving items of food, clothing, and rent might be dominated by the amounts for rent, generally a high percentage among these items. This last problem is overcome by using *relatives*, here price relatives, P_n/P_0.

The simple relatives price index is

$$I_n = \frac{\sum \left(\frac{P_n}{P_0}\right) \cdot 100}{k}, \quad \text{where}$$

Rule

$k =$ The number of items.

Reusing the numbers from Table 11 and the simple relatives form gives,

$I_0 = 100\%,$

$$I_1 = \frac{\left(\frac{.57}{.57} + \frac{1.67}{1.57} + \frac{.61}{.61} + \frac{.90}{.90} + \frac{1.18}{1.18}\right) \cdot 100}{5} = 101.3\%, \text{ and similarly}$$

$I_2 = 94.9\%.$

It should be apparent that the different procedures, aggregate versus relatives, lead to somewhat different index values. Moreover, the selection of items, the location and source of prices, errors in recording prices, selection of a base period, and numerous other factors affect an index. Kruskal [1] points out a number of interesting problems in attempting to make valid comparisons of food prices.

Index numbers are a part of all of our lives. Price movements as indicated by the stock and grain markets, by the CPI and other consumer indices, by Gross National Product figures, by unemployment figures, by prime interest rates, etc., guide decision-making not only in business and government, but for most individuals as well. For a more extensive study of this topic, including calculations with numerous weighted index forms, see [4].

REFERENCES

[1] William H. Kruskal, "The Cost of Eating" in *Statistics by Example, Exploring Data,* edited by Frederick Mosteller, et al. (Reading, Mass.: Addison-Wesley, 1973).
[2] Arthur Lockwood, *Diagrams* (New York: Watson-Guptill, 1969).
[3] *Statistical Abstract of the United States–1975* (Washington D.C.: U.S. Department of Commerce, 1976).
[4] Taro Yamane, *Statistics,* second edition (New York: Harper-Row, 1967).
[5] John Ingram, *Introductory Statistics* (Menlo Park, Ca.: Cummings, 1974).

Problems

1. Compute the grade-point average (on a four points = A scale) for the student whose course record is given:

Courses	Grade	Hours
Speech	B	3
History	B	4
Math	C	3
Science	C	4
Music	A	2

2. Compute your grade-point average for the past academic term. As an alternative use the grades from your last term in high school.

3. a. Determine the average rate of interest for these three loans: $50 at 12%, $100 at 18%, and $50 at 21%.

   b. What is the total dollar amount of interest charged for one year on these three loans?

4. Determine the mean interest rate (%) for these four loans: $100 at 12%, $40 at 16%, $50 at 18%, and $80 at 21%.

5. Give an index number showing the change in annual income for a person who earned $10,000 last year and $10,800 this year. Explain the meaning of your number. If the cost of living, as measured by the CPI, has gone up 8% during this same period, has his "buying power" increased or decreased? Explain.

6. Use the data in Table 10 and the idea of the "escalator clause" to determine the hourly wage in 1975 for a person who earned $4.08 per hour in 1974.

7. Convert Table 10 to an index with base = 1970 by dividing each entry by 116.3, then multiplying by 100. Explain the meaning of the new value for 1974.

8. Given the following basket of goods with prices:

   a. develop a simple aggregate price index (I_0, I_1, I_2).

b. develop a simple relatives price index.
c. interpret the meaning of your values for I_1, first for the aggregate price index and then for the relatives price index.

Item	P_0	P_1	P_2
Milk, half gal. 2%	.78	.82	.81
Eggs, 1 doz. large	.65	.69	.75
Flour, Gold Medal, 5 lb.	.55	.61	.69
Skippy Peanut Butter, 12 oz.	.41	.47	.47
Totals	2.39	2.59	2.72

9. Using the basket of foods given below:
   a. develop a simple aggregate price index (I_0, I_1, I_2).
   b. develop a simple relatives price index.
   c. explain the meaning of your values for I_2 for the aggregate price index and then for the relatives price index.

Item	P_0	P_1	P_2
Heinz Catsup, 14 oz.	.21	.26	.24
Hunts Tomato Sauce, 8 oz.	.13	.14	.11
Campbell's Pork and Beans, 1 lb.	.17	.18	.15

5/ Summary for Chapters 2 and 3

Descriptive tools include measures—averages and measures of variation—and graphical presentations. Raw score calculation of descriptive measures is appropriate for sets of about ten or fewer observations. Techniques were discussed for computing three averages—the arithmetic mean, the median, and the mode. Common measures of variation include the range and the standard deviation. The midrange and the range give quick, but coarse, checks on the descriptive measures.

In Chapter 2 we viewed graphical displays as data pictures. These include histograms, bar graphs, line charts (frequency polygons), pictograms, and pie charts. The emphasis was on reading and using data displays. Clear titles, scales, labels, and legends are necessary for usable data pictures.

Description from Grouped Data

The frequency distribution in Chapter 3 is a means for displaying larger sets of 20, 50, 100, or even more observations. Specific guides were given for constructing frequency distributions. The frequency distribution can serve as the basis for numerous graphical displays as well as a "grouped data" base for approximating values for the sample mean and the standard deviation. Answers to "How many?" and "What percent?" questions are forerunners to the probability techniques to be presented in Chapter 4.

The evaluation of standardized testing scores in the form of percentiles and percentile ranks use a number of descriptive tools. The concept of averages is extended to the weighted mean. Numerous averages, including the "aggregate" and "relatives" forms, are used in business indices such as the Consumer Price Index.

In Chapters 2 and 3 we have discussed statistical description. Two measures—the mean and the standard deviation—will be used in other work, and the frequency distribution will be expanded to include probability forms. Chapter 4 concerns the fundamentals of probability, including several specific types of probability experiments.

Review Problems

1. For the quiz scores 9, 13, 7, 16, 17, and 10, compute values for $\overline{X}$, the median, mode, s^2, and s. Make checks on your values for $\overline{X}$ and s.

2. Suppose you are the payroll manager for a firm that currently employs 600 people. From data for this same week of last year, the median wage was $236, the mean was $250, and the modal wage was $212. Assuming that positions and wages are somewhat stable, except for an overall wage increase of 5%, how much money is needed to pay these employees for this week?

3. The times required by a jogger to complete her usual run were 21.0, 18.0, 16.5, 17.2, and 17.3 minutes, respectively, for five consecutive days. Compute values for $\overline{X}$, the median, the range, s^2, and s.

4. Records from a computerized dating service gave the following number of repeated dates for computer-matched couples.

    a. Plot the data as a bar graph.

    b. Determine how many of the couples repeated dates at least one time? Three or more times?

c. Explain why an accurate value could not be obtained for the mean.

Number	Frequency
0	28
1	12
2	21
3	3
4 or more	36
	$n = 100$

5. For the observations -1.5, -2.3, -3.4, -1.8, and 0, compute values for $\overline{X}$, the median, the range, s^2, and s.

6. Two loans are each for $25 at 12% interest, while a third is for $50 at 15%. What is the average (mean) rate of interest paid on these three loans?

7. For the following frequency distribution, approximate $\overline{X}$, s^2, s, the median, and the mode:

Grade average	Frequency
1.8–2.1	4
2.2–2.5	6
2.6–2.9	4
3.0–3.3	8
3.4–3.7	3

8. For the following distributions of exam scores indicate which (A, B, neither, or both distributions) satisfies the statements:
   a. has the larger grade average
   b. has over 50% of the scores below 80
   c. has the smaller variance

Statistics	Distribution	
	A	B
Mean	83	80
Median	80	82
Standard deviation	4	6

Description from Grouped Data

9. For this frequency distribution, approximate $\overline{X}$, s^2, and s values. Estimate the mode:

Class	Frequency
16–21	15
22–27	16
28–33	5
34–39	5

10. The results of 50 tosses of a die (one of a pair of dice) appear below.
    a. Find $\overline{X}$, s^2, and s.
    b. Determine how many times an even number was observed. Next divide by the sum of the frequencies, and then multiply by 100. Finally, explain the meaning of the number you have just found.

Observation	1	2	3	4	5	6
Frequency	8	9	7	8	8	10

11. Following is a line graph intended to display sales over years. Explain how you would improve (change) the diagram. Redraw the diagram, incorporating your suggestions.

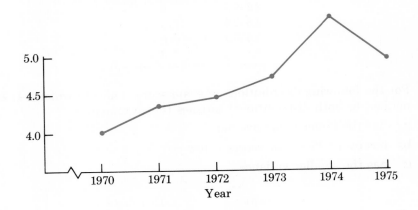

12. Following is the distribution of nightly earnings, including tips, for Mike, a waiter at Freddie's Steak House:
    a. Compute $\overline{X}$, s^2, s, and modal values. Based on the distribution (for three months equals 60 work days), approximately what total earnings might Mike expect for the year?

b. Determine on what percent of the days he made $27.50 or more?

Amount	Frequency (days)
$12.50–17.49	2
17.50–22.49	10
22.50–27.49	15
27.50–32.49	29
32.50–37.49	3
37.50–42.49	1
Total	60

13. Below is a frequency polygon showing a person's height over the years. Use it to approximate answers for the following. Answer the questions by inspection; you need not perform any calculations. (*Suggestion:* Use a ruler.)

   a. At what age was this person 40 inches tall?
   b. What was his height at age 6?
   c. What was his median height over the years 0 to 20?
   d. At what age was his median height attained?

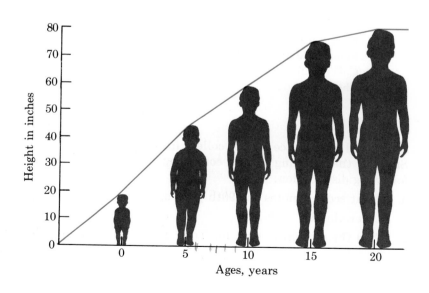

Description from Grouped Data

14. Suppose the Consumer Price Index rose from 150 to 155 between January and June. If you work under an "escalator clause" and were making $2.38 per hour in January, what should be your hourly wage in June?

15. Suppose a Wholesale Food Price Index lists the following:

Month	Value
January	180
April	186
August	192

    a. Convert this to an index with a base value of 100 in January.
    b. By how much (%) have food prices increased between January and April?

16. For the basket of goods given below, set a price index having a January base value = 100 using:
    a. simple aggregates.
    b. price relatives.
    c. Compare the resulting indices.

Item	January	February	March
Eggs, 1 doz. large	.55	.65	.69
Flour, Gold Medal, 5 lb.	.55	.59	.61
Coffee, Hills Bros., 1 lb.	.81	.82	.84
Cheerios, 10 oz.	.61	.63	.63

17. A national intelligence test for college freshmen is being standardized. Suppose the norming group, consisting of 500 students, gave the following distribution:
    a. What score denotes the 90th percentile?
    b. Determine $Q_3 = P_{75}$.
    c. Find the percentile rank for Tom, who scored 135 on this test.
    d. Mary's score is 162. Which quarter—1, 2, 3, or 4—includes her score?

Scores	Frequency
100–109	7
110–119	17
120–129	23
130–139	62
140–149	120
150–159	140
160–169	64
170–179	42
180–189	18
190–199	7
Total = 500	

18. Construct a "less than" cumulative frequency polygon (ogive) like the one in Figure 2, Section 3.3, but use the data given in Problem 17 above. Use your figure to check (approximately) each answer to Problem 17.

19. In amateur competition, a member of the U. S. Olympic team has run the 100 meter dash in 11.2, 11.3, 11.5, and 11.2 seconds. Compute the arithmetic mean for these times. Explain why the arithmetic mean would likely not be the measure used to describe her speed.

Probability Concepts

CHAPTER 4

Games, gods, and the laws of chance, Jimmy the Greek, Reno and Las Vegas—these are all names that bring to mind one thing—gambling. Most of us have spent very little time or money at the blackjack tables, playing roulette wheels, or shaking hands with one-armed bandits, but their mention brings an air of excitement. History has much evidence of man's interest in chance, in casting lots, in "horse trading," in the superstitions of some religions, and national beliefs. Today that interest is shown by the massive amounts spent at race tracks, casinos, on lotteries, and other games of chance. Unfortunately very few can consistently "beat the house" (see [1]), and most of us cannot expect to make a living by foretelling events of the future. Since probability goes far beyond games, our purpose will be to look at a few of the fundamental principles underlying the laws of chance.

Whether it's business, politics, the stock market, the weather, or just practical matters in our daily lives, we generally have less than complete information for making decisions. Frequently our decisions require an appraisal, either subjectively or objectively, of the possibilities of chance. And we need to be able to scrutinize the phrase, "The probability is . . . ." So we begin by defining *probability* and some basic properties of chance occurrences.

1/ Describing Experiments

We start with the basic units and procedures. For example, in order to add $2 + 4$ to get 6 one must first know the meaning of the symbols 2, 4, and 6

How can we determine our chances of getting the flavor of gum ball we want before putting a coin into the machine? If there are equal numbers of only two kinds, we are equally likely to get either flavor; but the mix rarely contains equal numbers!

and the meaning of the addition operator (+). Similarly, to discuss probability concepts we need to learn the basic elements and operators from which probabilities are formed. This is the objective of this section.

The vehicle for obtaining data is an experiment.

Definition   An EXPERIMENT is any well-defined process from which observations can be obtained.

The process of the experiment can be quite simple, such as matching coins, or quite complex, such as a science experiment requiring rigid levels of pressure, temperature, force, etc. The essential feature of a statistical experiment is that the outcome at any trial cannot be predetermined. An element of the unknown precludes fixed outcomes.

An experiment is defined by the possible outcomes called *simple events*. These will be symbolized by capital letters, $E_1, E_2, \ldots, E_n$.

Definition   A SIMPLE EVENT is an essential outcome resulting from a single trial of an experiment.

The following illustrates an experiment and some of its possible simple events.

Example   Mark, who is student teaching, is grading spelling papers for a class of 30 pupils. For his work the simple events are:
$E_1 =$ he grades Susy C's paper
$E_2 =$ he grades Billy J's paper
$E_3 =$ he grades Tommy Z's paper
.
.
.
$E_{30} =$ he grades Julie B's paper.

That is, $E_1, E_2, E_3, \ldots, E_{30}$ are the simplest elements in the experiment. These are the simple events. Usually we are interested in collections of these simple events. For example, $\{E_1, E_2\}$ is a collection that represents Mark grades Susy C's paper and that he grades Billy J's paper. The braces, { }, indicate a collection of one or more simple events. This is a *compound* event.

Statistical experiments require one more component. This is the chance of occurrence for the simple events. For example, in Mark's experiment it's possible that the papers are all mixed up, so the order of grading is essentially a random selection. If so, we could use statistics to determine the chance (probability) for each simple event, that is, the chance that any paper is graded. Then, with a few operational rules we could also compute probabilities for compound events.

Numerous questions can be asked about an experiment. Then it is most helpful to be able to describe all of the simple events. A complete list of simple events is described in a *sample space*.

A SAMPLE SPACE, S, is the collection of simple events for an experiment.   Definition

Of the several methods used to display sample spaces, we will discuss three: rectangular coordinates with ordered pairs, Venn diagrams, and tree diagrams.

A first display uses rectangular coordinates and ordered pairs as described in Appendix A, p. 363. This experiment relates times at bat and number of hits for one player; that is, these simple events are defined by two characteristics—times at bat and number of hits. See Figure 1. The experiment includes 15 possibilities expressed as simple events: $E_1, E_2, \ldots, E_{15}$. One description for these events requires a pair of *ordered* values, the first coming from the horizontal scale and the second from the vertical scale. For example the event $E_1 = (0, 0)$ indicates the player had zero (no) hits and no times at bat. Similarly $E_7 = (1, 2)$ indicates one hit in two times at bat. Let's assume that all possibilities for times at bat are described in Figure 1.*

Figure 1 / Describing a baseball experiment

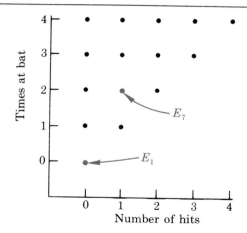

---

*It is conceivable that a player could have five or even more times at bat, but Figures 1 and 2 exclude those possibilities. We assume this player has at most four times at bat. Then Figure 1 indicates only those possibilities that are considered a part of this experiment.

Probability Concepts

Now numerous questions might occur to baseball fans. For example, how does one describe the event A this player bats four times, or the event B that he gets three or more hits, or the event C that his number of hits equals his times at bat? These are *compound events*, that is, collections of one or more simple events. See Figure 2. Compound events A, B, and C, can be described by the *ordered pairs*, as

$A = \{(0,4), (1,4), (2,4), (3,4), (4,4)\}$ the event he bats four times
$B = \{(3,3), (3,4), (4,4)\}$ the event he gets three or more hits
$C = \{(0,0), (1,1), (2,2), (3,3), (4,4)\}$ the event his hits equal his times at bat.

A compound event puts restrictions on the experiment, hence the space. For example, event A (four times at bat) restricts the outcomes to only five possibilities, those for which the second number in the ordered pair is 4. Observe how this is described in Figure 2.

Figure 2/ Graphical display of compound events

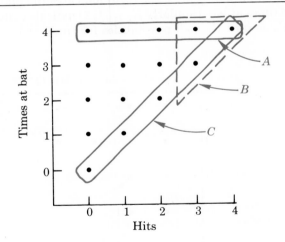

Now that we have defined the elements of our system—the simple events—we should identify event operations. Observe in Figure 2 that compound events can overlap. This commonality of shared points for two or more events is called their *intersection*. Thus A intersect B, A and B, includes only (3,4) and (4,4) from all the simple events in Figure 2.

Definition

The INTERSECTION of two events is an event that contains those simple events and only those which are in both.

Describing Experiments

The connective "and" identifies the intersection operation. Then A and B = {(3,4), (4,4)} is the event that this player bats four times *and* gets three or more hits. Another A and C = {(4,4)} is the event he bats four times *and* on each he gets a hit.

A second event operation, the *union,* signals the accumulation of simple events.

> The UNION of two events includes all simple events that belong to either one or both of the events.

Definition

The connective "or" indicates a union. Then from Figure 2, A union B, that is,

A or B = {(0,4), (1,4), (2,4), (3,4), (4,4), (3,3)}   while
B or C = {(3,3), (3,4), (4,4), (0,0), (1,1), (2,2)}.

Notice that a simple event, once included in the union, is not duplicated. For example consider (4,4) in A or B. The union specifies A *or* B *or* both, that is, at least one of the events. Here the player bats four times *or* gets three or more hits *or* does both.

One final event operation is the *complement.* The event that this player had other than four times at bat is the complement to event A and is symbolized A'. This includes possibilities of zero, one, two, or three times at bat. All are possibilities where the second number in the ordered pair is *not* 4. From Figure 2,

A' = {(0,0), (0,1), (0,2), (0,3), (1,1), (1,2), (1,3), (2,2), (2,3), (3,3)}.

A useful property of the complement is that an event A, in union with its complement, is the sample space $(A$ or $A') = S$.

> The COMPLEMENT of event A, symbolized A', is an event that contains all simple events in the space that are not in event A.

Definition

These, then, are the basic event operations—intersection, union, and the complement.

A second way to describe experiments uses Venn diagrams and the compound events. See Figure 3.

Figure 3 shows the total space as the area within the rectangle. Various compound events are described in a manner convenient for answering numerous probability questions. The space is separated into four *mutually exclusive* events, that is, into regions that do not overlap. In Figure 3 these are labeled A' and B', A and B', A and B, and A' and B. This separation or partition simplifies the process of enumerating the simple events that satisfy certain descriptions. Several examples follow.

Figure 3/ A general Venn diagram

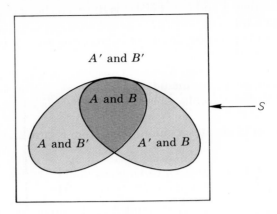

Example

For the baseball experiment with,

$A =$ the player bats 4 times, and
$B =$ he gets 3 or more hits   (as in Figure 2).

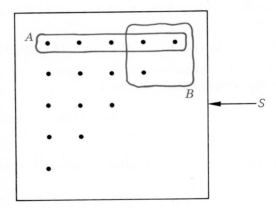

Now some "How many?" type questions can be answered by observation. See Figure 4. The answer to "How many simple events are shared by A *and* B?" is by count $N(A$ and $B) = 2$. Here $N(\ )$ denotes the number of simple events in the set. There are two possibilities whereby he might get three *or more* hits in four times at bat, that is, either three hits or four hits. As ordered pairs A and $B = \{(3,4), (4,4)\}$ so $N(A$ and $B) = 2$. Other counts include,

$N(A') = 10$ $N(A' \text{ and } B) = 1$
$N(B') = 12$ $N(A' \text{ and } B') = 9$
$N(A \text{ or } B) = 6$ $N(A \text{ and } B') = 3$

Check these using Figure 3 as a guide. Let each dot represent a simple event.

The disadvantage of the Venn diagram compared to a rectangular display is the loss of identity for simple events. This may be of little consequence if each has equal value, but can be a real loss if some simple events occur more often than others.

The full Venn display for the baseball experiment appears in Figure 4.

Figure 4/ A Venn diagram for the baseball experiment

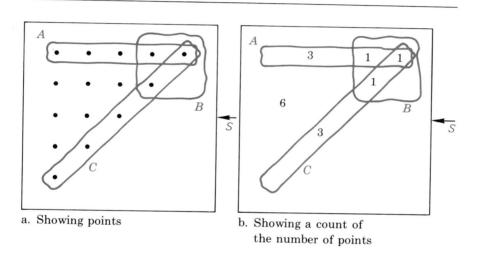

a. Showing points

b. Showing a count of the number of points

From this we get: $N(A \text{ and } B) = 2$, $N(A \text{ and } C) = 1$, $N(A \text{ or } B) = 6$, and $N(A \text{ and } C)' = 14$. The Venn diagram will be used for probability calculations where we count how many simple events are in a compound event.

The *tree diagram* gives us a third way of displaying the possible outcomes for an experiment. This display gets its name from its similarity to a tree with branches emanating from a single trunk.

> A rock group plans to schedule concerts in Chicago (C), Milwaukee (M), and St. Paul (S). What are the possibilities if we consider all possible orders of appearance? There are six. The branches display exclusive possibilities and are read beginning from the trunk (left) and moving along any branch to its tip. This tree has six branches,

Example

Probability Concepts

[tree diagram: First stop, Second stop, Third stop with branches C→M→S, C→S→M, M→C→S, M→S→C, S→C→M, S→M→C]

the first being Chicago to Milwaukee to St. Paul—C, M, S. The next is Chicago to St. Paul to Milwaukee—C, S, M. Each branch, from trunk to tip, describes a simple event.

The tree diagram is useful for describing experiments that require evaluation at regular intervals of time and for which an irregular pattern of action can occur. A time frame is set, then the branches describe a flow of action.

Example

In the preceding example, suppose the musicians decided to schedule Chicago as their second stop, even if their first stop is Chicago. Further, if the first and second stops are both Chicago, the third stop will be at Milwaukee. Otherwise a stop could be at any of the three cities. The tree diagram takes the pattern shown in the figure. There are seven possibilities. Observe the irregular pattern. A Venn diagram or rectangular description might not be so easily understood.

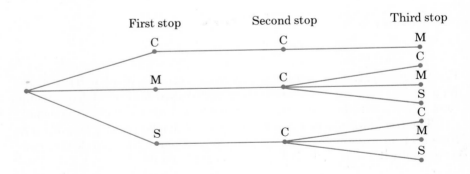

These devices—rectangular displays and ordered pairs, Venn diagrams, and tree diagrams—should be treated as alternative ways to visualize possible outcomes for an experiment. Frequently more than one can be used as visual aids to probability calculations.

or — union
and — intersection

Problems

1. Let $A = \{1\}$, $B = \{1, 2\}$, $C = \{2, 3, 4\}$, and S = sample space $= \{1, 2, 3, 4\}$. List all numbers (simple events) that satisfy the following:

   a. A or B b. A and B c. B or C d. B' e. $(A$ or $B)'$

2. For $S = \{1, 2, 3 \ldots, 10\}$, $A = \{1, 2, 3, 4, 5\}$, $B = \{3, 6, 9\}$, and $C = \{5, 10\}$, what is:

   a. A or B b. A and B c. B or C d. B' e. $(A$ or $B)'$

3. Following is a display of degrees offered for selected majors at State University:

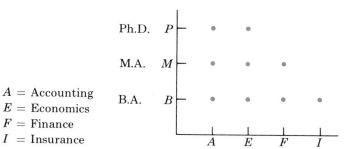

A = Accounting
E = Economics
F = Finance
I = Insurance

T = event the major is accounting = $\{(A, B), (A, M), (A, P)\}$.

   a. List ordered pairs for

   U = event the student is working on a master's degree
   V = event the student is an undergraduate.

   b. Describe each of the following by listing all ordered pairs that satisfy the event: $(T$ or $U)$, V', $(U$ or $V)'$, $(U$ or $V')$.

   c. Describe each of the following compound events using event operations and the symbols T, U, and V:

   $\{(A, M)\}$; $\{(A, B), (A, M), (A, P), (E, B), (F, B), (I, B)\}$; $\{(A, P), (E, P), (A, B), (E, B), (F, B), (I, B)\}$

   d. Are events T and U mutually exclusive? How about U and V?

Probability Concepts

4. An experiment consists of tossing two regular, four-sided objects (tetrahedra) and observing the two numbers, either 1, 2, 3, or 4, appearing on the down faces. One object is red, the other green.

a. Using the rectangular coordinate system as in Figure 1, describe the space. Record the numbers on the red object on the horizontal scale (the abscissa).

b. Describe each of the following events in words:

$A = \{(2, 4), (3, 3), (3, 4), (4, 2), (4, 3), (4, 4)\}$
$B = \{(1, 1), (1, 2), (1, 3), (1, 4), (3, 1), (3, 2), (3, 3), (3, 4)\}$
$C = \{(2, 2), (4, 4)\}.$

c. Describe each of the following by listing all ordered pairs that satisfy the compound event: $(A \text{ or } B)$; B'; A and C; A and C'.

d. Describe each of the events in part c in words.

e. Are events A, B mutually exclusive? What about B with C?

5. Given the Venn display: Determine $N(S)$, $N(A)$, $N(A \text{ and } B)$, $N(A \text{ or } B)$, $N(A \text{ and } B')$, $N(A' \text{ or } B')$, and $N(A \text{ or } B)'$.

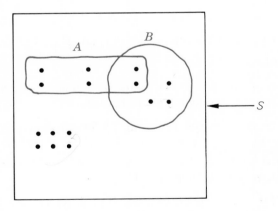

Problems

6. Find $N(S)$, $N(B)$, $N(A \text{ or } C)$, $N(A \text{ and } B \text{ and } C)$, and $N(B \text{ and } C')$ for the Venn display with the number of points indicated:

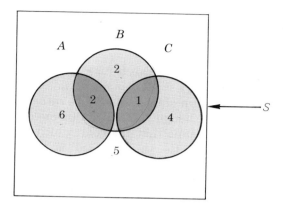

7. A sales representative wants to visit Atlanta, Chicago, Detroit, and New York during a one-week trip. Make a tree diagram displaying the 24 different ways that he could order his four stops. Assume he will visit a city no more than one time.

8. Give a statement in words that describes this tree diagram. Suppose the tree describes stops at Minneapolis (M), Cincinnati (C), and St. Louis (S), and each stop is a visit to a hospital in that city.

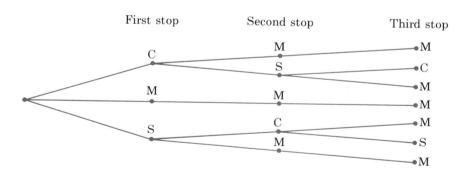

2/ Probability Rules

In statistics we seek patterns that can be used to describe chance experiments. For example by tossing a single coin many times, one probably would not observe an orderly pattern of heads and tails, rather they would occur more or less at random. Although a specific outcome cannot be foretold, it may be reasonable to assume that the coin, which has two sides, will show tails about one-half of the times. Based on this assumption, one might approximate the chance that the next toss would show tails as one-half. This illustrates the *equal likelihood definition of probability*.

Rule

If a chance experiment can give any of $N(S) = N$ equally likely and mutually exclusive simple events, and if $N(A)$ of these satisfy event A, then event A has probability

$$P(A) = \frac{N(A)}{N}.$$

As it is used here, "equally likely" means that for each simple event E, $P(E) = 1/N$. Intuitively then, this rule is suitable for "equal likelihood" games such as coin or dice tosses, lotteries, and even some casino games.

Example

The chance that a single toss of a "fair" coin shows tails is one-half, that any one of the six faces on a well-balanced die is the up-face is one-sixth, and that any one of n identical tickets in a lottery is chosen is $1/n$.

Imposing the equal likelihood definition of probability on the baseball experiment described earlier assigns a chance of 1/15 to each of the 15 simple events. Since A (event he bats four times) includes five simple events—$N(A) = 5$—then $P(A) = 5 \cdot (1/15) = 5/15$, or 1/3. Recalling that B is the event he gets three or more hits, while C requires that he hit each time that he bats, then

$$P(B) = \frac{N(B)}{N} = \frac{3}{15} = .20; \quad P(A \text{ or } B) = \frac{6}{15} = .40;$$

$$P(A \text{ or } C) = \frac{9}{15} = .60; \quad P(C') = \frac{N(C')}{N} = \frac{10}{15} = .67;$$

$$P(A \text{ and } B) = \frac{N(A \text{ and } B)}{N} = \frac{2}{15} = .13; \text{ and } P(A \text{ or } C') = \frac{11}{15} = .73.$$

Probability Rules

Figure 5/ The baseball experiment with equal likelihood probabilities

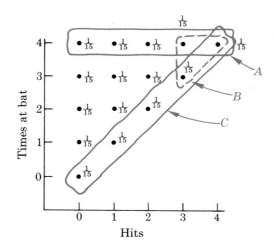

Check these results by counting the number of points in Figure 5 that satisfy each event, then multiply their number by 1/15. This gives the same value as adding $1/15 + 1/15 + \cdots + 1/15$ for the number of simple events in each compound description.

The equal likelihood definition is limited because most real experiments produce other than an equal chance for all outcomes. Under these circumstances we are obliged to derive the chance of occurrence by other means, commonly by experimentation or observation. Rather than assuming a coin is "fair," one approach is to repeatedly toss the coin and count the number of tails. This leads to a *relative frequency* estimation of probabilities.

Example

Repeated tossing of a coin showed the following ratio of tails to number of tosses: 6/10, then continuing 23/50, then 48/100, then 508/1,000, then 5,123/10,000, . . .

At some point after a sufficiently large number of tosses, a stable value is approached for the probability of tails. From the given sequence the empirical probability is about 51% that the next toss will be tails.

Rule

Empirical (relative frequency) probabilities are

$$P(A) = \frac{f_A}{n}$$

where

Probability Concepts

f_A = the frequency of occurrence of simple events in A
n = some large number of trials.

This procedure can be applied to the baseball experiment. In fact, it is *most* unlikely that the experience of any baseball player is that of Figure 5. Suppose the experience for one player, based on 400 games, is given in Figure 6. The probabilities are determined empirically from his experience by counting the number of times for each simple event, then dividing by 400. For example, $P\,[(1, 2)] = .06$ comes from the empirical information that in 24 different games this player had one hit in two times at bat so $P\,[(1, 2)] = 24/400 = .06$.

Figure 6/ The baseball experiment with empirical probabilities

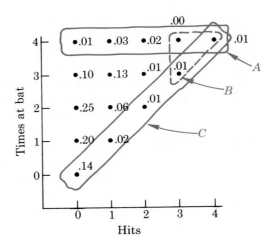

Using the empirical probabilities from Figure 6, and the events A = he bats 4 times, B = he gets 3 or more hits, and C = his number of hits equals his times at bat, $P(A) = P[(0, 4), (1, 4), (2, 4), (3, 4)\,(4, 4)] = .01 + .03 + .02 + .00 + .01 = .07$. And similarly in Figure 6, $P(B) = .02, P(C') = .81$, $P(A \text{ or } B) = .08$, $P(A \text{ and } B) = .01$, $P(A \text{ or } C) = .25$, and $P(A \text{ or } C') = .82$. The assignment of probabilities to the simple events is quite fundamental. Note the empirical assignment in Figure 6 is substantially different from the equal likelihood probabilities in Figure 5.

The equal likelihood and empirical probability rules are two ways of assigning probability values to the simple events. There are others. Subjective probabilities express the odds for and against an event. The language is common in racetrack betting.

Example

At post time during the 1975 Kentucky Derby, the odds on the winner, Foolish Pleasure, were four to one. That is, Foolish Pleasure was given an "odd" of four against winning the race to an "odd" of one for his winning the race. In other words, Foolish Pleasure was given an odd of one in $1 + 4 = 5$ for a $1/(1 + 4) = .20$, or 20% chance to win.

Subjective probabilities can be based on anything from long experience to hunches and guessing. There is no standard procedure for assigning probabilities to simple events.

Numerous procedures are used to assign theoretical probabilities to events in an experiment. Imagine a dart game where a record is made of the horizontal distance off-center for each dart. Conceivably, an expert dart thrower would hit near the center many times and fewer times for distances further removed from the center. Assuming this skilled person could toss forever, a theoretical probability pattern might appear similar to that in Figure 7. Since distances from the center might be associated with numbers on a real number line, the measure is continuous. So the "curve" in Figure 7 is smooth rather than a step-type function. This approximate pattern occurs quite frequently in real experience. It is called the *normal* or *bell curve* and will become central to our studies beginning in Chapter 6. Another game that illustrates theoretical probabilities appears in problem 4 of this section.

Figure 7 / A theoretical distribution of probabilities for distances off-center for dart tosses

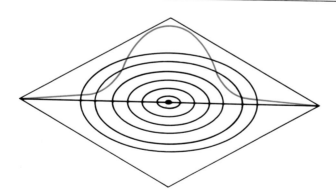

The point of all this is that "probability" can mean many different things. It is well worth one's time to find the basis for the probability assignment to simple events. For example, calculations for the baseball experiment from probabilities in Figure 5, and then in Figure 6 show that different assignments can lead to quite different conclusions.

Probability Concepts

Although some probabilities will be given, there is no guarantee that these will be given as precisely the values we need. So we use an algebra for probabilities.

The following are given as axioms, or assumptions:

AXIOM 1 $P(A) \geq 0$ for any event A defined in a sample space, S.
AXIOM 2 $P(S) = 1$.
AXIOM 3 $P(A \text{ or } B) = P(A) + P(B)$ for mutually exclusive events A and B.

An essential consequence of the first two axioms is that for any event A,
$$0 \leq P(A) \leq 1.$$

Rule

Probabilities *always* range between 0 and 1, inclusive.

Also, we have used Axiom 3 many times, for example, in finding $P(A) = P(E_1 \text{ or } E_2 \text{ or } E_3 \text{ or } E_4) = P(E_1) + \cdots + P(E_4)$. From the axioms it is possible to develop a general *rule of addition* for probabilities. However an intuitive approach is more appealing.

Rule

RULE OF ADDITION   For any events A, B, the probability that at least one of the events will occur is the probability of A plus the probability of B less the probability of their intersection:
$P(A \text{ or } B) = P(A) + P(B) - P(A \text{ and } B)$

Figure 8/ Venn display of the addition rule

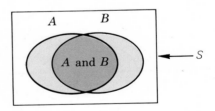

The addition rule is appropriate for finding the probability of event A or B or both. See Figure 8. Figure 8 shows that the intersection is a part of A; it is also a part of B. Then in the rule, $P(A) + P(B)$ counts this overlap two times. Once is enough, so the value of $P(A \text{ and } B)$ must be subtracted one time.

An example of problem solving follows. It uses the axioms and the addition rule. Be mindful that our rules are just expressions of logical reasoning, and in many problems just that—logic—will suffice.

Probability Rules

The probability that Mrs. Anderson will vote in a general election is 0.8, that Mr. Anderson will vote is 0.7. They have both voted in past elections 60% of the time. What is the probability that at least one member of this couple will vote in a forthcoming election?

Example

Let W = event the wife (Mrs. A) votes and
H = event the husband (Mr. A) votes.

Solution

This question asks for "at least one," that is, H or W (or both), so we want a value for $P(H$ or $W)$. What information is given? We know Mrs. Anderson votes 80% of the time, $P(W) = .80$. Mr. Anderson's voting record indicates $P(H) = .70$, and together $P(H$ and $W) = .60$. This information is sufficient for the addition rule, giving $P(W$ or $H)$ = $P(W) + P(H) - P(H$ and $W) = .80 + .70 - .60 = .90$. See the Venn diagram.

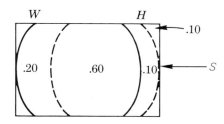

From this example we estimate that at least one of Mr. and Mrs. Anderson will vote ninety percent of the time. The approach here is an application of the procedures for working word problems from Appendix A, p. 366. In the preceding example, the first step was to identify each essential event, then attach relevant labels. Next the question was identified, and events expressed in probability notation. Taking care to correctly identify event relations, we recorded given probability values. Either (1) logical considerations or (2) application of probability rules to the "givens" leads to an answer. Frequently one approach can be used as a check for another approach. These steps lead to logical solutions for probability word problems. Try it on the several word problems that follow. The Venn diagram or other display can sometimes help one see event relations. Note that a probability value less than zero or greater than one indicates an error.

A second example using the addition rule also involves mutually exclusive events.

Events are said to be MUTUALLY EXCLUSIVE if they share *no* elements.

Definition

85

Probability Concepts

For mutually exclusive events A, B, we write $(A \text{ and } B) = \phi$ (the *null* or *empty event*); then $P(A \text{ and } B) = 0$. How this special relation affects the addition rule should be clear from the following example. Also observe how the steps for a logical solution (above) are carried out.

Example

The chance that Tom comes home for dinner is 0.7; the chance he eats dinner at a friend's home is 0.2; while the chance he eats elsewhere is 0.1. What is the chance that he will *not* eat dinner at home?

Solution

We want $P(H')$, where relevant events are:
$H = $ Tom eats dinner at home
$F = $ Tom eats dinner with a friend
$E = $ Tom eats dinner elsewhere.

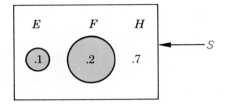

Consider a Venn display of the space of possibilities where the chance is recorded for each event. Here the events do not overlap; they are mutually exclusive. For example, $(E \text{ and } F) = \phi$. Then $P(E \text{ or } F) = P(E) + P(F) - P(E \text{ and } F) = .1 + .2 - 0 = .3$. But in fact $(E \text{ or } F)$ complements H on the space (collectively they fill up the space); that is, $(E \text{ or } F) = H'$. So the answer is $P(H') = .3$. As a check, $P(H') = 1 - P(H) = 1 - .7 = .3$, using Axioms 2 and 3.

Problems

1. Over the past ten years Professor Barr has missed only six class days. What is the probability that he will miss any one class day during this year? Assume there are 150 class days per year.

2. For $S = \{0, 1, 2, 3, \ldots, 9\}, A = \{2, 4, 6, 8\}, B = \{0\}, C = \{0, 1, 2, 3, 4\}$, and assuming the classical definition of probabilities, find:
   a. $P(S)$ b. $P(A \text{ or } B)$ c. $P(B \text{ or } C)$ d. $P(A' \text{ or } C)$ e. $P(C')$

Problems

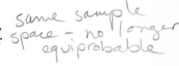

3. Using Problem 2 above, let $P(1) = P(3) = P(5) = P(7) = P(9) = 1/10$, $P(2) = P(4) = P(6) = P(8) = 1/20$, and $P(0) = 3/10$. Now answer parts a–e of Problem 2. Are any answers the same? Why or why not?

   Same sample space — no longer equiprobable

4. A game is played by spinning the arrow (see the figure below) and observing the number in the sector where the pointer stops. Spins where the counter stops on a line are respun. Assume the sectors have equal angles and the pointer stops in each sector about equally often.

For the compound event 1 = the pointer stops in a sector labeled 1, etc., find for any one spin:

a. $P(1)$
b. $P(2)$
c. $P(3)$
d. $P(2 \text{ or } 3)$
e. $P(2')$
f. $P(1 \text{ and } 3)$

5. Use the classical definition of probability and the results of Problem 4 in Section 1 of this chapter to evaluate each of the following:
   a. $P(A \text{ or } B)$
   b. $P(B')$
   c. $P(A \text{ and } C)$
   d. $P(S)$
   e. $P(A \text{ and } C')$

6. The following defy one or more of the probability axioms given on page 84. Tell which axiom(s) is violated.
   a. $P(A) = .3, P(B) = .8$ with A, B mutually exclusive.
   b. $P(C) = -.1, P(D) = .4$.
   c. For $S = \{E_1, E_2, E_3, E_4\}, P(E_1) = P(E_2) = P(E_3) = P(E_4) = 1/3$.
   d. $P(F) = .4, P(G) = .3, P(F \text{ or } G) = .8$ for F, G, mutually exclusive.

Probability Concepts

7. Using a Venn display, the equal likelihood definition, and the event descriptions in the problem, find probabilities for all events of Problem 5 in Section 1 of this chapter.

Use the systematic approach to solving word problems to answer the following questions.

8. The probability that Jon will buy his next sport coat at Burdines is .3, at the Toggery .5, and at the Pro Shop .2. What is the probability that he will buy at Burdines or the Pro Shop? What special event relation is used?

9. If of all marriages 10% result in a separation, 8% in divorce, and 4% in both separation and (later) divorce, what percentage of all married couples will have either a separation *or* a divorce?

3/ Probability Problems

In the last section we employed a systematic approach to problem solving. The process can be summarized by the following procedures:

1. Identify the essential descriptions and attach a relevant literal symbol to each. Write out the basic compound events, including the literal symbols, with a word description.

2. Find the statement of the problem. This is frequently the sentence that ends with a question mark. Using symbols and with appropriate event relations, express the question in probability form.

3. Reread the entire problem and select all information pertinent to the question. Quantify each piece of "given" information with proper event description and appropriate probability value. Write out the "givens" using probability notation. A visual display can be helpful.

4. Combine the "givens" either (1) by logical reasoning or (2) with probability rules and axioms to answer the questions.

5. Now look back through the solution. Check that you have:
   a. properly defined basic events
   b. correctly identified the question
   c. correctly interpreted the "given" information

d. correctly copied the given numeric values
e. taken a logical approach to solving the problem and
f. obtained a reasonable answer—remember that probability answers are between 0 and 1, inclusive.

A high percentage of the problems in this book are story or word problems. You should view this approach, spend some time analyzing each question, then follow through the steps for a solution. See if the visuals aid in your understanding. The discussion in Appendix A, p. 366 can be helpful, too.

The probability that Rex and Monica will eat at the Gridiron restaurant after a home football game is 0.90; for John and Lois the chance is 0.95; while both couples go there after 85% of the hometown games. If you go to the Gridiron after a hometown game, what is the chance you will see either one or both of these couples?

Example

The essential events are:
A = Rex and Monica eat at the Gridiron
B = John and Lois eat at the Gridiron.

Solution

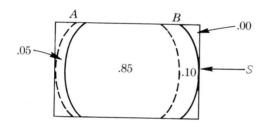

The statement of the problem (the last sentence) asks for "one or both," meaning "at least one." We want $P(A \text{ or } B)$. The "givens" include $P(A) = .90$, $P(B) = .95$, and $P(A \text{ and } B) = .85$. Again the addition rule or equivalent logic gives an answer,

$P(A \text{ or } B) = .90 + .95 - .85 = 1.0.$

You are certain to see at least one of these couples!

Another example involves the case of mutually exclusive events. Observe that realizing this condition simplifies our calculations.

In a psychology experiment guinea pigs are placed in a circular enclosure that contains three doors. Behind the first is a container of food, behind the second door is a container of water, and the third

Example

encloses an empty runway. Once behind a door the animal has no chance to return. Suppose for 400 guinea pigs that 210 went behind door 1, while 122 chose door 2, 56 chose door 3, and 12 did not leave the enclosure during the allotted time. What proportion got either food or water?

Solution

This time I have numbered the five steps for a systematic solution.

1. The basic events include:

   D_0 = guinea pig does not leave the enclosure

   D_1 = guinea pig enters door 1, food

   D_2 = guinea pig enters door 2, water

   D_3 = guinea pig enters door 3, empty.

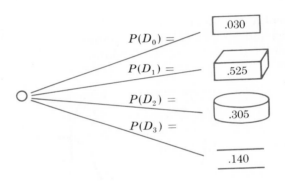

2. The question concerns those guinea pigs that receive food *or* water so we require $P(D_1 \text{ or } D_2)$.

3. Since we know the events D_0, D_1, D_2, D_3 are exclusive, then $P(D_1 \text{ or } D_2) = P(D_1) + P(D_2)$, using Axiom 3. We need only $P(D_1) = 210/400 = .525$ and $P(D_2) = 122/400 = .305$ using the given frequencies.

4. The solution is $P(D_1 \text{ or } D_2) = .525 + .305 = .83$.

5. This answer can be checked by evaluating the chance on $(D_0 \text{ or } D_3)'$, logically the chance for $(D_1 \text{ or } D_2)$. That is,

   $P(D_0 \text{ or } D_3) = 56/400 + 12/400 = 68/400 = .17$, so that

   $P(D_1 \text{ or } D_2) = 1 - P(D_0 \text{ or } D_3) = 1 - .17 = .83$ using Axioms 2 and 3.

Although the emphasis here is on probability calculations, you should realize the importance of having reasonable probability values for the basic events. That is, we can use correct logic and apply numerous probability rules, but an answer is worthless if the essential probabilities are substantially wrong.

Another widely used probability form concerns the intersection of two events. This is the *rule of multiplication*.

RULE OF MULTIPLICATION For two events A and B, the probability of their *intersection* (common occurrence) is the product

$$P(A \text{ and } B) = P(A) \cdot P(B/A),$$

with $P(B/A)$ = the conditional probability of B given A.

Or interchanging the roles of events A and B,

$$P(A \text{ and } B) = P(B) \cdot P(A/B).$$

Rule

The CONDITIONAL PROBABILITY of event B on the space of A, symbolized $P(B/A)$, is the probability of event B given that A has occurred. That is, event B is now confined to the subspace of S within event A.

Definition

The idea of *conditional* probabilities is described by an example. Consider an experiment in poker where each player is dealt five cards. If the 52 card deck is well-mixed, the chance at a first draw is 1/52 for each card, e.g., $P(\text{ace of clubs}) = 1/52$. Just suppose this is the first card dealt. Hoping to get a second ace (for a pair), each of the ace of diamonds, ace of hearts, or ace of spades has chance 1/51. Observe the reduction from 52 to 51 cards because the ace of clubs is gone. This change is described as $P(\text{ace of hearts/ace of clubs}) = 1/51$, where the slash (/) means *conditioned by* (or after) whatever follows.

A more general description of the conditional probability concept appears in this example.

Consider an experiment with equal likelihood for all simple events. We begin with a Venn display with $N(A) = 4, N(B) = 3, N(A \text{ and } B) = 1$,

Example

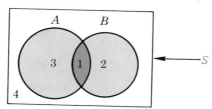

Probability Concepts

$N(S) = 10$. By count, $P(A) = 4/10$, $P(B) = 3/10$, $P(A \text{ and } B) = 1/10$, etc. For the conditional probability, $P(B/A)$, we consider the sub-experiment, hence the subspace, defined by A. Then B is described within A. Now $P(B/A)$ or the probability of B on the subspace of

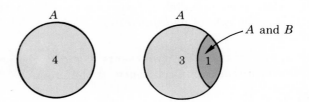

A is 1/4. That is, the conditional probability, here B conditioned by A, says the size of the experiment is reduced to the simple events in A. Of these four, one is shared with B. So the probability is 1/4.

One could take the posture that every event is conditioned or defined within some experiment, hence on some space. Then, for example, rather than writing $P(A)$, we would use $P(A/S)$. The former is preferred so long as it is clear that the chance for event A is in the context of a totally defined experiment. Otherwise restrictions that might affect the chance of event A happening should be indicated as a condition. What results is a conditional probability. Using the rule of multiplication requires our knowing one conditional probability value.

Example

There remain twenty gum balls in a penny gum machine. Of these, six are red, nine are yellow, and five are blue. Suppose Junior, who has 2¢, spends it all on this machine. Assuming he is paid off each time, what is the probability that both gum balls are red?

Solution

1. Let B_1 = event the first gum ball is red

   B_2 = event the second gum ball is red.

2. The question asks the chance that both are red, $P(B_1 \text{ and } B_2)$.

3. The information is given by displaying the possible contents of

| 6 red | 5 red |
| 14 other | 14 other |

20 balls at draw 1   19 balls at draw 2
(conditioned by B_1)

the machine. Assuming the classical definition, probabilities are obtained by counting $P(B_1) = 6/20$ and $P(B_2/B_1) = 5/19$.

4. Then by the rule of multiplication

$$P(B_1 \text{ and } B_2) = P(B_1) \cdot P(B_2/B_1)$$
$$= (6/20) \cdot (5/19) \doteq .30 \cdot .263 = .08.$$

5. Again, I check to be sure $0 \leq P(B_1 \text{ and } B_2) \leq 1$.

The next example, although more difficult, can be simplified with the aid of a tree diagram. Take care in reading the problem, then follow the steps for a systematic solution.

The incidence of color-blindness among males three to five years old in a certain population is 4 in 100, or .04. Of those males who are color-blind 99% will show positive (indicating color-blindness) on an initial screening test. While, of those who are not color-blind, one in fifty will show a positive test. What percentage of the boys (aged three to five) in this population will show a positive test?

Example

We let T = event of a positive test

C = event a boy is color-blind

C' = event a boy is not color-blind.

Solution

The given events with probabilities (in the order presented) are $P(C) = .04$, $P(T/C) = .99$, and $P(T/C') = 1/50 = .02$. $P(C')$ is $1 - P(C) = .96$.

The question asks $P(T) = ?$ A positive test can come from either of two sources: (1) those who are (2) those who are not color-blind. Symbolically $P(T) = P(T \text{ and } C) + P(T \text{ and } C')$. Notice that we are summing for two mutually exclusive events. These form two branches in a tree diagram; a positive test is described on each branch.

The question is answered by applying the multiplication rule twice, then applying the addition rule. That is,

For $C: P(C) \cdot P(T/C) = .04 \cdot .99 = .0396 = P(T \text{ and } C)$

For $C': P(C') \cdot P(T/C') = .96 \cdot .02 = \dfrac{.0192 = P(T \text{ and } C')}{.0588 = P(T)}$

Some .0588, or 5.88%, of all boys in this population who are aged three to five will show a positive initial screening test.

Conclusion

Probability Concepts

Several story problems are included in the section questions. The importance of logical checks cannot be overstressed. Remember that for probability questions an answer less than zero or more than one is *always* wrong! If encountered, check (1) your math, (2) assignment of event descriptions, and (3) that you are answering the right question.

Problems

1. Use the Venn diagram below and the equal likelihood definition to evaluate the following:

    a. $P(R)$
    b. $P(T')$
    c. $P(R \text{ or } T)$
    d. $P(R \text{ and } T)$
    e. $P(R/T)$
    f. $P(T/R)$

    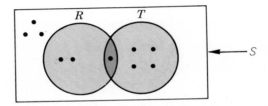

2. Given the mutually exclusive events A and B, for which $P(A) = 0.5$ and $P(B) = 0.3$, find:

    a. $P(A')$
    b. $P(A \text{ or } B)$
    c. $P(A' \text{ and } B)$
    d. $P(A/B)$
    e. $P(A' \text{ and } B')$
    f. $P(A \text{ and } B)'$

    (*Suggestion*: Use a Venn diagram and the rules of probability.)

3. Given $P(A) = .4$, $P(B) = .6$, and $P(A \text{ and } B) = .3$, indicate which of the following are true (T) and which are false (F):

    a. $P(A \text{ or } B) = 1$.
    b. A and B are mutually exclusive. That is, $P(A \text{ or } B) = P(A) + P(B)$.
    c. $P(B/A) = .75$.
    d. $P(A' \text{ and } B) = .10$.

4. Given $P(A) = .6$, and $P(B/A) = .5$, find $P(A \text{ and } B)$. Also find $P(A/B)$, given $P(B) = .5$.

For each of the following word problems use the systematic approach to problem-solving as outlined at the beginning of this section.

5. Of the 50 men in a certain club, 15 are under 21 years old and 37 work in salaried positions. Twelve are both under 21 *and* are salaried workers. What percent of this membership is under 21 *or* is salaried?

6. In a history class of 36 men and 24 women, 28 of the men and 20 of the women received grade C or better. If a student is chosen at random from this class, what is the probability that his or her grade is C or better, *or* is a woman?

7. A radio contains six transistors two of which are defective. If you replace two transistors, chosen by chance and without testing first, what is the probability that you will replace both defective ones?

8. Suppose the probability is 0.9 that your alarm goes off in the morning. If the alarm goes off, the probability is 0.6 that you will get up in time for your 8:00 class. If the alarm does not go off, the probability that you will get up in time is 0.1. What is the probability that you get up in time for your 8:00 class?

9. Art wants a new coat. Let

   A = event he buys at Ardan's wholesalers
   B = event he buys at Yonker's Men's Shop
   C = event he doesn't buy a new coat.

   Suppose $P(A) = .40$, $P(B) = .30$, and $P(C) = .30$, what is the chance Art will get a new coat? Check your answer by working the problem two ways.

REFERENCES

[1] Lester E. Dubins, *How to Gamble if You Must* (New York: McGraw-Hill, 1965).

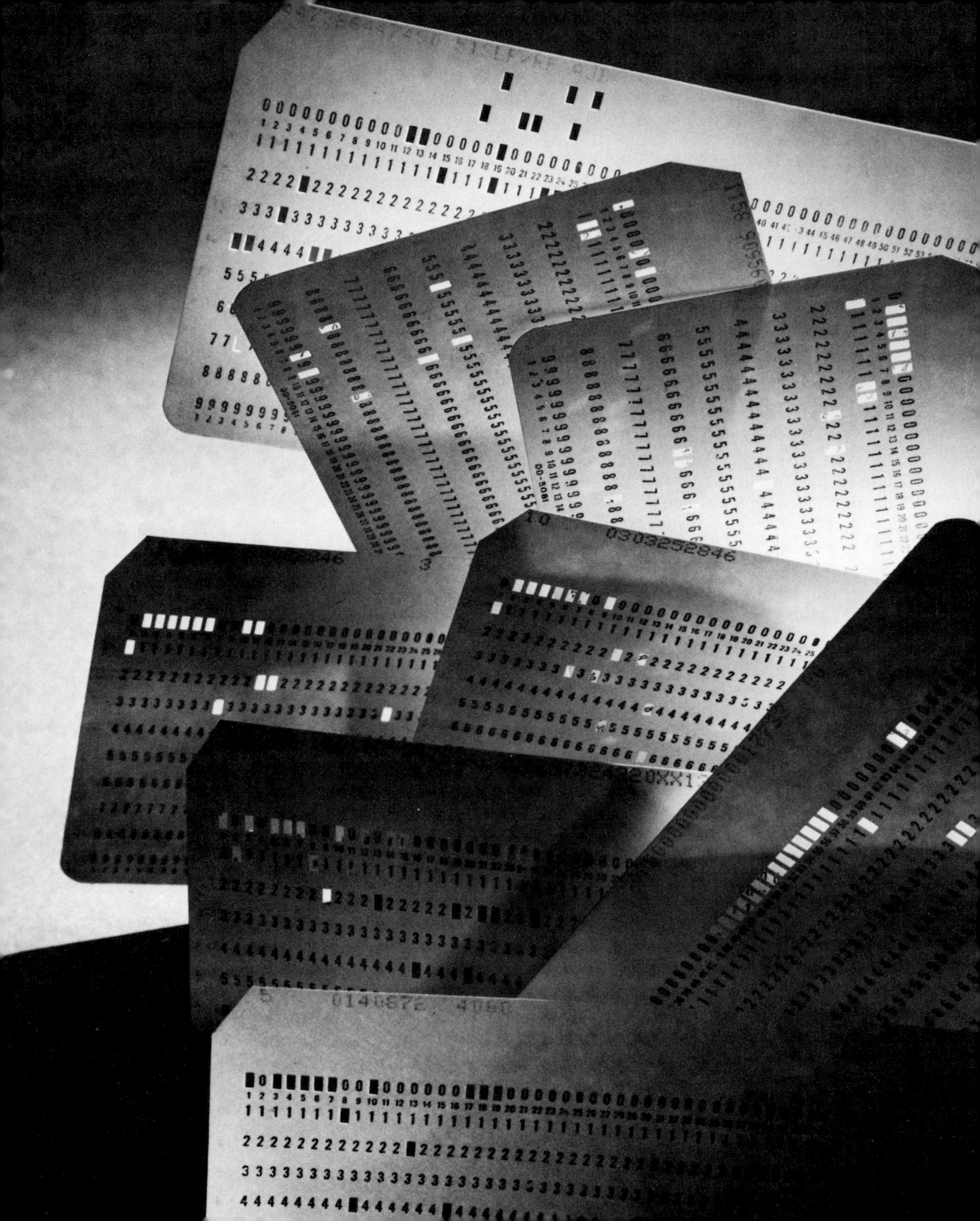

Binomial and Related Experiments

CHAPTER 5

We have developed some techniques for description. The concept of a frequency distribution was used to approximate averages and variation as well as to answer questions of "How many?" or "What percent?" More recently, in Chapter 4 we discussed experiments and ways of describing the chance of occurrence for numerous events. These concepts are interrelated and will be restudied as we look at several classes of experiments.

Here we consider experiments where all possibilities can be counted or otherwise enumerated. Any variable that describes such outcomes is *discrete*. This group includes binomial, hypergeometric, discrete uniform, and geometric variables.

But before discussing these we need one more event relation, *independence*. Consider an experiment of drawing two cards from a standard deck of 52 playing cards. Suppose someone asks "What is the chance for drawing first the ace and then the king of spades?" Of course we assume the cards are well-mixed. Then the answer depends on whether selection is made in sequence with the first card *not* replaced before the second is drawn, or if draws are made *with* replacement. Consider the first case, which is sampling *without* replacement. Here the ace of spades is one of 52 cards and so is drawn with chance 1/52. Then for a second draw, if the ace of spades is gone, there remain 51 cards, one being the king of spades. That is, after the ace the king of spades has chance 1/51. Using the notation of Chapter 4 where A = event ace of spades is drawn, and K = event king of spades is drawn, we get for ace and *then* king,

$$P(A \text{ and } K) = P(A) \cdot P(K/A) = 1/52 \cdot 1/51.$$

A binomial experiment can take only one of two possible values at any trial. These computer cards illustrate a binary system in a very simple way. The information given by the punched holes is interpreted by a computer and recorded internally by either the presence or absence of a pulse.

Next suppose the same experiment except that now after a first card is drawn, we replace it and the cards are again mixed before a second is drawn. This is sampling *with* replacement. Again the chance for the ace at the first draw, as well as for any other card, is 1/52. Specifically, the chance of drawing the king of spades at the first draw is also 1/52, i.e., $P(K) = 1/52$. The first card drawn is replaced, then at the second draw the chance for the king of spades is again 1 in 52. Here, for the order ace then king, $P(A \text{ and } K) = P(A) \cdot P(K/A) = 1/52 \cdot 1/52$. But observe that the conditional probability, $P(K/A)$, is now identical with the probability for a king on the first draw, $P(K) = 1/52$. That is, with replacement $P(K) = P(K/A)$. This somewhat common relation is called *independence*.

Definition — Two events are INDEPENDENT if the occurrence of one in no way affects the chance of occurrence for the other.

Replacing cards between draws then reshuffling gives fixed probabilities at successive draws, that is, independence. For sampling with replacement, the events ace and then king of spades are independent. More generally, we say the outcomes of draws are independent.

Similarly the upturned faces on two dice, the observed numbers on successive spins of a roulette wheel, and the birthdates for two unrelated persons can be considered independent events. This relation is essential in a first discrete variable experiment, the binomial.

1/ Identifying Binomial Experiments

One of the simplest experiments requires that each outcome take one of two possible values. In an electronic computer based on a binary system, each internal operation has two possibilities recorded by either the presence or the absence of a pulse. For the checkout of a manned space vehicle, each operating component is either set for a "go" or a "no go," and the total operation depends on a positive check for each part. Applications of this type are widespread and can lead to a binomial experiment.

The basis for binomial experiments is the *Bernoulli trial*. This requires that the outcome of a single trial is either a success or a nonsuccess and that the chance of success is the same for every trial. We are, however, generally interested in the outcomes of numerous Bernoulli trials, not just one. For example suppose you and a friend regularly "match" coins to decide who will buy coffee. For simplicity consider the outcomes for only four tries considering a match, m, a success and a nonmatch, n, a nonsuccess. The chance of a match stays the same for every toss, i.e., we have independence of outcomes and Bernoulli trials. The possibilities for four trials appear in Figure 1. If we assume the chance for a match is the same as the chance for a nonmatch, $P(m) = P(n) = 1/2$, then each of the 16 possibilities $E_1 = (n, n, n, n)$, $E_2 = (n, n, n, m)$, . . . , $E_{16} = (m, m, m, m)$, has probability 1/16.

Figure 1 / A tree diagram for a coin-matching experiment

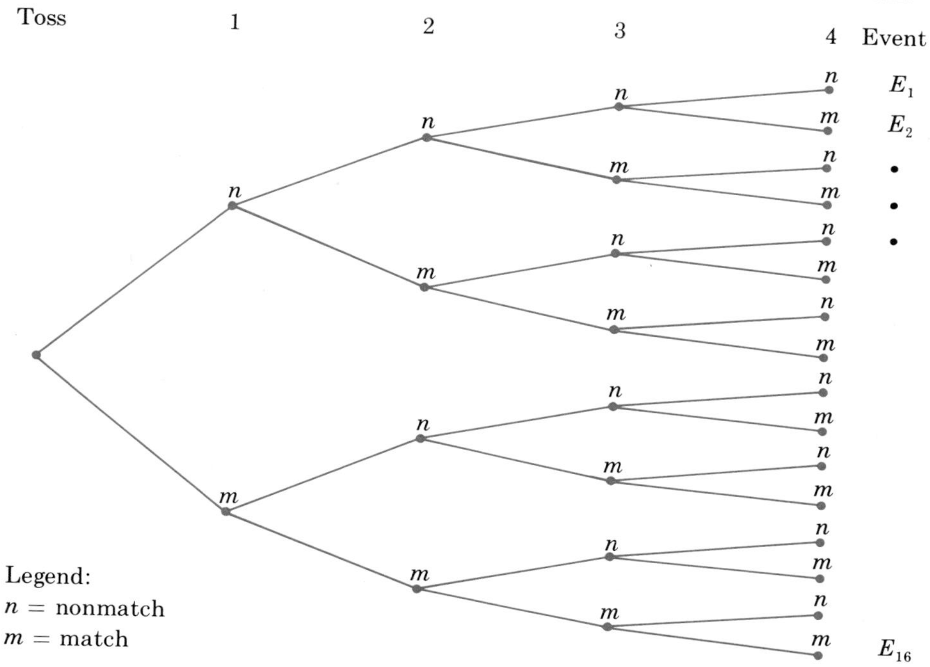

Legend:
n = nonmatch
m = match

Binomial and Related Experiments

Definition A RANDOM VARIABLE is a numerical real-valued function (rule) on a sample space.

In the coin-matching experiment above, we describe one random variable symbolized by "X." Here X names the possible number of matches in four tosses, either $X = 0$, $X = 1$, $X = 2$, $X = 3$, or $X = 4$. Now I want to assign probabilities to each of these values. This can be done by counting and with the aid of Figure 1. For example, Figure 1 displays only one possibility that gives no matches. This is described by the top branch (n, n, n, n), that is, E_1. Similarly, four branches each display one match, six show two matches, four three matches, and only one branch gives all four matches. The results, including all 16 possibilities, are summarized in Table 1. Thinking of the values $X = 0, 1, 2, 3, 4$ as names for compound events, we form a table of probabilities, Table 2, by comparing the number of ways for each event to the total, 16. The display in Table 2 is a *probability distribution*. The form is very much that of a frequency distribution, but with "probabilities" replacing "frequencies." This describes the total experiment or whole space of possibilities with an assignment of total probability, 1.

Table 1 / Possible Outcomes in a Coin-Matching Experiment

	Trial					Trial					Trial			
X	1	2	3	4	X	1	2	3	4	X	1	2	3	4
0	n	n	n	n		n	n	m	m		n	m	m	m
						n	m	n	m		m	n	m	m
	n	n	n	m	2	n	m	m	n	3	m	m	n	m
1	n	n	m	n		m	n	n	m		m	m	m	n
	n	m	n	n		m	n	m	n					
	m	n	n	n		m	m	n	n	4	m	m	m	m

Table 2 / A Binomial Probability Distribution

X	0	1	2	3	4	Total
$P(X)$	$\frac{1}{16}$	$\frac{4}{16}$	$\frac{6}{16}$	$\frac{4}{16}$	$\frac{1}{16}$	1

Definition A DISCRETE RANDOM VARIABLE is one which has at most a countable number of values.

Identifying Binomial Experiments

In the matching-coins experiment the random variable has possible values $X = 0, 1, 2, 3,$ or 4; their number is countable, being 5. Now consider another experiment, repeatedly tossing a coin until a heads appears. A heads could first appear on toss number one or on toss number two or on toss number three, etc. The number of possibilities becomes very large, so, to describe the possible outcomes, we define a random variable, $X =$ the number of tosses required to observe the first heads. Possibilities are $X = 1, 2, 3, 4, \ldots$. These numbers are the positive integers; their number is described as countably infinite.

There are some guides available for help in assigning probabilities to the values for a discrete random variable.

A probability distribution for a discrete random variable has — Rule

1. $P(X) \geq 0$ for any value X, and
2. $\Sigma P(X) = 1$ where the sum is over all possible values.

These conditions are, in fact, a reexpression of the three axioms of Section 4.2. Here events are described as the values taken by the random variable. Again the chance for any event must equal or exceed zero, and the total for exclusive events that define the experiment is one. Observe that Table 2 displays these properties. The values with nonzero probability are called *mass points*.

If the experiment contained not 16 possibilities, but 160 or 1,600, etc., would we need to display every conceivable outcome in order to answer probability questions? In this case the answer is no. Sometimes to avoid long and clumsy processes, we can develop a *probability function,* or *rule*. To illustrate a binomial probability rule for the coin-matching experiment, consider the event (outcome) of two matches in four tries. All of the orderings in Table 3 give two matches. There are six orderings. We have assumed a match as likely as a nonmatch; i.e., if both coins are fair,

Table 3/ Outcomes Yielding Two Matches

Order	Toss			
	1	2	3	4
1	n	n	m	m
2	n	m	n	m
3	n	m	m	n
4	m	n	n	m
5	m	n	m	n
6	m	m	n	n

Binomial and Related Experiments

(H, H) or (T, T) would give a match, while (H, T) or (T, H) would result in a nonmatch. The probability that any trial gives a match is assumed to be $1/2$. Then for the order, say (n, n, m, m), and assuming $P(m) = 1/2$ at each trial, we get $P(n \text{ and } n \text{ and } m \text{ and } m) = P(n) \cdot P(n) \cdot P(m) \cdot P(m) = 1/2 \cdot 1/2 \cdot 1/2 \cdot 1/2 = (1/2)^2 \cdot (1/2)^2 = 1/16$

With *independence* the chance of a match remains unchanged at $1/2$ whether the preceding outcome was a match or a nonmatch. In the calculation we have assumed a specific order (n, n, m, m), but the random variable describes the total number of matches. So to account for the six possible orderings of two matches in four tosses, we employ a combinatorial form, $\binom{n}{X}$. (See Appendix A, p. 369, also Problem 3 in this section). This gives $\binom{4}{2} = (4 \cdot 3)/(2 \cdot 1) = 6$ orders as displayed in Tables 1 and 3. Combining these results gives

$$P(2 \text{ matches}) = \binom{4}{2}\left(\frac{1}{2}\right)^2\left(\frac{1}{2}\right)^2 = \frac{6}{16}.$$

This agrees with Table 2 where six of 16 possibilities produce two matches. Similar procedures lead to the following rule.

Rule

The *binomial probability (function) rule* is

$$P(X) = \binom{n}{X} p^X q^{n-X}, \quad X = 0, 1, 2, \ldots, n$$

where

$X =$ the number of successes
$n =$ a fixed (specified) number of trials
$p =$ the fixed probability of success at each trial
$q = 1 - p =$ the probability of a nonsuccess at each trial.

$P(X)$ is the probability of exactly X successes in "n" repeated (Bernoulli) trials. The probability calculations for this experiment are given in Table 4. Compare these results with Table 2. The random variable and the probability rule are ways of expressing possibilities and probabilities respectively for a statistical experiment.

The following conditions specify the binomial experiment:

1. The outcome of each trial can be described as either a success or a nonsuccess.

Identifying Binomial Experiments

Table 4 / Calculations for a Binomial Probability Distribution, $n = 4$, $p = 1/2$.

Number of Matches, X	Probability, $P(X)$
0	$P(0) = \binom{4}{0}\left(\frac{1}{2}\right)^0\left(\frac{1}{2}\right)^4 = \frac{1}{16}$
1	$P(1) = \binom{4}{1}\left(\frac{1}{2}\right)^1\left(\frac{1}{2}\right)^3 = \frac{4}{16}$
2	$P(2) = \binom{4}{2}\left(\frac{1}{2}\right)^2\left(\frac{1}{2}\right)^2 = \frac{6}{16}$
3	$P(3) = \binom{4}{3}\left(\frac{1}{2}\right)^3\left(\frac{1}{2}\right)^1 = \frac{4}{16}$
4	$P(4) = \binom{4}{4}\left(\frac{1}{2}\right)^4\left(\frac{1}{2}\right)^0 = \frac{1}{16}$
	Total $\frac{16}{16} = 1$

2. The chance of a success is the same at every trial.
3. Trials are independent.
4. The random variable describes the number of successes in "n," a fixed number of Bernoulli trials.
5. The order in which the successes occur is not important.

As another example of a binomial experiment, consider the births, by sex, at a single hospital. If in the past, 60% of the babies have been boys, what is the chance that of the next five babies born there at least two will be boys? The binomial function is appropriate to determine the probability distribution for $X =$ the number of boy babies in the next five births. For example, for $X = 2$ with $n = 5$, and, if we can assume the future is like the past, $p = P(b) = .6$, and assuming independence for births, then

$$P(2 \text{ boys}) = \binom{5}{2}\frac{.6 .6 .4 .4 .4}{b\ b\ g\ g\ g}^* = \binom{5}{2} \cdot (.6)^2(.4)^3 = .2304.$$

But the answer to "at least two" requires probabilities for $X = 2, 3, 4$, then 5. Since we need most of the probabilities, Table 5 displays the total distribution. See Figure 2 for graphic presentations.

---

*$bbggg$ is one ordering of two boys and three girls. $\binom{5}{2}$ considers all orders.

$(.05)^6 = 6/09.05$

Binomial and Related Experiments

Figure 2 / Probability distribution for number of boy babies, $n = 5$, $p = .6$

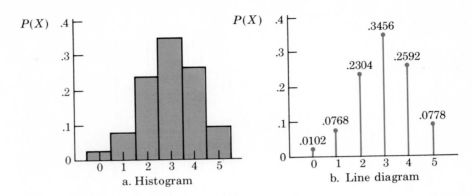

a. Histogram b. Line diagram

The two diagrams in Figure 2 illustrate alternative ways to display the probability distribution for a discrete random variable. Figure 2a, a probability histogram, centers a block of mass about each mass point. This graph has the advantage of showing probability as an area. Figure

Table 5 / Probability Distribution for a Binomial Experiment with $n = 5$, $p = .6$*

Number of boys, X	Probability measure, $P(X)$
0	$P(0) = \binom{5}{0} \cdot (.6)^0 \cdot (.4)^5 \doteq .0102$
1	$P(1) = \binom{5}{1} \cdot (.6)^1 \cdot (.4)^4 = .0768$
2	$P(2) = \binom{5}{2} \cdot (.6)^2 \cdot (.4)^3 = .2304$
3	$P(3) = \binom{5}{3} \cdot (.6)^3 \cdot (.4)^2 = .3456$
4	$P(4) = \binom{5}{4} \cdot (.6)^4 \cdot (.4)^1 = .2592$
5	$P(5) = \binom{5}{5} \cdot (.6)^5 \cdot (.4)^0 \doteq .0778$
	Total 1.0000

*Binomial point probabilities appear in Table V, Appendix B. See the Student Workbook for directions for using this table.

Identifying Binomial Experiments

2b better demonstrates the fact that for a discrete random variable the probability, or *mass*, exists at only a countable number of points. So the probability (mass) over any interval consists entirely of the mass at the mass points.

We are now equipped to compute the probability that *at least* two of the next five babies will be boys. Realizing that "at least two" means $X = 2, 3, 4,$ or 5, gives (using Table 5 or Figure 2):

$$P(X \geq 2) = P(2) + P(3) + P(4) + P(5)$$
$$= .2304 + .3456 + .2592 + .0778 = .9130.$$

This result is quite likely.

Our description of the distribution of boy babies in Table 5 holds answers to many other probability questions including,

$$P(0 \leq X \leq 2) = P(X = 0 \text{ or } 1 \text{ or } 2) = P(0) + P(1) + P(2)$$
$$= .0102 + .0768 + .2304 = .3174.$$
$$P(0 < X \leq 2) = P(X = 1 \text{ or } 2) = P(1) + P(2)$$
$$= .0768 + .2304 = .3072$$
$$P(X \leq 1.5) = P(0 \leq X \leq 1.5) = P(X = 0 \text{ or } 1) = P(0) + P(1)$$
$$= .0102 + .0768 = .0870.$$
$$P(2 < X < 3) = P(\phi) = 0 \text{ (again, } \phi \text{ is the null or empty event)}.$$

To read each probability question, substitute the phrase, "the number of boy babies," for X. Use Figure 2b to check the answers. (*Note:* These solutions follow the systematic approach to problem solving shown in Section 4.3. Knowing the random variable and the probability distribution reduces the process to (1) identifying the question and (2) determining the particular values of X included in the question. Appropriate probabilities are summed.)

Questions about discrete random variables require computing the probability associated with certain mass points. Often one can enumerate the mass points, then add the probabilities (mass).

In Figure 2 there appears a "welling-up" of the probability about the value $X = 3$. Since the random variable describes all possible values, and the distribution defines total probability, this value is a population average. $X = 3$ is the mode. We shall see that it is also the distribution mean. Section 3 concerns determination of values for the population mean, variance, and standard deviation.

Problems

1. Which of the conditions required of a probability distribution for a random variable are *not* met in each of the following?

   a.

X	P(X)
0	1/2
1	1/3
2	1/4

   b.

X	P(X)
1	1/4
2	1/2
3	3/2

2. Complete the following table to make a probability distribution, and then evaluate a–c:
   a. $P(0 \leq X \leq 2)$ b. $P(0 < X \leq 2)$ c. $P(0 < X < 2)$.

X	P(X)
0	1/8
1	2/8
2	3/8
3	2/8
Total	1

3. The coefficients of the terms in the expansion of the binomial theorem are combination counts. Using Table VI in Appendix B as a check, determine:

   a. $\binom{1}{0}, \binom{5}{0}, \binom{10}{0}, \binom{1}{1}, \binom{5}{5}, \binom{10}{10}$

   b. $\binom{3}{1}, \binom{4}{1}, \binom{20}{1}, \binom{3}{2}, \binom{4}{3}, \binom{20}{19}$

   c. $\binom{6}{0}, \binom{6}{6}, \binom{6}{1}, \binom{6}{5}, \binom{6}{2}, \binom{6}{4}$,

   Give generalizations for values ($n \geq 1$) of:

   d. $\binom{n}{0}$ and $\binom{n}{n}$

e. $\binom{n}{1}$ and $\binom{n}{n-1}$

f. Use $\binom{n}{X} = \binom{n}{n-X}$ for $n \geq X$ and Table VII

to find $\binom{12}{11}, \binom{15}{15}, \binom{18}{16}$ and $\binom{20}{15}$.

4. Assuming that the chance for a successful business investment is 3/4 and that investments can be considered independent events, complete the probability distribution for the number of successful business investments among four.

X	P(X)
0	$P(0) = \binom{4}{0}\left(\frac{3}{4}\right)^0\left(\frac{1}{4}\right)^4 = 1/256 = .0039$
1	
2	
3	
4	

5. a. If the probability that the jury in a certain court will contain no women is 0.1, set up the probability distribution for $X =$ the number of juries that contain no women. Let $n =$ four trials and assume each jury is chosen independently of all others. (*Suggestion:* Use Table V, Appendix B.)

b. Use the table you just constructed to determine $P(X \geq 2)$.

6. Assuming a binomial experiment, use Table V, Appendix B to find:
   a. $P(3 \leq X \leq 6)$, for $n = 20$, $p = .2$
   b. $P(3.6 < X \leq 4)$, for $n = 20$, $p = .7$
   c. $P(X > 0)$, for $n = 25$, $p = .25$

7. The probability that any one of five duck hunters will bag his limit on a hunting trip is .2. Assume none of them go together and that their experiences are independent. What is the probability that:
   a. at least four of these hunters will return with their limits?
   b. exactly three will return home with their limit (and the others will not)?
   c. the first three (hunters 1, 2, and 3) each return with their limit and the others do not? (*Hint:* This is similar to, but is *not* binomial.)

Binomial and Related Experiments

8. Recently oil companies have gone to offshore drilling for wells. Assume that along a certain offshore area the chance is 0.1 that any well will be profitable. Also assume independence.
   a. What is the chance that *at least* one of four wells will be profitable?
   b. If five wells are drilled, what is the chance that only one well will be profitable?

9. For the following (1) give a number answer and (2) indicate if event independence was used:
   a. Compute the probability for a pair of ones (snake eyes) on a pair of dice.
   b. The card is replaced, and cards are reshuffled after each draw. What is the chance of drawing the ace of spades and the ace of clubs, not necessarily in that order, in two draws from a standard deck of 52 playing cards?
   c. Same as part b above except that the cards are not replaced.
   d. One percent of all dishwashers of a certain make have defective timers. Of those with defective timers, ten percent have a defective cabinet light. Overall one and one-half percent of the dishwashers have defective cabinet lights. What is the probability that any dishwasher will have both a defective light *and* a defective timer?

2/ Related Experiments

The binomial is by no means the only discrete random variable. It is, however, quite common. In hopes of emphasizing the fact that "binomial" is *not* synonymous with "discrete random variable," three other discrete types are introduced. These are the *hypergeometric,* the *discrete uniform,* and the *geometric* random variables. Since these are not as central to our work as is the binomial, only a brief discussion follows.

Conditions that identify the hypergeometric distribution are:

1. Items of two or more types can be observed at any trial.
2. The number of items of each type is fixed. Also the number of trials, the sample size, is fixed.

3. The random variable describes the number of items for a single type in the sample.
4. The chance for selection of any type changes with succeeding trials; that is, outcomes are not independent.

Although Condition 1 is different in that "over two" possibilities may exist at each trial, Condition 4 generally distinguishes this from the binomial. The lack of independence for outcomes resulting from draws, e.g., of cards, lottery tickets, etc., is often evident when sampling is *without replacement* (recall Section 5.1). Problem 8 in this section is a good chance for you to explore the difference between binomial and hypergeometric probability calculations.

The *hypergeometric probability (function) rule* is   Rule

$$P(X) = \frac{\binom{a}{X}\binom{b}{n-X}}{\binom{a+b}{n}}, \quad \text{with } X = 0, 1, 2 \ldots, a \text{ and where}$$

a = the total number of items of a first type
b = the total number of items of all other types
n = sample size
X = the number of items of a first type in the sample
$P(X)$ = the probability of X (number of) items of a first type being in the sample.

The characteristic that distinguishes this from a binomial experiment is the lack of independence, called *dependence*.

Suppose there remain twenty gum balls in a penny gum machine. Of these, six are red, nine are yellow, and five are blue. The return is one ball per penny. If Junior, who has 2¢, "blows it all" on bubble gum, what is the probability that he will get (1) both red or (2) exactly one red? We will also make (3) the probability distribution.   Example

Here the random variable is X = the number of red gum balls in a sample of n = two gum balls. This number (X) is either 0, 1, or 2. Then $a = 6$, the number of red balls, so $b = 9 + 5 = 14$ other balls. The experiment has $\binom{a+b}{n} = \binom{20}{2} = 190$ possibilities for two gum balls (see Table VI, Appendix B). That is, the experiment is comprised of 190 simple events.   Solution

Binomial and Related Experiments

1. The first question requires the evaluation $P(X = 2)$. Since $\binom{a}{X} = \binom{6}{2} = 15$ and $\binom{b}{n-X} = \binom{14}{0} = 1$,

$$P(2) = \frac{\binom{6}{2}\binom{14}{0}}{\binom{20}{2}} = \frac{15 \cdot 1}{190} \doteq .08.$$

2. For the second question, $X = 1$ red and subsequently $n - X = 2 - 1 = 1$ other ball, giving

$$P(1) = \frac{\binom{6}{1}\binom{14}{1}}{190} = \frac{6 \cdot 14}{190} \doteq .44.$$

3. The remaining mass is at $X = 0$ so

$$P(0) = \frac{\binom{6}{0}\binom{14}{2}}{\binom{20}{2}} = \frac{1 \cdot 91}{190} = .48.$$

These results for $P(0)$, $P(1)$, and $P(2)$ give a probability distribution:

X	$P(X)$
0	.48
1	.44
2	.08
	Total = 1.00

Since nonzero mass exists for only three values, the last one computed can be checked as a complement, $P(0) = 1 - [P(1) + P(2)] = 1 - [.08 + .44] = .48$. In many cases the hypergeometric experiment is identified only by the realization that the sampling is *without replacement*. The result, as we saw in the example on page 92, is varying probabilities. So the probability of getting a red gum ball depends on which draw we are talking about. Outcomes for the draws are *not* independent.

Another variable closely related to the binomial is the *geometric random variable*. As for the binomial, this requires independent Bernoulli trials—again two possibilities each having fixed chance. Now the random variable $X =$ the number of trials *to a first success*.

Rule

The *geometric probability (function) rule* is

$$P(X) = q^{X-1} \cdot p \qquad X = 1, 2, 3 \ldots$$

where

X = the number of that trial which results in the first success
p = the fixed chance of a success at each trial
$q = 1 - p$ = the constant chance of a nonsuccess
$P(X)$ = the probability that the first success occurs on trial number X.

A specific order is required: nonsuccess, then a nonsuccess, etc., for $X - 1$ trials and then at trial number X, there is a success. Subsequently no evaluation of possible orders, i.e., combinatorial count, is necessary.

Example

Five percent of those who come to a teller station at State Bank will make a business account deposit (including batches of coins, etc.). If this teller has just completed one such transaction, what is the probability that the next will occur on or before the fifth customer?

Solution

The conditions of a binomial experiment are met, *except* X = the number of customers to the first business account. That is, this is a geometric random variable. Probabilities essential to answer this question include, for $p = .05$:

X	$P(X)$
1	$(.95)^0 \cdot .05 = .05000$
2	$(.95)^1 \cdot .05 = .04750$
3	$(.95)^2 \cdot .05 \doteq .04512$
4	$(.95)^3 \cdot .05 \doteq .04287$
5	$(.95)^4 \cdot .05 \doteq .04073$
6	$(.95)^5 \cdot .05 \doteq .03869$
7	$(.95)^6 \cdot .05 \doteq .03675$
.	.
.	.

The distribution is left incomplete because, theoretically, this can go on, 8, 9, 10, . . . , to a countably infinite number. The question is answered by

$$P(X = 1 \text{ or } 2 \ldots \text{ or } 5) = .05 + .04750 + \ldots + .04073 \doteq .23.$$

The geometric form is common in "game" problems.

The *discrete uniform random variable* has an extremely simple probability distribution. Figure 3 displays one such distribution for $n = 4$

Figure 3/ Discrete uniform probability distribution, $n = 4$

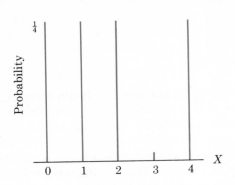

possible values. The number of mass points, n, and their individual values are needed for the probability rule.

Rule

The *discrete uniform probability (function) rule* is

$$P(X) = \frac{1}{n} \quad \text{for } X = X_1, X_2, \ldots, X_n$$

with

X_i = the individual values
$P(X)$ = the probability for the discrete value X.

In this experiment each possible outcome has equal chance.

Example

For the distribution displayed in Figure 3 the probability rule is
$P(X) = 1/4, \quad X = 0, 1, 2, \text{ or } 4.$

Then, $P(X = 0 \text{ or } 1) = 2/4$
$P(X \neq 0) = P(X = 1 \text{ or } 2 \text{ or } 4) = 3/4$
$P(X > 2) = P(4) = 1/4.$

There are numerous other discrete random variables, but these will suffice for our work.

Problems

1. At the end of the season, a nursery puts nine apple trees on sale from which the tags have fallen. It is known that four of the trees are of the "red delicious" variety and the rest are "grimes golden." If a customer purchases three of the trees:
    a. what is the probability that two are red delicious and one is grimes golden? Assume the types are indistinguishable. (*Note:* The entire distribution is *not* needed to answer this question.)
    b. construct the probability distribution for $X =$ the number of red delicious trees in the sample.

2. In a class of 20 kindergarten children, four have some form of visual malfunction. None are recognized to have a visual problem. If ten of the children are tested, what is the chance that at least two will be found who have visual problems? Do not make the entire probability distribution. Assume independence.

3. A geometric experiment is defined by $P(X) = (.8)^{X-1} \cdot (.2)$. $X = 1, 2, 3,$ .... Construct the probability distribution table through $X = 6$, then find $P(X > 6)$. Use four-decimal accuracy.

4. Of all the customers at a supermarket, ten percent will purchase five or fewer items. What is the probability that the fourth person in a checkout line will be the first to purchase five or fewer items?

5. Given $P(X) = 1/5$, $X = -4, -2, 0, 1,$ and 2, compute:
    a. $P(X > 0)$
    c. $P(X < -2 \text{ or } X > 1)$
    b. $P(X \neq 0)$
    d. $P(0 < X < 1)$

6. Of the cars passing a U. S. Department of Agriculture (USDA) checkpoint, five percent are carrying contraband fruit or vegetables. What is the probability that:
    a. of the next ten cars, at most one is carrying some contraband fruit or vegetables?
    b. the third car passing the USDA checkpoint is the first one transporting contraband fruit or vegetables?

7. Ten colored beads are placed in a closed container. One is green, four are white, three are red, and two are yellow. If someone is stringing these beads in essentially random order, what is the chance that:

Binomial and Related Experiments

   a. two of the first six placed are yellow?
   b. the last bead placed is red? (*Hint:* How many red would be placed among the first nine?)
8. A standard deck of playing cards contains two red suits, totalling 26 cards, and two black suits, which also include 26 cards. If two cards are picked at random, what is the probability that:
   a. for draws with replacement, both will be black?
   b. for draws without replacement, both will be black?
   c. without replacement, both will have the same color? (*Hint:* This includes possibilities of two red *or* of two black.)

3/ Describing Distributions

Earlier we discussed the mean for a collection of observations. For larger samples the data was grouped and the computations structured using a frequency distribution. The latter procedure, slightly modified, is used to describe the mean for a probability distribution. We consider only random variables with a reasonable number of values like the binomial, hypergeometric, and discrete uniform random variables.

Rule

For X, a discrete random variable with probability (function) rule $P(X)$, the *true mean* is $\mu = \sum [X \cdot P(X)]$.

The change from the grouped data sample calculation is essentially replacing long-run relative frequencies—f/n—in $\overline{X} = \sum [X(f/n)]$ by probabilities, $P(X)$. The following examples illustrate the computations.

Example

Suppose that in the past Rollins College has defeated Stetson in 0, 1, or 2 of their in-season soccer matches 10%, 70%, and 20% of the time, respectively. Based on this record, how many games should Rollins *expect* to win against Stetson in the regular season play?

Solution

The probability distribution for $X =$ number of games that Rollins wins is:

Describing Distributions

X	P(X)		[X · P(X)]
0	.10		0
1	.70		.70
2	.20		.40

then

$\mu = 1.10$

Using the rule for the population mean;

$\mu = 0 \cdot .10 + 1 \cdot .70 + 2 \cdot .20 = 0 + .70 + .40 = 1.1$ games.

This illustrates a rather important point. In a given year Rollins will win either 0, 1, or 2 of the games. The mean is not used to predict the outcome of a single event. But "in the long run" or "on the average" Rollins is expected to win 1.1 out of every two games of soccer that they play against Stetson. Another name for the true mean is the *expected value*. In any season Rollins will most likely win one game; 1 is the mode.

The next example concerns a binomial experiment. The general rule, just discussed, is applied. Then we introduce an equivalent, but shorter rule.

In the coin-matching experiment, let X = the number of matches in $n = 4$ attempts. The probability distribution is extended for calculation of the mean value. By observation, this particular probability distribution is symmetrical about the value two (see the table) giving mean = median = mode = 2. The calculations confirm $\mu = 2$. (*Note:* For hand calculation of the mean, variance, etc., from probabilities given as ratio fractions, the least work and greatest accuracy (both) result from calculations using ratios rather than decimal fractions.)

Example

X	P(X)	[X · P(X)]
0	1/16	0
1	4/16	4/16
2	6/16	12/16
3	4/16	12/16
4	1/16	4/16
		32/16 = 2 = μ

The rule for evaluating the true mean is quite general and requires only that the random variable be discrete. A shortcoming is that many calculations are required for even a modest number of possibilities. Consequently,

Binomial and Related Experiments

equivalent but shorter rules have been developed. We will consider only the binomial case here.

Rule

The *mean* for the probability distribution of a binomial random variable is computed as $\mu = np$.

Example

In the coin-matching experiment with $n = 4$ tosses, and $p = 1/2$ (chance of a match), $\mu = 4 \cdot 1/2 = 2$. This agrees with the earlier result.

Although simplified rules exist for finding the mean in other discrete distributions, remember that one procedure works for all of them, that being the "definition" form,

$$\mu = \sum [X \cdot P(X)].$$

A second analogy relates calculation of the variance, σ^2, to s^2 for grouped observations. The essential change, again, is replacing long-run relative frequencies by probabilities.

Rule

For any discrete random variable with probability (function) rule $P(X)$, the *true variance*, σ^2, is the expected value of the squared differences taken about the distribution mean:

$$\sigma^2 = \sum [(X - \mu)^2 \cdot P(X)]$$

Procedures change very little from those for calculating the values for a sample variance, s^2, for grouped sample data. Recall $s^2 = \sum [(X - \mu)^2 \cdot f/(n-1)]$. For sufficient trials, meaning for quite large "n," $(f/(n-1))$ becomes very much like (f/n), and approximates $P(X)$, hence the preceding rule.

Example

Again the coin-matching experiment provides straightforward calculations. Recalling that $\mu = 2$, the calculation of a value for σ^2 follows:

X	$P(X)$	$(X - 2)$	$(X - 2)^2$	$[(X - 2)^2 \cdot P(X)]$
0	1/16	$0 - 2 = -2$	4	4/16
1	4/16	-1	1	4/16
2	6/16	0	0	0
3	4/16	1	1	4/16
4	1/16	2	4	4/16
Total	1			$1 = \sigma^2$

Notice the calculation actually requires one less step than computing values for s^2 because "division" has already been made in the values for $P(X)$. As before the standard deviation is the square root of the variance, here $\sigma = \sqrt{\sigma^2} = \sqrt{1} = 1$.

A mathematically equivalent form exists for computing the variance in binomial experiments.

Rule

For binomial random variables the *variance* for the probability distribution is $\sigma^2 = npq$, with the *standard deviation* $\sigma = \sqrt{npq}$. Again $q = 1 - p =$ the chance of a nonsuccess.

For the coin-matching experiment with $n = 4$, $p = 1/2$, the variance is $\sigma^2 = 4 \cdot 1/2 \cdot 1/2 = 1$. This is the simpler, therefore preferred, computational form for binomial experiments.

As a final example calculations are made of mean, variance, and standard deviation values for a discrete uniform distribution.

Example

Let $X = 0, 1, 2,$ or 4 with $P(X) = 1/4$ for each. See the probability distribution and calculations below. Do you follow all of the steps? By general rules $\mu = 7/4$ and $\sigma^2 = 35/16$, so that $\sigma = \sqrt{35/16} = \sqrt{35}/4 \doteq 1.5$.

X	$P(X)$	$[X \cdot P(X)]$	$\left(X - \dfrac{7}{4}\right)$	$\left(X - \dfrac{7}{4}\right)^2$	$\left[\left(X - \dfrac{7}{4}\right)^2 \cdot P(X)\right]$
0	$\dfrac{1}{4}$	0	$-\dfrac{7}{4}$	$\dfrac{49}{16}$	$\dfrac{49}{64}$
1	$\dfrac{1}{4}$	$\dfrac{1}{4}$	$-\dfrac{3}{4}$	$\dfrac{9}{16}$	$\dfrac{9}{64}$
2	$\dfrac{1}{4}$	$\dfrac{2}{4}$	$\dfrac{1}{4}$	$\dfrac{1}{16}$	$\dfrac{1}{64}$
4	$\dfrac{1}{4}$	$\dfrac{4}{4}$	$\dfrac{9}{4}$	$\dfrac{81}{16}$	$\dfrac{81}{64}$
	Total $\mu = 7/4$				$\sigma^2 = \dfrac{140}{64} = \dfrac{35}{16}$

The true mean, μ, and the standard deviation, σ, describe the probability distribution in much the same way that $\overline{X}$ and s describe a frequency distribution. In Chapter 6 we will show one way that these values—μ and σ—can be used to describe a probability distribution.

Problems

1. Let X be a binomial variable with $n = 5$, $p = .4$.
   a. Find the mean, mode, variance, and standard deviation. Use the short forms.
   b. Check the values for the mean and variance using the longer, general forms.

2. Records at a certain hospital show that 50% of the births have been boy babies. For $n = $ six births and assuming no multiple births, etc., set down the probability distribution; use this to find μ, σ^2, σ, and the mode. See Table V in Appendix B. If possible use a second method to find the values for μ and σ^2.

3. Complete the following table for a binomial experiment and give values for n and p:

X	$P(X)$	$[X \cdot P(X)]$	$(X - .80)$	$(X - .80)^2$	$[(X - .80)^2 \cdot P(X)]$
0	.36	0	$-.8$	.64	.2304
1	.48	.48	.2	.04	.0192
2	.16	.32	1.2	1.44	.2304
Total	1	$.80 = \mu$			$.4800 = \sigma^2$

4. Use the probability rule $P(X) = 1/5$ for $X = -4, -2, 0, 1, 2$.
   a. Construct a probability distribution.
   b. Using the general forms, find values for μ and σ^2.

5. Using the general forms, compute values for population mean and variance for the probability distribution of Problem 1 in Section 2.

6. Explain the difference between the procedures for s^2 and those for σ^2.

7. Assuming a discrete distribution for $X = $ the number of tests given in your classes and using the table:

   a. find values for μ, σ^2, and σ.
   b. give $P(X \leq 3)$, $P(X \neq 0)$, and $P(1.3 < X \leq 4)$.

X	P(X)
1	.10
2	.20
3	.30
4	.30
5	.10

8. Evaluate each of the following:
   a. For X binomial, $n = 5, p = .2$, find μ, σ, $P(X = 4)$.
   b. For X hypergeometric, $a = 2, b = 4, n = 3$, find μ, $P(X = 1)$, $P(X \geq 2)$.
   c. For X uniform, $P(X) = 1/3, X = -1, 0, 2$, find μ, σ^2, $P(X = 0)$.

4/ Essay Example

Probabilities are a vital concern in a number of fields. The failure rate, or some similar criterion, is a gauge of the academic difficulty for our colleges and universities. In industrial production continued effort is given to maintain a low percentage of unacceptable items. Watchwords in medicine include recovery rate and percent immunity. Professionals seek high reliability in testing automobile, aircraft, and other machine parts. The insurance industry bases their rates, the premiums that you and I pay, on probabilities determined from their insurance experience. All of these areas depend on extensive information as the basis for determining the chance of certain events.

The role of probabilities in reliability work is exemplary for many other areas. *Reliability* concerns the chance that a device will perform adequately under usual conditions for a specified time. Reliability has been a major concern in the recent Apollo missions that landed our astronauts on the moon. Lieberman [2] depicts a simplified model for an Apollo system in Figure 4.

The system described in Figure 4 might be simplified to a schematic.

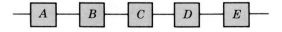

Figure 4 / A model for an Apollo system

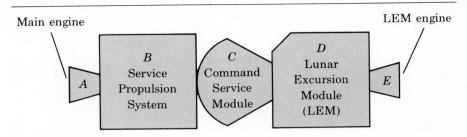

This configuration is called a *series* system. The schematic shows that if any one component fails, the total system will fail. For example, a failure in the performance of the command module would lead to failure for the entire system. Graphically this might appear as a break, e.g., failure of

the command module to allow the LEM to dock. In order for the system to function the first component must operate properly *and* so must the second *and* so on to the last component. Thus evaluating the reliability of the system requires knowing the probability that each component will function properly. Typically these probabilities are approximated through mock tests on scale models or by repeated computer evaluations, called simulations. Then if we assume the reliability for each component is unaffected by the performance of all other components (i.e., if we can assume independence), the system reliability for a series system is a product. For a series system with three components, and with reliabilities of $P(A) = .98$, $P(B) = .95$, and $P(C) = .99$, the system reliability is

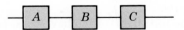

$P(A \text{ and } B \text{ and } C) = P(A) \cdot P(B) \cdot P(C) = .98 \cdot .95 \cdot .99 = .92.$

Another, possibly more functional, arrangement is the *parallel system*.

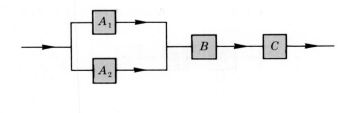

Essay Example

A simple schematic shows two components in parallel with two others in series. This could represent A_1 and A_2 as two main engines, one being a backup. The advantage here is that proper performance of either A_1 or A_2 (or of both) will allow the system to function to B. Increased reliability can result from the parallel configuration. This gain is illustrated using the component probabilities of the preceding system. The new system's reliability is:

$$P[(A_1 \text{ or } A_2) \text{ and } B \text{ and } C] = [P(A_1) + P(A_2) - P(A_1 \text{ and } A_2)] \cdot P(B) \cdot P(C)$$
$$= (.98 + .98 - .98 \cdot .98) \cdot .95 \cdot .99 = .94.$$

Again we have assumed that each component is unaffected by the performance of all others. In the end a potential gain in reliability must be weighed against the additional cost for duplicating a component.

Numerous combinations of parallel and series configurations lead to other reliability (probability) evaluations. Several other possibilities are presented in Review Problem 24 in Chapter 6.

REFERENCE

[1] Gerald J. Lieberman, "Striving for Reliability" in Judith M. Tanur, Frederick Mosteller, and others, *Statistics: A Guide to the Unknown* (San Francisco: Holden-Day, 1972).

The Normal Distribution

CHAPTER 6

Theoretically a continuous random variable admits more values than could ever be counted. Numerous physical dimensions for persons including height, weight, arm length, and characteristics such as age and I.Q. are commonly recorded for only a limited number of values. These are, however, characteristics that can be imagined on a continuous interval scale. Thus to expedite matters we say Jim is thirteen years old and Beth is 48 inches tall, rather than to say Jim is thirteen years, two months, six days, nine hours and three minutes, etc., and Beth is 48.267 . . . inches tall. These figures are subject to the restrictions of precision for our measuring instruments. Moreover, exacting measurement may be unnecessary for getting the message across.

The primary distinction here is that for continuous random variables more exacting measuring devices yield more precise values. Commonly the result is viewed as greater decimal accuracy in numbers. For discrete variables, accuracy beyond a certain level has little effect on probability descriptions. For example, counting the number of children in a family as 2 serves just as well as would 2.0 or 2.00; the variable is discrete. That is, $P(X = 2) = P(X = 2.0) = P(X = 2.00)$. Yet one might perceive quite different results from many repeated measurements of the length of an object, say a table, depending on the level of precision; the measure is continuous. A visual comparison of the consequence of "shrinking the scale" is given in Figure 1 for (1) a discrete and (2) a continuous random variable. Thus for the discrete variable, the appearance of the distribution remains unchanged with increased precision. But for a continuous random variable,

If you buy a pearl clam, what is the probability that you will pick one with a value greater than $10? By using the normal approximation, you can estimate your chances before you buy.

Figure 1 / Discrete versus continuous random variables

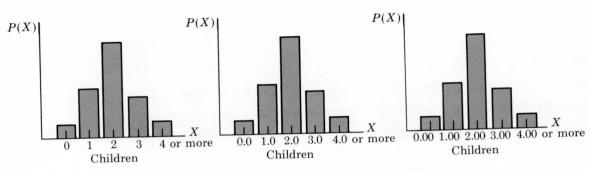

a. X: Number of children in a family, a discrete variable

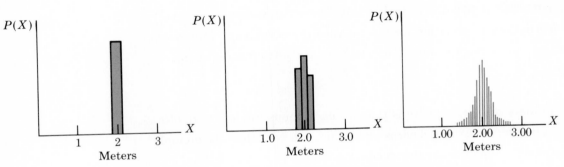

b. X: Measures of length in meters for a fixed object, a continuous variable

the probability distribution can be quite distorted if imprecise measurements are taken. Here two-decimal accuracy was required for a reasonably clear approximation of the probability distribution. This pattern, quite common in ordinary experience, approximates a bell, or normal, curve.

Continuous random variables take real values and a probability curve has a nonnegative height. Consequently, the idea of probability is in this case analogous to area. However, the total area is restricted to a size of one.

1/ Continuous Experiments— The Uniform Distribution

The uniform distribution describes one of the simplest experiments arising from continuous measurements. More than a countable number of values are admitted, and the random variable has zero probability of taking any single value. Here, as for other continuous variables, probability is defined only over intervals of nonzero length.

> For a continuous random variable X, the PROBABILITY FUNCTION is a rule that defines the probability over intervals of values of X.

Definition

We begin with a probability function for the uniform distribution.

> The *continuous uniform random variable* has the probability (function) rule
>
> $$f(x) = \frac{1}{b-a}$$
>
> for $a < X < b$, where $b > a$. ($f(X)$ will denote probability functions for continuous random variables; $P(X)$ is for discrete variables.)

Rule

A graph of the function shows that the distribution is *uniform* throughout the interval (a, b) (see Figure 2). That is, the area is the same for subintervals of equal length anywhere between points a and b in Figure 2.

Figure 2/ The continuous uniform distribution on (a, b)

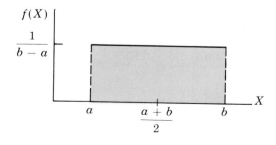

The Normal Distribution

Several essential features can be observed from the figure. First, because the distribution is uniform, the mean and median have the same value, $(a + b)/2$. Moreover, probability calculations can be associated with the area within the rectangle of height $1/(b - a)$ over the interval (a, b). Total area gives total probability,

$$[1/(b - a)] \cdot [(b - a)] = 1.$$

(Recall that the area for a rectangle is determined by multiplying length times width.) This same procedure can be used to find the area, hence the probability, over any interval within $a < X < b$. Yet for any interval entirely below point a or entirely to the right of point b there is zero area and consequently zero probability.

The conditions required to define a probability distribution for continuous random variables are very similar to those for discrete variables. They include:

1. The probability over any interval must take a value greater than or equal to zero.

2. Total probability is one.

The next example displays these properties for one continuous uniform distribution. Here probabilities are just the areas for rectangles, with the total area being one. The restrictions on the random variable define the width of the interval of interest. The height, a constant, is determined by the uniform rule.

Example

The continuous uniform distribution over the interval $a = 0.0$ and $b = 4.0$ has mean = median = $(0.0 + 4.0)/2 = 2.0$. The probability distribution is displayed in Figure 3. Several probability evaluations are:

a. $P(X < 1.0) = \dfrac{1.0 - 0.0}{4} = \dfrac{1}{4}$, or .25

b. $P(1.0 < X < 3.0) = \dfrac{3.0 - 1.0}{4} = \dfrac{2}{4}$, or .5

c. $P(X > .50) = \dfrac{4.0 - .50}{4} = \dfrac{3.50}{4} = .825$, or $\dfrac{7}{8}$

d. $P(X < 4.0) = \dfrac{4.0 - 0.0}{4} = 1.$

Continuous Experiments—The Uniform Distribution

You can check these answers by observing the shaded areas in Figure 3. Question d includes all possibilities, and so gives an area equal to one.

Figure 3/ The continuous uniform distribution on (0, 4)

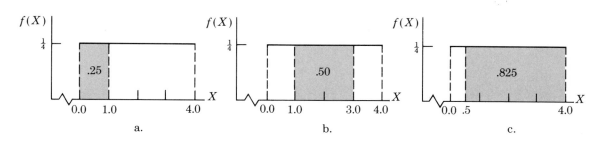

For discrete probability distributions the only measure is at a countable number of *mass points*. Yet it is quite the opposite for a continuous random variable. Here the probability is zero at any single point (X — value). To illustrate, consider the last example where $f(X) = 1/4$, $0 < X < 4$. What is the probability that X will equal 2, i.e., $P(X = 2) = $? Thinking of probability as area requires computing the area of the line in Figure 4.

Figure 4/ Probability at a point for a continuous variable

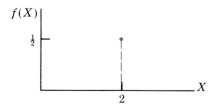

We begin by approximating this area. Suppose an accuracy of 0.1. Then "1," as 1.0 on a meter stick, would be recorded for any observed measurement between 0.95 and 1.05 meters giving area $= (1.05 - 0.95)/2 = 0.1/2 = .05$. Yet with increased accuracy (to the nearest part of one-hundred) the area or probability for the reading 1.00 is reduced by a factor of ten, giving area $= (1.005 - .995)/2 = .01/2 = .005$. It should seem reasonable

The Normal Distribution

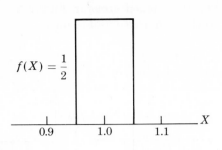 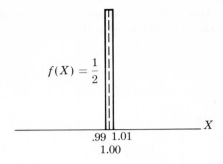

that with even greater accuracy the area is further reduced. In fact as the interval width goes to zero, so too must this area go to zero. But this is logical because, in the furthest reduction, we are trying to find the area for a line. But a line, defined as the distance between two points, has only one dimension, length. A line has zero width. Consequently the area over any point is zero and the probability is zero. The conclusion holds true for all continuous probability distributions.

Rule

For discrete random variables nonzero probability exists only at specific mass points, hence the probability over any interval is the sum of probabilities for the mass points on that interval. For continuous random variables nonzero probability exists only over intervals; the probability at a point is zero. The probability on an interval is the area on that interval with the total area being one.

This means for example that for continuous random variables, probability over the interval, say $(1/2 \leq X \leq 1)$, would have the same value as on the interval $(1/2 < X \leq 1)$ since $P(X = 1/2) = 0$. For discrete random variables the probabilities are the same only if $X = 1/2$ is *not* a mass point.

The mean and variance forms for the continuous uniform distribution are as follows:

Rule

The mean and variance for a continuous uniform distribution are

$$\mu = \frac{a + b}{2} \quad \text{and} \quad \sigma^2 = \frac{(b - a)^2}{12}.$$

The computation of mean, variance, and standard deviation values for continuous probability distributions is determined by integral calculus. However, knowledge of calculus is not assumed here, so the derivation of these forms is left as an exercise for those with the appropriate background.

A second example illustrates calculations for a particular continuous uniform experiment.

Example

The time in minutes that a certain young man waits for his girlfriend while she gets ready for their date is described by $f(X) = 1/15$, $5 < X < 20$. What is the probability that he will have to wait (a) at most 10 minutes? (b) exactly 10 minutes? (c) from 7.5 to 15 minutes? (d) On the average, how long must he wait?

Solutions

The uniform rule gives,

a. $P(X \leq 10.0) = (10 - 5)/15 = 5/15 = 1/3$
b. $P(X = 10.0) = 0$
c. $P(7.5 < X < 15) = (15 - 7.5)/15 = 1/2$
d. Mean $= (a + b)/2 = (5 + 20)/2 = 12.5$ minutes.

The uniform probabilty form is unusually simple, yet it allows us to visualize probability calculations by equating probability to the area beneath a "curve" and defined over some interval.

Problems

1. For $f(X) = 1/10$, and $-5 < X < 5$, (a) find μ, σ^2, and σ, (b) compute $P(X < 0)$, $P(X = 1)$, and $P(-2 < X \leq 3)$.

2. For $f(X) = 1/12$, and $-6 < X < 6$, find μ, σ^2, and σ.

3. Consider a discrete uniform probability (function) rule $P(X) = 1/10$, $X = -5, -4, -3, -2, -1, 1, 2, 3, 4, 5$.
   a. Display the probability distribution as a line diagram (see p. 104).
   b. Compute $P(X < 0)$, $P(X = 1)$, and $P(-2 < X \leq 3)$
   c. Compare the results of parts a and b in this problem with the results of Problem 1.

4. Given the probability distribution below find:
   a. the probability function
   b. values for μ, σ^2, and σ.
   c. $P(X \leq 2)$, $P(X < 2)$, and $P(X = 2)$

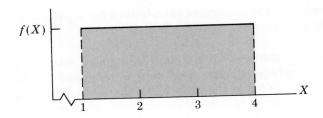

5. The light at a certain intersection is set in the following manner. It is green starting at 12:00 and lasting for one minute. It does not turn green again for two minutes. The cycle is repeated throughout the day.
   a. If you come to this intersection at 8:02, will the light be green?
   b. What percentage of the time during any given hour will it be green?
   c. If your arrival time at this intersection is uniform over the interval 7:59 to 8:06, what is the probability that you will *not* have to stop for this light?

2/ The Standard Normal or Z-Distribution

The normal distribution is symmetric and has a concentration of probability near its central value with less probability concentrated in the extremes. The central value is the mean and the median and the mode. By its symmetry one half of the area is to the left and one half is to the right of the center. There are many variations of this pattern. See Figure 5 for three examples. For example, the distribution in Figure 5 centered at *a* theoretically might be of heights for men in Japan, that centered at *b* for

Figure 5/ Several normal distributions

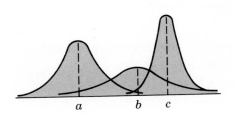

The Standard Normal or Z–Distribution

heights of men in the United States, and that centered at c for the men of Ghana. All are characterized by a familiar bell shape, but each is distinguished by its central value, the mean, and by the amount of variation. That is, calculation of normal probabilities requires values for two population characteristics, the two parameters μ and σ.

Because the normal distribution is not a regular polygonal figure (i.e., a rectangle, triangle, etc.), more complicated procedures are required to compute probabilities. Fortunately one can avoid new computations for normal distributions having different values for the parameters by using the *standard normal, or Z-form*.

> The STANDARD NORMAL or Z-DISTRIBUTION has zero mean and a standard deviation of one, $N(0, 1^2)$. (*Note:* This notation, which is quite common, indicates a normal distribution having mean zero and standard deviation squared (variance) of 1^2.)

Definition

The standard or Z-score is a coding device for converting any normal distribution to the Z-form. We will employ this tool in the next section, but first we will explore the standard normal distribution.

The probability associated with the standard normal distribution is recorded in Table I, Appendix B. Certain benchmarks or guides, if remembered, can give useful checks on your probability answers. See Table 1. The table of normal curve areas gives probability (or area) on an interval from the center, $Z =$ zero (0), to a given Z-value. Although the table indicates positive Z-values, symmetry dictates equal areas for equal, but negative Z-scores. This is the basis for doubling probabilities, like $.3413 \cdot 2 = .6826$, or 68.26% in Table 1. This means that over 68% (just over 2/3) of the area is within one Z (standard deviation) unit of distance of the mean. Half of this amount is between $Z = 0$ and $Z = 1$. The other half is between $Z = -1$ and $Z = 0$. Further, the interval $Z = -2$ to $Z = +2$ includes 95.44% of the total, and ± three standard deviation units, or Z-scores, contain almost all (99.74%) of the area. See Table 1.

Table 1 / Benchmarks for the Standard Normal Distribution.

Z-Score	Tabled Measure	Area between $-Z$ and $+Z$
1	.3413	68.26%
2	.4772	95.44%
3	.4987	99.74%

The Normal Distribution

Use of Table 1 for Z-distribution problems is simplified by remembering that the Z-values in Table I, Appendix B, and on a (Z-) curve diagram are located on the margins. Areas (probabilities) are located in the body of each figure. See Figure 6. Reading Table I, Appendix B, for normal curve areas and Z-scores is illustrated in six examples. You should check the numbers that I have taken from Table I.

Figure 6/ A visual relation between the Z-curve and the standard normal (Z) probability table (Table 1, Appendix B)

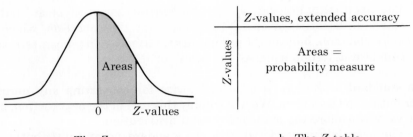

a. The Z-curve

b. The Z-table

Example

What percent of the area is below $Z = +1.00$?

Solution

From Table I the area from the mean to $Z = 1.00$ is .3413. By symmetry .5000 of the area is below the mean.

So $P(Z < 1.00) = .5000 + .3413 = .8413.$

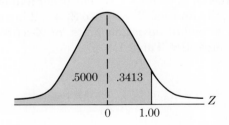

A visual display gives a guide for the inclusion or exclusion of certain areas and should be made. The column headings in Table I indicate accuracy for the Z-value to parts of one hundred. In the problems and examples all Z-scores will be recorded to this accuracy.

Example

What is the probability for a reading between $Z = 0$ and $Z = -1.35$?

The Standard Normal or Z–Distribution

Solution

The area for $Z = 1.35$ is .4115. So by symmetry for positive and negative Z-values, the area from $Z = -1.35$ to $Z = 0$ is .4115; that is, $P(-1.35 < Z < 0) = .4115$. We said earlier that for continuous variables, $P(Z = a) = 0$ for any real number a. So in fact, $P(-1.35 < Z < 0) = P(-1.35 < Z \leq 0) = P(-1.35 \leq Z < 0) = P(-1.35 \leq Z \leq 0) = .4115$.

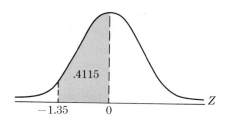

What fraction of the total area is to the right of $Z = 2.486$?

Example

We begin by rounding to the nearest part of one hundred, $Z = 2.486 \doteq$ to $Z = 2.49$. Table I shows the area .4936 for $Z = 2.49$. However, we seek the area to the right of this point, that is, the remainder of the 50% of area that is right of center. So $P(Z > 2.486) \doteq .5000 - .4936 = .0064$. This is less than one percent, i.e., 0.64%.

Solution

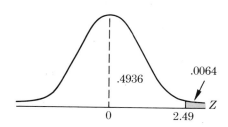

Now we reverse the process by assuming areas (probability measure) and find the associated Z-values. The following can be answered by going to the body of Table I, making a trial-and-error search for the value, and then "realizing" its position to the margins to determine the Z-value. Several examples should clarify the process.

Find the Z-value for which 19.50% of the area is between $Z = 0$ and the Z-value.

Example

Going to Table I, the value .1950 is in the sixth row in the third column. Reading to the left margin gives Z accurate to the nearest tenth, i.e.,

Solution

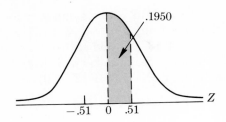

0.5. Reading up to the column heading gives extended accuracy to .01. Then $Z = 0.51$. Since the question makes no requirement on direction, either above or below $Z = 0$, the answer can also be $Z = -0.51$.

Example

We seek a Z-value for which 75% of the area is to its left. That is we seek the third quartile value, Q_3.

Solution

Since the area to the left (.7500) exceeds 50%, this Z-value is positive. The tabled area of interest is $.7500 - .5000 = .2500$. However, no such value appears in Table I. The value we seek, .2500, is between .2486 and .2517. The nearer value is .2486 (It is "off" by -14 parts, while .2517 is "off" by $+17$ parts), so we read the Z-value for area .2486.

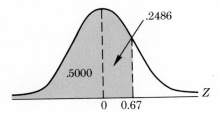

This is $Z = 0.67$. That is, $P(Z < .67) = .5000 + .2486 = .7486, \doteq 75\%$. Although it would be nice to get an answer closer to the exact value, to do so requires either interpolation techniques or a more extensive table. For our needs this approximation should be close enough.

Example

Find the value for which the area between $+Z$ and $-Z$ is 0.3400.

Solution

Again symmetry is used. The area of interest is centered at $Z = 0$, so that $.3400/2 = .1700$, or 17% resides between 0 and the unknown (positive) Z-value. Table I displays .1700 as the area to $Z = 0.44$. This is the answer. (*Note:* This type of question will be asked again in Chapter 8 when we develop *confidence intervals*.)

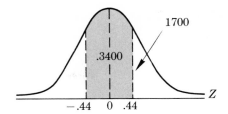

All normal probability questions are treated by conversion to the standard or Z-form. Table I will be used often in our work. Be sure you have mastered it before going on.

Problems

Draw a bell curve showing a display of pertinent Z-values and areas as part of your solution for each of the following.

1. Given that Z is the standard normal variable, find:
   a. $P(-1.00 < Z < 0.00)$
   b. $P(-1.96 < Z < 1.96)$
   c. $P(1.00 < Z < 2.33)$
   d. $P(Z < -1.64)$
   e. $P(Z > 1.96)$
   f. $P(Z < 1.00 \text{ or } Z > 2.00)$

2. Given the following probabilities (areas), find Z_0 for which:
   a. $P(-Z_0 < Z < Z_0) = .9802$
   b. $P(Z < Z_0) = .0505$
   c. $P(Z_0 < Z < 0) = .3810$
   d. $P(Z < Z_0) = .4960$
   e. $P(Z > Z_0) = .0038$
   f. $P(Z < Z_0) = .6772$

3. For the following find the nearest Z-value that will satisfy the statement:
   a. 9.68% of the measure is below this Z.
   b. .4192 is the area between $Z = 0$ and this value.
   c. This is (approximately) the first quartile value.
   d. 92.22% of the area resides above this value.
   e. 99.80% of the measure falls between $\pm Z$.

The Normal Distribution

4. Find areas (probabilities) for:
   a. $Z < 1.326$
   b. $0 < Z < 1.095$
   c. $Z \geq -.163$
   d. $-1.364 < Z < 2.41$
   e. $Z < 4.00$
   f. $Z > 3.09$

5. Find the following, which intermix the processes. Use Table I.
   a. Find Z_0 such that the area to the left of Z_0 is .5040.
   b. Compute $P(Z > -3.20)$.
   c. Find the area for $(-2.065 < Z)$.
   d. If $P(.81 < Z < Z_0) = .1932$, find Z_0.
   e. Find the area between $Z = -1.56$ and $Z = -.56$.
   f. Which is larger, P_{60} (60th percentile value on Z-scale) or $Z = 0.30$? Justify your choice.

3/ The Normal Distribution with Applications

Obtaining *standard (Z) scores* is a coding technique which converts values for any normal distribution to comparable values on the standard normal scale where the mean is 0 and the standard deviation is 1.

Rule

To convert from a normal distribution with known mean, μ, and variance, σ^2, $N(\mu,\sigma^2)$, to the standard normal form, $N(0,1^2)$, use standard or Z-scores:

$$Z = \frac{X - \mu}{\sigma}.$$

This conversion allows all normal distributions to be related to one form, that for which probabilities (areas) appear in Table I. Figure 7 symbolizes the conversion to this common form. The Z-form holds a special relation to the standard deviation. That is, one unit on the scale of Z equals the distance spanned by one standard deviation unit. Numerically when $X - \mu = \sigma$, then $Z = \sigma/\sigma = 1$. Thus Z-values record differences where the basic unit is the standard deviation. The standard normal distribution provides one pattern, a quite common one, for assigning probability based on standard deviation units.

Figure 7 / Conversion to the standard normal disdribution

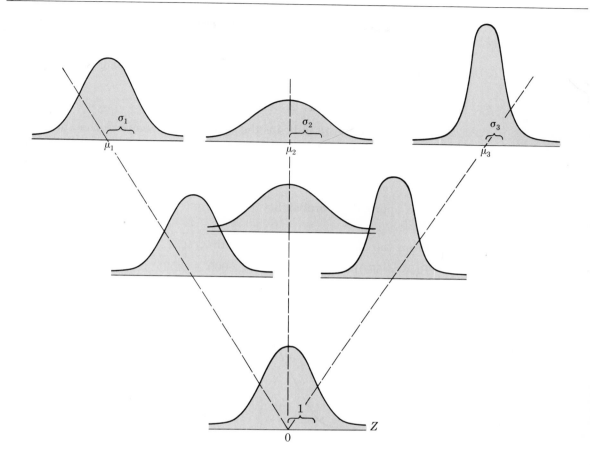

We are now equipped to use the normal distribution to answer probability questions. The procedure adds one step—conversion to Z-scores—to those of the preceding section. Also, the solution is a slight modification of the systematic approach to problem solving as given in Section 4.3. The primary procedural changes are (1) a drawing replaces the systematic listing of "givens" and (2) the solution requires the logical use of Z-scores and Table I to determine the probability answer. A number of examples follow. $N(\mu, \sigma^2)$ indicates a normal distribution with mean μ, and variance σ^2.

Example

Suppose the Minnesota Highway Patrol has clocked many cars in a marked speed zone outside of Little Falls and has found that the mean

The Normal Distribution

speed is 50.0 miles per hour with a standard deviation of 3.0 miles per hour. If the distribution of auto speeds is normal and if any car exceeding 55.0 miles per hour is speeding, what percent of the cars that travel this stretch of road are speeding?

Solution

The variable is $X =$ the speed for cars on this stretch of road. The probability distribution is $N(50.0, 3.0^2)$.

1. The question asks us to find $P(X > 55.0)$.

2. Using the figure to indicate the "givens," we compute the appropriate Z-scores:

$$Z = \frac{X - \mu}{\sigma} = \frac{55.0 - 50.0}{3.0} = \frac{5.0}{3.0} = 1.67$$

3. The appropriate probability (area) from Table I, Appendix B, is .4525.

4. From the standard normal curve diagram,

$$P(X > 55.0) = P(Z > 1.67) = .5000 - .4525 = .0475 \text{ or } 4.75\%.$$

We conclude that 4.75% of those who travel this stretch of road are speeding.

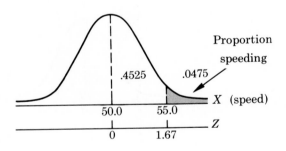

The display gives a visual check to determine the "area" under question. Procedures and computations then become logical.

Example

Union negotiators for the Brotherhood of Diesel Engineers are asking for a $1.45 per hour raise. Assuming that past union increases have averaged $1.00 with a standard deviation of $.30 per hour, what is the probability that their increase will be less than $1.45? Assume a normal distribution.

Solution

The variable is $X =$ the wage increase, with $N[\$1.00, (\$.30)^2]$ distribution.

1. The question is $P(X < \$1.45)$.
2. The value in question, $\$1.45$, equates to
$$Z = \frac{1.45 - 1.00}{.30} = \frac{.45}{.30} = 1.50.$$
3. From Table I, $P(0 < Z < 1.50) = .4332$
4. Logical determination gives:
$$P(X < \$1.45) = .4332 + .5000 = .8332.$$

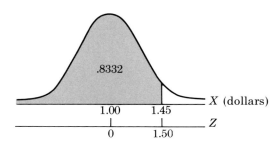

There is an 83.3% chance that this group will receive something less than a $1.45 per hour raise.

The next example differs only in that we are required to determine two probabilities. A diagram, if drawn properly, can assist in the solution.

Example

Suppose that in the preceding situation the union expects to get something less than their request, but hopes for a raise no less than $.90 per hour. What is the chance they will receive a raise within this interval, $.90–$1.45?

Solution

The random variable remains X = the wage increase, with $N[\$1.00, (\$.30)^2]$ distribution.

1. This question asks for $P(\$.90 < X < \$1.45)$.
2. Equated to Z-values,
$$Z_1 = \frac{1.45 - 1.00}{.30} = 1.50$$
$$Z_2 = \frac{.90 - 1.00}{.30} = \frac{-.10}{.30} = -.33$$

3. Table I gives, $\quad P(-.33 < Z < 0) \quad = .1293$
$\qquad\qquad\qquad\qquad P(0 < Z < 1.50) \quad = .4332$
4. We seek the sum $P(-.33 < Z < 1.50) = .5625$

There is a 56.25% chance that they will receive a raise of between $.90 to $1.45 per hour.

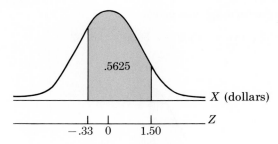

Although many normal probability questions can be answered using the straight Z-distribution, those relating to percentiles and percentile ranks are more easily treated using a cumulative distribution. Consider a large population of persons whose scores on a standardized test can be assumed to follow a normal distribution. Questions like "What value is at the 90th percentile?" and "What is the percentile rank for Jay who achieved 480?" are more easily answered from a distribution that accumulates normal probabilities.

The cumulative distribution function for Z is symbolized $\phi(Z)$— *phi* of Z. The cumulative distribution is described in Figure 8.

Figure 8/ The cumulative (Z) distribution function

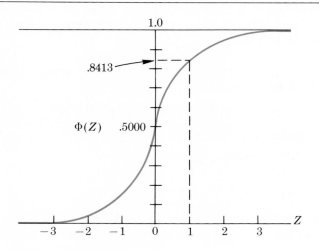

The Normal Distribution with Applications

The cumulative probability distribution extends the idea of cumulative frequencies (see Section 3.3). The symmetry of the Z-curve appears in Figure 8 in that the right side of the $\phi(Z)$-curve is reflected diagonally about the value .5000 for $Z = 0$. This is the basis for determining values of $\phi(Z)$.

For values of Z greater than zero, $\phi(Z)$ = the probability value in Table I added to .5000.

Rule

Thus $\phi(1) = .3413 + .5000 = .8413$. See Figure 8. Through the symmetry of the curve, readings for Z-values less than zero, i.e., $\phi(-Z)$, require subtracting: $1 - \phi(Z)$. See Figure 9.

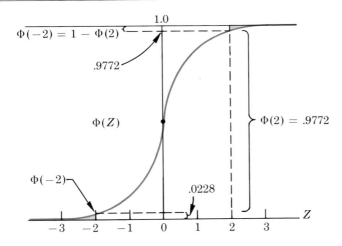

Figure 9 / Cumulative probabilities for negative Z-values

For values of Z less than zero, the cumulative probability is $\phi(-Z) = 1 - \phi(Z) = 1 -$ (the probability value for Z in Table I added to .5000).

Rule

The shaded regions in Figure 9 demonstrate the symmetry about $Z = 0$. Then, since the total probability is one, $\phi(Z) + \phi(-Z) = 1$ so that $\phi(-Z) = 1 - \phi(Z)$. For example, Figure 9 shows $\phi(-2) = 1 - \phi(2) = 1 - (.4772 + .5000) = .0228$. $Z = 0$ is somewhat unique being neither positive or negative; $\phi(0) = .5000$ by symmetry.

Cumulative normal probabilities are commonly used in the interpretation of standardized test results. Determination of percentile ranks and percentiles (raw scores) are described in the following examples.

141

The Normal Distribution

Example

The distribution of aptitude scores on a standardized test had a mean of 432 with a standard deviation of 42 points and can be considered normal. Jay scored 480. Let's determine his percentile rank.

Solution

The variable X = individual aptitude scores is normal, $N[432,(42)^2]$.

1. We seek the percentile rank for a score $X = 480$.
2. This raw score is converted to a Z-value,

$$Z = \frac{480 - 432}{42} = \frac{48}{42} = 1.14.$$

3. Since this Z-value exceeds zero, then from Table I

$$\phi(1.14) = .3729 + .5000 = .8729.$$

The score 480 ranks at the 87th percentile (rounded) on this normal distribution. Jay has achieved at a level above 87% of those who took this test.

The only distinction for determining percentile ranks for scores below the mean is the calculation of $\phi(-Z)$.

The processes of determining the raw score associated with any percentile reverses what we have just done. It uses the same process as in Section 2 where we knew the probability and wanted the associated Z-value. An additional step is necessary to convert from Z to raw-score (X) form.

Example

Let's find the value at the 40th percentile for the distribution given in the last example.

Solution

As an initial check this value is less than 432 = the mean = median = P_{50}.

1. P_{40} for a normal distribution means that the accumulation $\phi(-Z)$ = .40. Note that the percentage being less than 50% indicates a negative Z-value.
2. Thus $\phi(-Z) = 1 - \phi(Z) = .40$, or $\phi(Z) = 1 - .40 = .6000$.
3. To use Table I we must subtract .5000 so we seek the Z-value that gives area, .6000 − .5000 = .1000. This is approximately the area for $Z = .25$.
4. But P_{40} is below the mean, so by symmetry $-Z = -.25$.
5. The remaining step involves decoding the Z-form:

$$-.25 = \frac{X - 432}{42}, \text{ or } X = 432 - (.25 \cdot 42) = 421.5.^*$$

Then $P_{40} = 421.5$. A score of 421 is just below the 40th percentile; 422 is just above the 40th percentile.

The preceding is a quite common application of Z-scores in education and psychology testing. A normal distribution is assumed in the guides to interpreting the tests. However there are a countable number of possible scores, so the normal distribution is only an approximating probability pattern. It does, however, give good approximations for many standardized tests. The error in the normal approximation is ignored here, but will be the primary topic of the next section.

Problems

1. Given the distribution of X is normal with:
   a. $\mu = 4$, and $\sigma = 1$, find $P(X < 2)$
   b. $\mu = -60$, and $\sigma = 6$, find $P(X \leq -54)$
   c. $\mu = 10$, and $\sigma = .8$, find $P(X \leq 5)$
   d. $\mu = 2$, and $\sigma = 2$, find $P(-2 < X < 3)$
   e. $\mu = -1$, and $\sigma = 2$, find $P(X < -1.5 \text{ or } X > 0)$

2. For a normal distribution with mean 6.00 and standard deviation 0.50, that is, for X being $N[6.00, (.50)^2]$, find:
   a. $P(X < 6.75)$
   b. the area to the left of 6.30
   c. the measure above $X = 6.30$
   d. $P(0 < X < 6.00)$
   e. the median for the probability distribution

3. Suppose the average serving of roast beef in a cafeteria line is 5.0 ounces with a standard deviation of 0.5 ounces. If, unknown to the management, a standards inspector comes through the line, what is the probability that his serving will weigh less than 4.0 ounces? Assume normal distribution.

---

*Page 365 illustrates similar procedures.

The Normal Distribution

4. One-pound boxes of sugar are known to contain an average of 1.03 pounds of sugar with a standard deviation of 0.02 pounds. If the weights of the boxes have a normal distribution, what percent will weigh less than 1.0 pound?

5. The Soapy Suds Corporation owns a machine that fills boxes of laundry flakes. The machine is set for a 49.0 ounce fill. Fills have a standard deviation of 0.6 ounces and are normally distributed.

   a. What proportion of the boxes will have a net weight differing from the mean weight of 49.0 ounces by more than 1.0 ounce?

   b. If refilling each reject by hand costs the company an average of 8¢, what is the expected cost of refills per 100 boxes produced? Assume fills between 48.0 and 50.0 ounces are acceptable and that all rejects are refilled by hand.

6. The average time required for students to finish a test is 40.5 minutes with a standard deviation of 9.6 minutes. If the distribution of times is normal, what percent of the students will finish the exam in 60.0 minutes?

7. A standardized exam has a mean of 386 with a standard deviation of 20. Assuming the distribution of scores is normal:

   a. what is the percentile rank for a score of 400?

   b. find the value of the 60th percentile, P_{60}.

   c. find the values for P_{25} and P_{75}.

   d. which quarter of the distribution includes the score 396?

   e. determine the percentile rank for a score of 370.

8. In a very large class of Humanities I, the final 200-point exam had a mean of 140 with a standard deviation of 18.2 points. A normal distribution was a reasonable approximation to the actual dispersement of scores. The instructor decided to "grade on the curve" and assigned 8% As, 24% Bs, 36% Cs, 24% Ds and 8% Fs.

   a. Determine the value that separates A and B grades.

   b. What letter grade would be assigned a score of 165?

   c. What is the lowest passing score (lowest D)?

   d. What percent of the class achieved below Kathy, whose raw score was 185?

4/ Normal Approximations

Binomial variables were discussed at length in Chapter 5. Without indicating so, we restricted the probability questions to cases in which the sample size, n, was 25 or smaller. This allowed use of Table V. For larger samples we will use a *normal approximation* to binomial probabilities.

Suppose we want the probability of nine *or fewer* successes in a binomial experiment of $n = 50$ trials with the fixed chance of success being 0.2. The *exact* probability is

$$P(X = 0 \text{ or } 1 \text{ or } \ldots \text{ or } 9)$$
$$= \binom{50}{0}(.2)^0(.8)^{50} + \cdots + \binom{50}{9}(.2)^9(.8)^{41} = .4437.$$

Unfortunately the calculation by hand is a formidable chore. We obtained it from a table [1]. As an alternative the normal distribution can provide an approximation.

> The *normal approximation* to probabilities for binomial experiments gives reasonable values if both np and nq are five or more.

Rule

The normal approximation is quite good for p close to 1/2, even for small samples like $n = 15$ or 20, but is not so good if p approaches either 0 or 1. For any value of p, the approximation improves as n increases.

The normal curve approximates the area bounded beneath the binomial probability distribution with the area over an appropriate interval bounded by the normal curve. See Figure 10 for a graphic example of the approximation. The histogram represents the exact binomial probabilities.

Figure 10/ Normal approximation to binomial probabilities

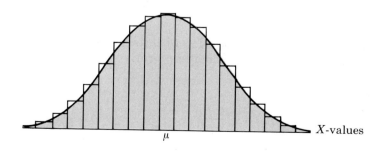

The Normal Distribution

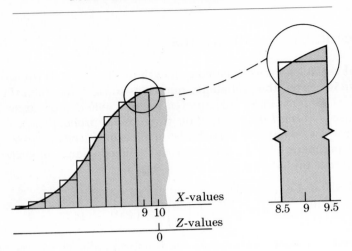

Figure 11 / The normal approximation at $X = 9$ for a binomial distribution with $n = 50$ and $p = .20$

The normal (smooth) curve approximates this "area." Figure 11 includes a portion of the exact representation for the binomial distribution with $n = 50$ and $p = .20$. The error, due to the *normal approximation,* for this interval of one unit width is the difference in the area of the two shaded regions (see cutout).

Steps for the normal approximation include:

1. Determine the interval of X-values of interest. Indicate the integer values that are included in the event.

2. Decide whether to include the extreme value(s) with others in the event. This will involve either adding .5 to or subtracting .5 from the extreme integer value to compute a Z-score. See Figure 11; the .5 is a *continuity correction.*

3. Compute the mean, $\mu = np$, and compute the standard deviation, $\sigma = \sqrt{np(1-p)}$.

4. Compute Z-score(s) and proceed as with the exact normal distribution, but using the normal approximation form:

$$Z \doteq \frac{(X \pm .5) - np}{\sqrt{np(1-p)}}.$$

The question at the beginning of this section asks for $P(X \leq 9)$ so the values of interest include $X = 0, 1, 2, \ldots,$ or 9. The extreme integer value is 9. This is to be included in the count 0, 1, 2, . . . , 8, 9 so the *continuity correction* is added giving $(9 + .5)$. Then for $n = 50, p = 0.2$, the mean is $\mu = 50 \cdot 0.2 = 10$, and $\sigma = \sqrt{50 \cdot .2 \cdot .8} = \sqrt{8} = 2.83$. The normal approximation form gives

$$Z = \frac{(9 + .5) - 10}{2.83} = \frac{-.50}{2.83} = -0.18.$$

Using Table I leads to $P(X \leq 9) \doteq P(Z < -0.18) = .5000 - .0714 = .4286$. The difference between this and the exact value (.4437) is .0151, which is considered small. Observe that by the rule we are "legal" for a normal approximation; "np" equals 10. This approximation was reasonably close to the exact value.

In the examples that follow you should concentrate on (1) recognition of binomial experiments and (2) the necessity for an approximation, especially when n exceeds 25.

A critical point in solving normal approximation problems is deciding whether to include or to exclude (cutoff) extreme values. Check your thinking with the following examples.

Example

In Boone eighty percent of the families with telephones have at least one member at home between 5 and 7 P.M. on weekdays. If a random sample of 100 residential numbers is selected and calls are made during this time period, what is the chance that *75 or more* in the sample will have someone at home?

Solution

We let $X =$ the number of families in Boone with some member at home during this time period. The conditions for a binomial experiment are met (see Section 5.1). Moreover, the normal approximation is appropriate for $np = 100 \cdot .8 = 80$ and $nq = 100 \cdot .2 = 20$.

1. The interval of X-values of interest is *75 or more*, so $P(X \geq 75) = ?$
2. Consequently the continuity correction is subtracted to include 75 with those that are larger: 75, 76, ... The extreme value becomes $(75 - .5)$.
3. The mean is $\mu = 100 \cdot .8 = 80$ with standard deviation $\sqrt{100 \cdot .8 \cdot .2} = 4$.
4. The appropriate Z-value is:

$$Z = \frac{(75 - .5) - 80}{4} = -1.38$$

The Normal Distribution

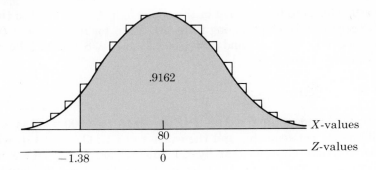

giving *probability (area)* = .4162. Then

$$P(X \geq 75) \doteq P(Z \geq -1.38) = .4162 + .5000 = .9162.$$

There is about a 91.6% chance that someone would be available to answer the phone in each of *75 or more* of 100 homes in this sample.

Again we meet one of those pointed phrases "or more," "more than," etc. Converting this to a correct numerical value is central to normal approximation solutions.

The next example is somewhat unique in that we seek a point probability, that is, the mass at a point, with the aid of the normal approximation.

Example

Pearl clams are sold at a marine attraction and opened in the presence of the buyer. If overall, one in ten clams contain pearls with a value in excess of $10.00, what is the probability that *exactly 10* from a batch of 100 clams would contain such a high-valued pearl? (Every clam is guaranteed to have a pearl).

Solution

The variable is X = the number of clams whose pearls exceed $10.00 in value. This is a discrete binomial experiment with $n = 100, p = 0.10$. The normal approximation is appropriate ($np = 10$, $nq = 90$).

1. A point probability, $P(X = 10)$, is requested.

2. This approximation requires both subtraction and addition of the correction factor, subsequently two Z-values, to form an interval of approximation. That is, the approximated area is that on the interval from 9.5 to 10.5. The normal approximation is shaded in the figure.

Normal Approximations

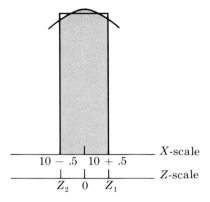

3. The mean is 10, and the standard deviation, $\sigma = \sqrt{100 \cdot .1 \cdot .9} = 3$.
4. The solution follows:

$$Z_1 = \frac{(10 + .5) - 10}{3} = \frac{.5}{3} \doteq .17 \quad \text{Area} \quad .0675$$

$$Z_2 = \frac{(10 - .5) - 10}{3} = -\frac{.5}{3} \doteq -.17 \quad .0675$$

So that $P(X = 10) \doteq P(9.5 \leq Z \leq 10.5) = .1350$.

There is approximately a $13\frac{1}{2}\%$ chance that a selection of 100 pearl clams from this population will produce *exactly* ten with pearls valued over $10.00.

The normal approximation has wide use and it can be used to approximate the probabilities for numerous discrete distributions. However, it does not give reasonable approximations to all discrete distributions and should not be used if the exact distribution is substantially different from a bell pattern.

Problems

1. For a binomial experiment with $n = 25$ and $p = .2$, calculate $P(7 \leq X \leq 9)$ using:
   a. the exact binomial probabilities, Table V.
   b. the normal approximation with the aid of Table I. Compare answers.

2. Answer both parts of Problem 1, assuming a binomial experiment with $n = 25$ and $p = .3$. Which approximation ($1 - b$ or $2 - b$) was closer to the exact value? Why?

3. Good-By-Bug is designed for an 80% kill of Florida mosquitos upon contact. Assuming that this value is correct, what is the probability that a controlled experiment results in a kill of <u>over</u> 85 of a swarm of 100 Florida mosquitos?

4. If 40% of the customers of a department store use credit cards for their purchases, what is the probability that in a random sample of 100 customers, <u>45 or more</u> use credit cards?

5. An auctioneer has found that the probability is about .20 that the buyer of his first item will buy more (in dollar value) than any other single buyer. If the auctioneer handles 100 auctions, what is the probability that *at most* 20 of the first item buyers will buy the most? Assume X, the number of first buyers, follows a binomial probability distribution.

6. From past experience 72 percent of the pupils in Grade 5 in a school system have passed a standardized English exam on sentence structure. What is the probability that 150 of 200 fifth-graders in this system will pass this exam?

7. At the O.K. Cafe, experience shows that one third of the customers choose the "Special." If on a given day the cooks prepared enough food for 160 "Special" lunches and there are a total of 450 customers, what is the probability that the demand for the "Special" will exceed the supply?

5/ Summary for Chapters 4, 5, and 6

These chapters concern probability and probability experiments. Experiments are real situations that involve the occurrence of chance events. They form a framework for describing possible outcomes. Techniques for displaying the space of possibilities include ordered pairs, rectangular coordinates, and Venn and tree diagrams.

Probability is defined by a classical (equal-likelihood) definition, the relative frequency definition, and several other forms. The relative frequency definition states that after many trials of an experiment a stable relative frequency of events is observed. This long-run relative frequency becomes the probability.

An arithmetic of probabilities includes the axioms and addition and multiplication rules. Special relations exist for events that are (1) mutually exclusive or that are (2) independent. A five-step systematic approach provides a method for answering probability word problems.

Chapter 5, on discrete variables, includes the binomial, hypergeometric, geometric, and discrete uniform types. These experiments have few enough outcomes to be enumerated. Answers to probability questions for discrete variables are most easily found using several of the Tables in Appendix B.

Mathematical expectation concerns evaluating distribution values—the mean or expected value, the variance, and the standard deviation. Calculation of values for the population mean, μ, and the standard deviation, σ, are very similar to earlier grouped-data calculations on the sample mean, $\overline{X}$, and the sample standard deviation, s, respectively.

Continuous (random) variables, as discussed in Chapter 6, have zero probability at a point, and have nonzero probability only over intervals. In contrast, for discrete distributions the only nonzero probability occurs at a countable number of mass points. Then the only nonzero probability on an interval is for the mass points within the interval.

The continuous uniform distribution was used to relate continuous probability measure to area, with total area restricted to size one. Now the parameters, or numerical characterizations, become central to evaluating questions about the probability distribution. For the normal distribution, values for μ and σ are required in scaling to standard or Z-form. The latter allows one to use Table I for all normal probability questions. This is the first of many applications using the standard deviation.

The normal distribution serves to answer questions about many experiments. Among the varied applications are determination of percentiles

The Normal Distribution

and percentile ranks for scores on standardized tests. The normal distribution can be used to approximate answers to probability questions about some discrete variable experiments, too. The normal distribution gives reasonable approximations to binomial probability questions so long as both $np \geq 5$ and $nq \geq 5$.

REFERENCE

[1] Edwin B. Cox, Editor, *Basic Tables in Business and Economics* (New York: McGraw-Hill, 1967).

Review Problems

1. Assume a normal distribution with the given mean and standard deviation:
   a. $\mu = 0, \sigma = 1$, find $P(Z > 1.06)$.
   b. $\mu = -2, \sigma = 2.5$, find $P(-1 < X \leq 0.5)$.
   c. $N(0,1)$, find Z_0 such that $P(Z < Z_0) = .9821$.
   d. $N(0,1)$, find Z_0 such that $P(-Z_0 < Z < Z_0) = .2358$.
   e. $\mu = 1{,}010, \sigma = 200$, evaluate $P(X > 1300)$.
   f. $\mu = 0, \sigma = 2$, find the 70th percentile value.

2. For $P(A) = 0.2$, $P(B) = 0.4$, and $P(A/B) = 0.5$, find:
   a. $P(A \text{ and } B)$ d. $P(A' \text{ or } B')$
   b. $P(A \text{ or } B)$ e. $P(B/A)$
   c. $P(B')$ f. $P(A \text{ and } B')$

   (*Suggestion:* Use a Venn diagram.)

3. A binomial experiment consists of 100 trials each having chance 0.2 of resulting in a success. Find the probability that the number of successes exceeds 25. Use the normal curve approximation.

4. For $S = \{0, 1, 2, \ldots, 99\}$, $A = \{0, 2, 4, 6, \ldots, 98\}$, $B = \{12, 24, 36, \ldots, 96\}$, $C = \{0, 5, 10, 15, \ldots, 95\}$ and assuming the equal-likelihood definition for probabilities, determine:
   a. $P(A \text{ or } B)$.

b. all numbers that belong to A *and* B.
c. whether events B and C are mutually exclusive.
d. $P(A')$.
e. $P[(A \text{ and } C)']$.

5. For $P(X) = \frac{1}{6}$, $X = -6, -4, -1, 0, 2, 3$:
   a. compute μ, σ^2, and σ.
   b. find $P(X \geq -1)$.

6. In a certain state five percent of all the land parcels is used for recreation purposes such as parks. What is the probability that among five land parcels selected at random two will be used for recreation purposes? (*Hint:* Use Table V).

7. Suppose there are two "duds" (shells that will not fire) in a box of nine shells.
   a. How many possible samples of three shells could be chosen?
   b. How many of the samples will contain exactly one dud (and consequently two good shells)?
   c. What is the probability that a hunter who picks three shells from such a box will get exactly one dud?

8. Apples from a large shipment average 8.0 ounces in weight with a standard deviation of 1.6 ounces. What proportion of the fruit weighs between 6.6 and 9.4 ounces? Assume normal distribution of weights.

9. A box contains three white, three black, and four green balls. What is the probability that if two balls are drawn from the closed box:
   a. without replacement, both will be green?
   b. with replacement, both will be green?
   c. without replacement, both will have the same color?

10. A father and son are playing one-on-one basketball. Usually the father hits three times out of four and the son makes four times out of seven. If the father makes a shot, the chance that his son then scores is 50%. What is the probability that if both shoot once, both make their shot?

11. The average score on a nationwide 200-point test was 170.0 with a standard deviation of 10.0. The distribution of scores is considered normal.
    a. What is the percentile rank for a score of 175?
    b. Determine the value for the 45th percentile.

The Normal Distribution

12. Of five college men who are single, the probability is 0.1 that any one of them *will not* get married within the next five years. What is the probability that at least three *will* get married within the next five years?

13. For the members of a union, income is normally distributed with $\mu = \$12,000.00$ and $\sigma = \$1,000.00$. Find the proportion of members that make under $\$10,500.00$?

14. In a particular region, the distribution of the four basic blood groups is O, 45%; A, 40%; B, 10%; AB, 5%. What is the probability for a randomly selected couple that:
    a. both are blood type A?
    b. neither is type O?
    c. the wife is type A and the husband is type B?
    d. they have the same blood type?
    e. the wife is type O and the husband is either O or A?

    You should assume that blood type is independent for a husband and wife.

15. If 30% of the patients afflicted with a certain illness have to be hospitalized, what is the probability that:
    a. two or fewer in a group of five who contract this illness will have to be hospitalized?
    b. at most 100 in a group of 300 who have this illness will have to be hospitalized?

16. If 10% of the power units for one make of Citizens Band radio burn out before their guarantee has expired, what is the probability that a merchant who has sold 100 such C.B. radios might be asked to replace *at least* 20 of them?

17. A "fair" die is repeatedly tossed until the side with six spots turns up. What is the probability that at most three tosses will be required? Assume the chance for each face at each toss is 1/6.

18. A production process is "out of control" and checked for repair if in a sample of twenty successive units, five or more are defective. If 10% of its output is defective, what is the chance that the process will be declared "out of control" after a sample of twenty is checked?

19. In tossing a pair of dice there are 36 possible outcomes. In half of these the total number of spots on the up-faces of the two die is an even number. Given that the total is even, what is the probability that the total is six?

Review Problems

20. For the wheel pictured below approximate the probability that fewer than 12 spins in 100 will show a one? Assume independence of spins, and equal size for the sectors. The pointer is respun if it stops on a line.

21. The manufacturer of a new drink, *Slim Cola,* is conducting a taste-test against their own *Diet Cola.* If there is essentially no difference in taste, what is the probability that at most 60 of 100 tasters would prefer the new cola? (*Hint:* Use $p = 1/2$.)

22. Assume the heights of men applying for Air Force pilot training are normally distributed with a mean of 68.2 inches and $\sigma = 1.5$ inches. Past experience indicates that 20% will be classed as short, 55% as medium height, 19% as tall, and 6% as very tall. What is the tallest height in the medium class? How would Bob, who is 5'9" tall, be classified?

23. The arrival time of the Northern Europe flight to Miami International Airport is uniform over the time interval of 8:05 to 8:15 P.M. What is the latest time at which you should arrive at the receiving gate to be 80% sure that you will arrive ahead of the flight passengers?

24. Using procedures from the essay example of Section 5.4 compute the system reliability for each of the following. Assume independence and use $P(A_1) = .93, P(A_2) = .98, P(B) = .97,$ and $P(C) = P(D) = .95$.

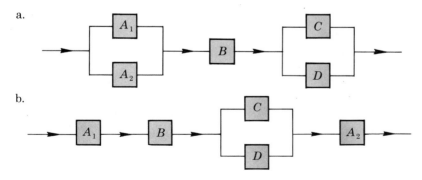

25. If events A and B are mutually exclusive, are A' and B' also exclusive? (*Suggestion:* Use a Venn diagram.)

155

Sampling and the Central Limit Theorem

CHAPTER 7

Sampling plays a central role in statistics and data analysis. Even the everyday processes that we use in consumer buying include some visual investigation or sampling. The regular use of sampling is also seen in voter polls, hourly weather readings, the use of meters to make traffic counts, and in medicine, where periodic readings are made of heartbeat, temperature, and blood pressure. Also recall the A. C. Nielsen example in Chapter 1 concerning television program ratings. The purpose of sampling is to observe less than the population of values thus saving time and cost, but still to get a good description of population characteristics.

Sometimes sampling provides the only reasonable alternative. Consider a manufacturer of automobile horns. One test center has a soundproof room where the walls are filled with horns; one to a hundred horns may be blowing at one time. In the process of testing for quality of sound, strength of materials, and so forth, these horns can be literally blown to pieces—they're not much good after being tested. Obviously only a small percentage, a sample, of the total production could be observed; if all were tested, none would remain to be sold. Destructive sampling is an extreme case. It is a common practice, though, for manufacturers of soaps, sprays, deodorants, or similar items, to disperse quantities of their products to some, but not all, households as a means of introducing the product into the consumer market. Costs, in time and money, are paramount in determining the sample size.

The Central Limit Theorem states that as sample size increases, a distribution approaches the bell-shaped curve of the standard normal distribution. This can help us answer probability questions about the mean of a sample.

1/ Sampling in Experiments

Since cost is a central factor, an important consideration in deciding whether to sample or to take a complete census is the population size. An example is the fact that our United States population census is taken only once each ten years.

Population sizes are classified as either finite or infinite. The population of our United States is finite, but large. We say a population is finite as long as its elements can be counted or labeled, 1, 2, 3, . . . , N, for some positive integer, N, of specified value. Otherwise, it is infinite.*

A reason for drawing samples is to describe, approximately, a probability distribution. Generally we project from a sample value or *statistic* to some population value, a *parameter*.

Definition
: A STATISTIC is a numerical characteristic of a sample. A PARAMETER is a numerical characteristic of a population.

The primary difference is that since a parameter is determined from the total probability distribution, it has a fixed value; whereas the value for a statistic depends on the chance occurrence of items selected in one or more samples. In statistical inference one objective is to make reasonable projections from a statistic to a related parameter. Statistics used in this way are called *estimators*. Common estimators include $\overline{X}$ for μ, s^2 for σ^2, and X/n (relative frequency) for the population proportion, p, in binomial experiments.

Definition
: An ESTIMATE is the value obtained when an estimator is applied to a sample.

---

*For example, suppose $N = 1,000$. Then a finite population of size 1,000 is one for which the elements can be assigned labels, 1, 2, 3, . . . , through 1,000. The smallest population size that is considered infinite is that which is countably infinite. For this the elements can be given labels 1, 2, 3, . . . , 1,000, . . . , 1,000,000, . . . , ∞ (∞ means infinity). This means that although we might envision assigning a counting label to each element, their number is so great that it would be physically impossible, say in your lifetime, to assign a value to each one. An example is the collection of positive integers. A biggest value is not specified so people often use the symbol ∞ to describe it—1, 2, 3, . . . , ∞. A population of infinite size has as many or more elements than the number of positive integers.

For example, in a political poll an estimator for the true proportion of voters who favor an issue is the relative frequency of preference in a sample of voters. The estimator is X/n for $p = $ the true proportion of voters who favor the issue. If $X = 62$ voters favor the issue of $n = 100$ voters in the sample, then the point estimate of the true proportion in favor is 62/100, or 0.62.

One purpose for sampling is to make reasonable projections. For example, suppose we sample one hundred potential voters from a single precinct as a voter poll. As an alternative, we might select ten voters from each of ten precincts in this population. Which sample should give more meaningful projections? As you might suspect, the answer depends in part on how the individuals are picked within the groups. Having defined a population, reasonable estimation also requires selection of a representative, or reasonable, sample. One procedure that assures representativeness is *random sampling*. In fact, this is the foundation of much of classical statistics.

> RANDOM SAMPLING is any procedure that guarantees an equal chance of inclusion for every item in the population at a first selection, and equal chance for all remaining items at each succeeding selection.

Definition

With finite populations this condition can be visualized. For example in a lottery with N tickets, each carries a distinct label. If these tickets are placed in a container, thoroughly mixed, and a single ticket drawn, the resulting selection gives a random sample of one. The chance for every ticket is $1/N$. The remaining tickets could again be mixed, and then a second chosen. By repeating this process n times, a random sample of n tickets is selected. This process might be used, for example, if $n = 5$ tickets were to be chosen from $N = 100$. This illustrates sampling *without replacement* and without regard to order so that a sample of $n = 5$ from a population of 100 could be chosen in any one of

$$\binom{100}{5} = \frac{100 \cdot 99 \cdot 98 \cdot 97 \cdot 96}{5!} = 75{,}287{,}520 \text{ ways.}$$

Since a specific set of five tickets is one of 75,287,520 possibilities,

$$P(\text{any random sample of five from 100}) = 1/\binom{100}{5} = .0000000133.$$

> The probability for any specific sample of n items, selected at random and without replacement, from a finite population of size N is $P(\text{any sample of } n) = 1/\binom{N}{n}$.

Rule

Notice that when $n = 1$, the rule gives

$$P(\text{any sample of 1}) = \frac{1}{\binom{N}{1}} = \frac{1}{N}.$$

Combination counts are the basis for probability calculations with random sampling from finite populations. Samples are generally drawn without replacement. Consequently the condition of randomness does not depend on the probability distribution, but rather on the process by which sample items are drawn. To insure randomness, we must sample by procedures that guarantee all units an equal chance for selection. For samples of $n < N$, random sampling guarantees representativeness in the sense that, in the long run, every sample of n has an equal chance to be chosen.

For infinite populations, a random sample is chosen by independent selection of items from the distribution. For example, a random sample of two is obtained for a coin toss (binomial) experiment: a coin is shaken (mixed), tossed, and the face observed. It is then reshaken, retossed, and again, a face is observed. The process yields two observations, independently determined, from the distribution of possible outcomes for an experiment in coin tosses. The space is infinite because we can imagine more than a physically countable number of tosses. This consequence of random sampling applies only to infinite populations. That is, for infinite populations, probabilities are based on the original distribution and the condition of independence.

Problems

1. Let a population of size 4 be defined by the values $-1, 2, 3$, and 4. Assuming random samples of $n = 2$, there are $6 = \binom{4}{2}$, or the pairs $(-1, 2), (-1, 3), (-1, 4), (2, 3), (2, 4)$, and $(3, 4)$. Determine, but do not singly identify, the number of distinct samples of $n = 2$ (a) from $N = 40$, and (b) from $N = 400$.

2. For a certain infinite population, 20% of the probability resides on the interval 2.0 to 3.0. What is the probability that a sample of two chosen at random from this population will give both values on this interval? At least one on this interval?

3. A recent survey indicates that in a certain precinct 400 persons voted in the last election. There were 642 voters registered, and the mean age of those who voted was 28.4 years.

a. For the population of all voters in this precinct, identify and estimate as many parameters as you can.

b. Could the sample of those who voted be considered a random sample of all the voters in this precinct? Explain.

4. Suppose that you wanted to make a survey of the safety condition of cars in your county. If there are 150,000 cars registered in the county, how many random samples of size 100 are possible? Don't compute a final answer.

5. Mr. Baker's television is not operating properly and, since he has just moved to this locality, he is uninformed as to which repair shops give satisfactory service. Unknown to him, the Better Business Bureau has rated the firms in this area as follows: (S = satisfactory, etc.)

Firm	Service	Firm	Service
1	S	9	U
2	U	10	S
3	S	11	S
4	U	12	U
5	S	13	U
6	S	14	S
7	S	15	S
8	U		

a. If Mr. Baker selects a firm at random from the telephone directory, what is the chance that the firm selected is rated satisfactory by the Better Business Bureau?

b. If he narrows his choice by selecting a simple random sample of three from which to obtain repair estimates, what is the probability that *at least one* will be rated satisfactory?

6. A child takes an intelligence test in which he or she is given six cards, each containing one of the letters b-i-s-h-o-p, and asked to find as many two-letter words as are possible. There are six—is, hi, pi, so, ho, and oh—that have meaning in the English language.

a. Assuming that a card can be used only once in any two-letter word, display all 30 possibilities. Consider all orders, e.g., both "is" and "si."

b. Using random selection, what is the chance any child would "guess" at least one of the meaningful two-letter words?

2/ Sampling Distributions

Up until now we have considered taking only one sample for an experiment. Yet in the last section we observed that from a (finite) population of size N, there exist $\binom{N}{n}$ possible samples, each of size n. It seems logical to ask "What kinds of values would we observe for statistics from these different samples?" and "Is it possible to determine some regularity of chance of occurrence for the various values taken by a statistic?" We will answer these questions for the case in which the statistic is the sample mean. But this first requires an understanding of the probability distribution of sample means, the *sampling distribution*.

Definition

A SAMPLING DISTRIBUTION is the probability distribution for a statistic.

We are now describing the sample mean as a chance quantity, that is, a random variable that can take many different values. Then two or more distinct samples *can* give distinct values. Consider a population with values 1, 2, 3, 4, 5, and 6. Random samples of three would include those shown in the table, and more.

Sample	Sample values		
1:	1	2	3
2:	1	3	6
3:	4	5	6

In any case, these samples give $\overline{X}_1 = 2$, $\overline{X}_2 = 3\frac{1}{3}$, and $\overline{X}_3 = 5$, respectively. *This is an entirely new concept for we are no longer thinking of the sample mean as a single value, but rather as a random variable that can take many values.* With random sampling, we describe a probability distribution for $\overline{X}$. Sampling distributions also exist for the median, the mode, s^2, X/n, etc. We will discuss the distribution for $\overline{X}$ because it is the most common in our use.

Explaining the concept of a sampling distribution requires going back to the parent distribution. Suppose the original (unit) experiment gives rise to a discrete uniform distribution,

$$P(X) = 1/6, \quad X = 1, 2, 3, 4, 5, 6$$

and that we draw random samples of size $n = 3$. For example, the experiment might be a small-scale lottery with the first ticket drawn from among those numbered 1 to 6 inclusive, and then, *without replacement,* a second draw, then a third. For the parent distribution, $\mu = 3.5$ and the standard deviation is 1.7. The probability distribution appears in Figure 1. Our

Figure 1 / Probability distribution for a discrete uniform random variable

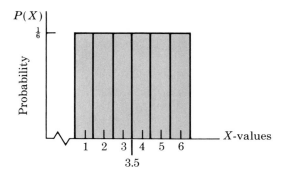

interest is not, however, in this distribution, but rather in the distribution for the $\binom{6}{3} = 20$ means associated with the 20 random samples, each of $n = 3$ observations. The samples and their means appear in Table 1. Each takes probability 1/20. This distribution is displayed in Figure 2. It contains several properties that indicate $\overline{X}$ is a suitable statistic for estimating μ. First, by observation the mean of the sample $\overline{X}$-values is 3.5. That is, the mean of the $\overline{X}$-distribution is the same as the mean of the X-distribution. This property is *unbiasedness*. The distribution of sample means is unbiased under random sampling. That is, it has the same mean as the parent distribution. Also the distribution of $\overline{X}$ values (see Figure 2) compared to

Figure 2 / Sampling distribution for the mean, $n = 3$, $N = 6$

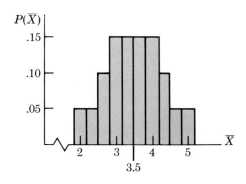

the parent distribution (see Figure 1) shows a marked clustering of values near $\mu = 3.5$. Thus, even though a single sample will give a value for $\overline{X}$ other than 3.5, most samples have mean ($\overline{X}$) value reasonably close to 3.5. This property, which concerns the closeness of $\overline{X}$-values to the true mean, needs more discussion.

The concentration of mean values about $\mu = 3.5$ is heightened with increased sample size. For example, for the same parent distribution, there are $\binom{6}{4} = 15$ distinct random samples of size $n = 4$. Each has probability 1/15. See Table 2.

Table 1 / The Sampling Distribution of the Mean, $n = 3, N = 6$

Samples	$\overline{X}$	$P(\overline{X})$
(1,2,3)	2	$\frac{1}{20} = .05$
(1,2,4)	$2\frac{1}{3}$	.05
(1,2,5), (1,3,4)	$2\frac{2}{3}$	.10
(1,2,6), (1,3,5), (2,3,4)	3	.15
(1,3,6), (1,4,5), (2,3,5)	$3\frac{1}{3}$	.15
(1,4,6), (2,3,6), (2,4,5)	$3\frac{2}{3}$	.15
(1,5,6), (2,4,6), (3,4,5)	4	.15
(2,5,6), (3,4,6)	$4\frac{1}{3}$	.10
(3,5,6)	$4\frac{2}{3}$	.05
(4,5,6)	5	.05
	Total	1.00

To compare the effect of increased sample size, this distribution ($n = 4$) and the distribution for samples of three (see Figure 2) are presented together in Figure 3.

Both distributions center about the true mean, $\mu = 3.5$. This is a characteristic of random sampling and is not altered by sample size. How-

Sampling Distributions

ever, the distribution of $\overline{X}$-values for $n = 4$ is more closely clustered around the true mean. Larger samples generally produce closer estimates. This

Table 2/ The Sampling Distribution of the Mean, $n = 4, N = 6$

Samples	$\overline{X}$	$P(\overline{X})$
(1,2,3,4)	2.50	$\frac{1}{15}$
(1,2,3,5)	2.75	$\frac{1}{15}$
(1,2,3,6), (1,2,4,5)	3.00	$\frac{2}{15}$
(1,2,4,6), (1,3,4,5)	3.25	$\frac{2}{15}$
(1,2,5,6), (1,3,4,6), (2,3,4,5)	3.50	$\frac{3}{15}$
(1,3,5,6), (2,3,4,6)	3.75	$\frac{2}{15}$
(1,4,5,6), (2,3,5,6)	4.00	$\frac{2}{15}$
(2,4,5,6)	4.25	$\frac{1}{15}$
(3,4,5,6)	4.50	$\frac{1}{15}$
	Total	1

Figure 3/ Sampling distributions for the mean

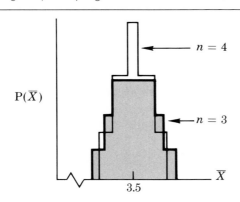

has been extended to a *law of large numbers** which, loosely interpreted, *says that for a quite large sample size the sample mean has a high probability of taking a value near the true mean.* Note that the present example is not a large sample, but Figure 3 indicates how the bigger sample size resulted in a greater cluster of mean values near the true mean. Here the standard deviation of the sampling distribution, called the *standard error of the mean,* decreases with increased sample size. Operationally, evaluation of the standard error of the mean for a discrete random variable uses the same processes as the calculation for σ except that the basic variable is changed to $\overline{X}$-values.

Rule

The *mean* and the *standard deviation* for the sampling distribution of the mean are

$$\mu_{\overline{X}} = \sum_{\text{all } \overline{X}} [\overline{X} \cdot P(\overline{X})]$$

$$\sigma_{\overline{X}} = \sqrt{\sum [(\overline{X} - \mu_{\overline{X}})^2 \cdot P(\overline{X})]}$$

(*Note:* $\sigma_{\overline{X}}$ is also called the standard error of the mean.)

The mean and the standard deviation of the mean are computed for the distribution in Table 2 where $n = 4$.

$$\mu_{\overline{X}} = \left[2.50 \cdot \frac{1}{15}\right] + \left[2.75 \cdot \frac{1}{15}\right] + \cdots + \left[4.50 \cdot \frac{1}{15}\right] = \frac{52.50}{15} = 3.50$$

$$\sigma_{\overline{X}}^2 = \left[(2.50 - 3.50)^2 \cdot \frac{1}{15}\right] + \left[(2.75 - 3.50)^2 \cdot \frac{1}{15}\right]$$

$$+ \cdots + \left[(4.50 - 3.50)^2 \cdot \frac{1}{15}\right] = \frac{4.3750}{15} \doteq .2917$$

$$\sigma_{\overline{X}} = \sqrt{.2917} \doteq .54.$$

n	$\sigma_{\overline{X}}$
3	.76
4	.54
5	.37

The mean and the standard error (deviation) for the sampling distribution in Table 1 ($n = 3$) are $\mu_{\overline{X}} = 3.50$ and $\sigma_{\overline{X}} = .76$. For random samples of $n = 5$ the values are $\mu_{\overline{X}} = 3.5$, $\sigma_{\overline{X}} \doteq .37$ (calculations and sampling distribution are not shown). The relation of sample size and standard error shows, in this case, that for larger sample size the $\overline{X}$-values are more closely clustered. Calculations are simplified according to the following rule.

Rule

For finite populations with $n/N \leq 0.05$ (a 5% sample at most):

$$\mu_{\overline{X}} = \mu, \text{ and}$$

$$\sigma_{\overline{X}} \doteq \sigma/\sqrt{n}$$

---

*A more rigorous statement and development of this law can be found in [1].

where μ and σ are, respectively, the mean and the standard deviation for the original probability distribution. In sampling from an infinite population, the last equation is also exact.

This rule gives us the capability to find the *mean* and *standard error* for the distribution of sample means provided we know the values for three things—μ, σ, and n. Further, it points out that as n increases, $\sigma_{\bar{X}} = \sigma/\sqrt{n}$ decreases. With increased random sample size, estimates of the mean generally will be closer to the correct value.

Problems

1. Assuming a random sample of size 100 from a very large population with a mean $\mu = 50$ meters and a standard deviation $\sigma = 20$ meters, find values for the mean and standard deviation for the distribution of the sample mean. That is, determine values for $\mu_{\bar{X}}$ and $\sigma_{\bar{X}}$.

2. We say that under random sampling $\bar{X}$, the sample mean, is an "unbiased estimator of the population mean." What is meant by this statement?

3. What is the difference between $\sigma = \sigma_X$ and $\sigma_{\bar{X}}$ in sampling theory? (*Hint:* To what distribution of values does each refer?)

4. The wages earned at "The store" are a population of $N = 6$ with the discrete uniform distribution given in the table below:

Employee	$X =$ hourly wages	$P(X)$
Abel, A	$1	$\frac{1}{6}$
Bean, B	2	$\frac{1}{6}$
Copp, C	3	$\frac{1}{6}$
Dilts, D	3	$\frac{1}{6}$
Elan, E	4	$\frac{1}{6}$
Foss, F	5	$\frac{1}{6}$

a. Find the mean of all possible random samples of two (wages of two workers) assuming sampling is without replacement.
b. Find the mean of the sample means of Part a.
c. Compute the "standard error of the mean" under random sampling.
d. What percentage of the individual wages are between $2.50 and $4.50, inclusive? What percent of the mean wages are in this same pay interval?

5. Work Parts a, b, c of Problem 4, but use $n = 3$. Compare your results.

3/ The Central Limit Theorem with Applications

We have seen that the sampling distribution for $\overline{X}$-values has the same mean as its parent and that its standard deviation is, for a 5% or smaller sample, the deviation of the population distribution divided by the square root of the sample size. In order to perform probability calculations and inference we have only to determine a probability form. A clue to an appropriate form is hinted at in Figure 3. Although these sampling distributions are both discrete and therefore are drawn with "steps," there appears to be a tendency toward a concentration of values near the center. This "welling up" of values about the mean is not unique to any one experiment, but in fact with sufficiently large samples, occurs for the distribution of means from any parent distribution. As the sample size increases, the distribution of the sample mean approaches the shape of the normal distribution, no matter what the distribution of the original variable. This is explained in a fundamental theorem of statistics—the Central Limit Theorem.

Theorem

CENTRAL LIMIT THEOREM For random samples from *any* probability distribution with finite mean and variance, as n becomes increasingly large, the sampling distribution of the sample mean is approximately normally distributed.*
(See the *next* rule for a computational form.)

This theorem may not seem unusual as we saw numerous normal-like

---

*In the case of a census (100% sample), the sampling distribution of the sample means is a degenerate distribution. That is, with total sample, $\overline{X}$ goes to a single value, μ, the true mean.

The Central Limit Theorem with Applications

distributions in Chapter 6 including wage distributions (Brotherhood of Diesel Engineers), automobile speeds, some test scores, and physical characteristics (such as heights), etc.

It would seem just as natural as kittens coming from cats that the distribution of $\overline{X}$-values is normal if the original distribution of (X-) values is itself normal. And possibly there is an indication toward normality for some binomial experiments when n is large (see Figure 9, p. 141), but it is somewhat surprising that for an X-distribution like the uniform, the distribution of $\overline{X}$-values also approaches the normal. See Figure 4. For sufficiently large random samples, the sampling distribution for $\overline{X}$-values will be normal with mean and standard deviation values determined by the next rule.

Figure 4 / Varied distributions with sufficiently large n have normal distribution for $\overline{X}$

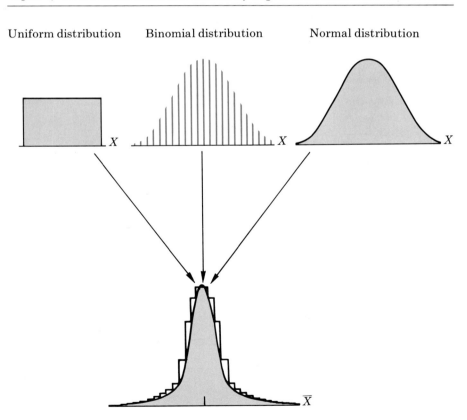

Sampling and the Central Limit Theorem

Rule

Calculation of Z-values for the sampling distribution of the mean for samples of more than 30 observations uses

$$Z = \frac{\overline{X} - \mu}{\frac{\sigma}{\sqrt{n}}}.$$

That is, this standardized form is essentially $N(0, 1)$. This form, like the original Z, compares the deviation (here of $\overline{X}$ values from μ) to the size of the standard unit of deviation on the scale of $\overline{X}$-values, i.e., $\sigma/\sqrt{n}$. It is used when the question asks for probabilities about "average" values. The examples following show a selection of diverse applications.

Example

Suppose that the average starting salary for spring term graduates in accounting (U.S.) was $9,100.00 with a standard deviation of $450.00. What is the probability that a random sample of 100 would have mean salary of $9,000.00 or more?

Solution

The distribution of individual salaries is unknown. It is not essential, but we do need its mean and standard deviation. The question concerns $\overline{X}$ = mean for samples of 100 incomes so:

1. We seek $P(\overline{X} \geq \$9,000.00)$.

2. Using the rule,

$$Z = \frac{\overline{X} - \mu}{\sigma/\sqrt{n}} = \frac{\$9,000.00 - \$9,100.00}{\$450.00/\sqrt{100}} = \frac{-\$100}{\$450/10} = \frac{-100}{45} = -2.22.$$

3. Table I (p. 376) gives $P(-2.22) < Z < 0) = .4868$.

4. An appropriate resolution of area using the figure gives
$P(\overline{X} \geq \$9,000.00) = .4868 + .5000 = .9868$.

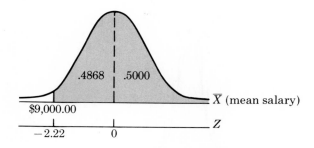

There is a 98.68% chance that the mean salary for a random sample of 100 accounting graduates would exceed $9,000.00.

The Central Limit Theorem with Applications

In this case the standard deviation unit (standard error) is $45.00; that is, $450.00/$\sqrt{100}$ = $45.00 (not $450.00). A common error is to forget, or to ignore, the division by the square root of the sample size.

The next example points out that sizing the standard deviation unit is a central consideration in working normal probability problems.

Example

The experience of a telephone company has been that the average amount owed on overdue accounts is $12.00 with a standard deviation of $1.50. The distribution of these amounts is essentially normal. What is the probability that (1) the mean for a random sample of 100 overdue accounts exceeds $12.50, and (2) a single overdue account will show a value in excess of $12.50?

Solution

(1) The variable is $\overline{X}$ = mean (or average) for samples of 100 accounts.

1. We seek $P(\overline{X} > \$12.50)$

2. $Z = \dfrac{\overline{X} - \mu}{\sigma/\sqrt{n}} = \dfrac{\$12.50 - \$12.00}{\$1.50/\sqrt{100}} = \dfrac{\$.50}{\$.15} = 3.33$

3. $P(\overline{X} > \$12.50) = P(Z > 3.33) \doteq 0.0000$

   because this exceeds the largest entry in Table I.

4. There is essentially zero chance for a mean value in excess of $12.50 as long as random samples of $n = 100$ are used. The size of a deviation unit for this experiment is $1.50/$\sqrt{100}$ = $.15. See Figure 5a.

Figure 5/ Comparison of deviation units when the variable is $\overline{X}$ or X

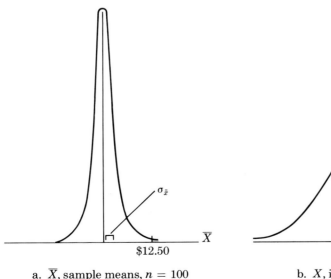

a. $\overline{X}$, sample means, $n = 100$ b. X, individuals, $n = 1$

Solution

(2) The variable is X-individual amounts owed. This becomes a normal distribution problem as shown in Chapter 6.

1. We want $P(X > \$12.50)$ for X normally distributed
2. The first Z-form (from Chapter 6) gives
$$Z = \frac{X - \mu}{\sigma} = \frac{\$12.50 - \$12.00}{\$1.50} = \frac{\$.50}{\$1.50} = .33$$
3. $P(0 < Z < .33) = .1293$ from Table I
4. $P(X > 12.50) = .5000 - .1293 = .3707$ (see Figure 5b).

Over thirty-seven percent of the individuals with overdue accounts owe more than $12.50. Observe that the deviation unit, here $1.50, is $\sqrt{100} = 10$ times that of the $\overline{X}$-distribution, $.15.

This example shows that it is important, even critical, to determine which variable—$\overline{X}$ or X—is under question. The first question asks for the "mean or average," the latter about the "individual." This choice can greatly alter the probability answer as we've just seen in the solutions to (1) and (2).

Now for a final example using the new Z-rule.

Example

If Whisperjet flights between Atlanta and Chicago average 120 passengers with a standard deviation of 12, what is the probability that 64 flights (considered a random sample) would average 120 to 125 passengers, inclusive?

Solution

The variable is $\overline{X}$ = average number of passengers. The question is $P(120 \leq \overline{X} \leq 125)$, then
$$Z = \frac{(125 - 120)}{12/\sqrt{64}} = \frac{5}{12/8} = \frac{5}{1.5} \doteq 3.33$$

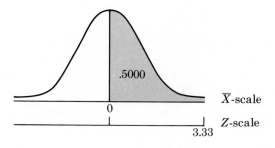

Since $\mu = 120$, $P(120 \leq \overline{X} \leq 125) \doteq P(0 \leq Z \leq 3.33) \doteq .5000$ from Table I.

The preceding examples are just a few applications of the central limit theorem. Don't be surprised when it appears again, many times, in the remaining chapters. It is likely the most widely used rule in statistics.

Problems

1. On the average, containers of a frozen vegetable lose 6.5 grams in weight during storage with a standard deviation of 1.8 grams. What is the probability that a case of 36 containers, assumed to be a random sample, will have a mean loss of weight in excess of 7.0 grams during storage?

2. Flashlight bulbs manufactured by a batch process have a mean lifetime of 600.0 hours with a standard deviation of 45.0 hours. Samples of 36 bulbs are taken from a batch. What percent of these samples will have a mean lifetime between 585.0 and 620.0 hours?

3. Given that X is normally distributed with a mean of 30 and standard deviation of 8, calculate the probability that the sample mean, $\overline{X}$, based on a random sample of size 30, will be less than 32.

4. At a savings and loan association, the average amount for short-term loans to university students is $225.00 with a standard deviation of $12.50. The distribution of amounts loaned is normal.
   a. What percent of the (individual) loans are for $200.00 or less?
   b. What is the probability that 25 loans, considered a random sample, will have a mean over $220.00? Here the central limit theorem applies even though n is less than 30 because we started from a normal parent distribution.

5. Suppose that individuals eating at the Pancake House spend, on the average, $1.24 for breakfast with a standard deviation of $.15.
   a. What percentage of the customers spend over $1.30 for their breakfast? Assume normal distribution of amounts spent.
   b. What is the probability that a sample of 100 customers will spend an average of between $1.20 and $1.30?

6. The mean height for a population of 10,000 college men is 69.2 inches with a standard deviation of 2.5 inches.

a. Find the probability that in a random sample of 100 college men, the mean height will be greater than 69.9 inches.

b. How many of these men are over 6 feet tall? Assume a normal distribution for heights.

4/ Essay Example

Statistical sampling is an accepted practice in most quarters of business and industry. Since World War II, statistical quality control procedures have been used in food processing and in the inspection of manufactured goods. For example, you have probably purchased a coat, sweater, or other garment recently that contained a label "Inspected by No.___." Today sampling techniques are used in tax auditing, inventory processes, ordering supplies for large organizations, and, most recently, for passenger load counts on commercial airlines.

One instance of the application of statistical sampling comes from an experience where the author directed the development of a sample inventory (counting) of the parts holdings for a road equipment sales dealership [2]. A week prior to taking the inventory, the usual computer listing was prepared in a special format—listing all parts by price classes rather than by the usual part number sequencing. Using this information, a second computer analysis produced initial sampling numbers, including means and standard deviations. At this time inventory standards were set. These included an allowable error of 3%, meaning that our findings might indicate anywhere from $97 to $103 for every $100 worth of actual inventory, and a 95% probability level. These initial numbers were treated with appropriate sampling rules to determine part-sample requirements. Of the 37,324 part numbers, 11,277—30.2%—were sampled.

A random numbers generator program was used to select the specific sample of parts. This computer program assured that every part in each part class had equal chance to be included for inspection. The computer output was a deck of punched cards punched by part number, one card per part. The quantity for each of these parts was to be counted.

The cards were then taken to the parts storage bins where a physical count was made. Usual inventory procedures were followed in making the bin counts which were then recorded in pencil on each card. The bin count, the quantity of that part found in storage, was keypunched onto each card. Next the weekly computer listing was observed and from it the

computer counts were also punched onto the sample cards. Unit prices were recorded at this time.

The "value" for an individual part was defined as the unit price multiplied by the quantity on hand. Separate determinations were made of the (1) bin (actual) value and (2) the computer listing value. These determinations projected inventory totals of $905,339 and $893,085, respectively. This was a discrepancy of 1.5%, well within our 3% limit and sufficiently close so that the computer listing was accepted as the inventory standard. Any discrepancies of sampled items were corrected on the computer listing to agree with the actual bin count. The following week, the overall computer listing was taken as the inventory value.

The inventory was successful to the extent that a precision of within 2% was attained—again, 3% was our original goal. Moreover, the procedures were physically workable in that it reduced the total time for taking the inventory from 1,350 man-hours for a full count the previous year to 280 man-hours using this sampling plan. Additionally, the computer listing, which was used daily for parts service, was updated and correct for over 30% of the parts holdings.

Although this example is quite specific, these techniques could be applied to other inventories, e.g., large department stores, or wholesale warehouses, etc. The greatest saving is in the reduction of man-hours of counting low-cost items that are stocked in large quantities. The savings can be one of worker morale also, especially if (as was our case) the inventory is taken at year's end.

One part of this project that is easily overlooked is the extensive task of preparing a *population list,* in this example, all parts listed by price. Our job consisted of preparing a computer program to convert a list of 37,324 parts (including about five descriptive entries per part) from a number-sequence list to a price-ordered list. In this case the cost was several weeks of time and several thousand dollars. Yet once prepared the procedure could be reused for many years. Generally speaking, preparing a suitable list is probably the most difficult task in a sampling experiment.

Although random sampling was the basic procedure, this project used a more complicated sampling design that involved grouping parts by their unit value. However, the process for the computerized "random number generator program" was random sampling. The techniques were computer oriented, yet we can understand the basic processes with the aid of Table VII, Appendix B. Suppose, for illustration, a population of $N = 90$ machine parts from which we are to select a random sample of $n = 15$ (i.e., a $16\frac{2}{3}\%$ random sample). Each part could be assigned a unique two-digit pair or label from 01, 02, . . . , through 90. Then parts could be chosen for the sample by selection of two-digit pairs from those randomly set in Table VII. Numbers could be selected from those listed in columns 1 and 2 or those in 3 and 4, etc. in the table. The starting point is an arbitrary choice so we begin at the top row with the two left-most columns and move down the rows in these columns. Using this procedure, the first fifteen labels are 13, 59, 72, 88, 70, *45*, 87, 40, 14, 63, $\boxed{45}$, 71, 00, 51, 12. However, the representation 00 is not assigned to any member of the population. This or any other of 91, 92, . . . , 99, inclusive is ignored. Also, the second 45 (see box) is a repeat and is also deleted. This is sampling without replacement. We need two more responses. Continuing on in Table VII gives 85 and 89. Should we need more sample, additional units would be obtained by continuing through columns 1 and 2, then to columns 3 and 4 and so forth. The parts associated with the labels 12, 13, 14, 40, 45, 51, 59, 63, 70, 71, 72, 85, 87, 88, and 89 constitute the random sample. Again, the processes used in the "random numbers generator program" are more complicated, but gave equivalent results.

REFERENCES

[1] Paul G. Hoel, Sidney Port, and Charles Stone, *Introduction to Probability Theory* (Boston: Houghton-Mifflin Company, 1971).

[2] John Ingram, "A Sampling Inventory for Machine Parts" (Unpublished work done for Gibbs-Cook Caterpillar Corp., Des Moines, Iowa, 1969).

[3] Taro Yamane, *Elementary Sampling Theory* (Englewood Cliffs, New Jersey: Prentice Hall, Inc., 1967).

Estimation and Sample Size

CHAPTER 8

The life process continually requires decisions—what clothes to wear today, to drive or walk to the nearest postal service box to mail a letter, should I get groceries today or tomorrow, and so on. Although we may not associate the use of statistics with our day-by-day decision making, Dr. Thomas Harris in *I'm OK—You're OK* [2] considers that we might use "probability estimating" in our psychological processes of decision making,

> . . . One of the realities of the human predicament is that we frequently have to make decisions before all the facts are in. This is true of any commitment. It is true of marriage. It is true of voting. It is true of signing a petition. It is true of the establishment of priorities. It is true of those values we embrace independently—that is, with the Adult . . . the capacity for probability estimating can be increased by conscious effort. Like a muscle in the body, the Adult grows and increases in efficiency through training and use. If the Adult is alert to the possibility of trouble, through probability estimating, it can also devise solutions to meet the trouble if and when it comes.

In other words, probabilities, although not formalized as much as we will do here, *may* be a part of the processes we use in making decisions in our daily lives. If you can accept this premise, it may be easier to visualize the approaches to certain quantifiable decisions as presented here.

Our concern is restricted to decisions where the action can be stated in quantitative terms. Principally this relates to projections about averages or percentages. Here inference is divided into two major categories.

The first, estimation, concerns projections about the size of a parameter. For example, most college students use their cumulative grade point

The fully representative sample shown at the bottom of this photograph is every statistician's dream. Using statistical tools, we can establish procedures that bring us close to obtaining "perfect" samples.

average, gpa, as a personal gauge about how well their college work is going. A gpa of 2.8 (on a 4.0 scale) at the end of the junior year is likely to be sufficient for an individual to estimate his chances of eventually graduating. One probably feels that with this many hours behind, "I can achieve at least a 2.0 overall gpa." In estimation we commonly want to develop probable intervals or spans of values. For example, recall the statement in the first chapter "If the Nielsen sample . . . produces a rating of 20% for a number of programs, the *true* rating lies somewhere between 18.8 and 21.3 for two out of three programs." We are now equipped to understand the meaning of such a statement.

A second major decision area is hypothesis testing. In statistical testing we attempt to substantiate a tentative theory about the value of some parameter. Politicians typically employ polls to discern their voter strength. If sample evidence supports a majority of people in favor of him, the candidate may take a somewhat relaxed campaign approach. If, on the other hand, the polls show him to be the underdog, favored by less than a majority, he may decide to take quite different action. A systematic approach to hypothesis testing and making decisions based on sample evidence is a central theme in Chapter 9.

In this chapter we discuss estimation and sample size relations. Processes are outlined for the estimation of two parameters—the mean, μ, and the binomial population proportion, p. Sample size is central to determination of proper probability forms and here large sample techniques are discussed. These relate to the standard normal distribution as described by the central limit theorem. Later chapters will concern smaller sample cases.

1/ Estimation of the Mean—Large Samples

Estimation is a process for sizing parameters. Initially, parameter size can be approximated through a point estimator, that is, by a statistical rule. Thus the grade point average you have achieved in coursework so far gives a value to approximate your eventual grade point average. Suppose that for one year of work your grade average is 3.1. Then $\overline{X} = 3.1$ is one point estimate for the final grade average. Here the parameter is μ, the final cumulative grade point average, the estimator is $\overline{X}$. The estimate, 3.1, is based on your grades for one year.

> A POINT ESTIMATE is the single value taken by an estimator (statistic) when it is applied to a specific sample.

Definition

This value—3.1—may sound fine, but what does it mean in terms of grade security? For example, had you taken a different selection of courses would the value be higher or less? Will the value likely go up or down in subsequent terms? Other than for comparison to absolutes (like 2.0 and 4.0), a point estimate affords limited information. Also, point estimation gives no way of assessing the chance that the value is reasonably close to the eventual average. Of course there is something better that overcomes these shortcomings. This is *interval estimation*.

In interval estimation we seek a span of values that "should" include the true value. Here an interval of values is developed about the point estimate. The process is developed first by example.

> In an attempt to estimate the time required to commute to an 8:00 A.M. class, a student has timed her trips for 36 class days (assumed a random sample). The trips have averaged 24.2 minutes with a standard deviation 4.2 minutes. She wants an interval on the average time needed to reach her 8:00 A.M. class. Suppose a .95 probability value.

Example

> We seek two numbers, for now l and u, that will contain a span of possible values for "average" time of travel. The sample of 36 is sufficiently large to allow use of the central limit theorem. The problem is one of scaling from the Z back to the distribution of sample means. A 95% centered probability interval on the scale of Z, where $\sigma_Z = 1$, spans -1.96 to 1.96. We seek a comparable span on the scale of $\overline{X}$-values, centered at the value for the point estimate. This is the observed

Solution

Estimation and Sample Size

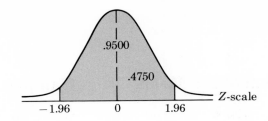

value $\bar{X}_0$, or 24.2 minutes. From this as the central value, we record a distance Z(a number of) deviation units in either direction. On the scale of $\bar{X}$-values, the deviation unit is $\sigma_{\bar{X}} = \sigma/\sqrt{n}$. Hence the forms

$$L = \bar{X}_0 - \left(\frac{Z\sigma}{\sqrt{n}}\right) \quad \text{and} \quad U = \bar{X}_0 + \left(\frac{Z\sigma}{\sqrt{n}}\right)$$

delineate the span. Treated with sample values, these forms give specific values l and u respectively. Here

$$l = 24.2 - \left(\frac{1.96 \cdot 4.2}{\sqrt{36}}\right) = 24.2 - 1.4 = 22.8 \text{ minutes}$$

$$u = 24.2 + \left(\frac{1.96 \cdot 4.2}{\sqrt{36}}\right) \doteq 24.2 + 1.4 = 25.6 \text{ minutes.}$$

Conclusion

This person can be fairly sure that her trips will average from 22.8 minutes to 25.6 minutes. Her travel could be planned accordingly.

The procedures shown in the example allow us to formulate the interval estimation problem in general terms, and then to resolve it in the specific case.

Definition

In INTERVAL ESTIMATION of the mean, μ, we seek a pair of statistics, say L and U, such that $P(L < \mu < U)$ is some preassigned value between 0 and 1, inclusive.

In the example the probability level, also called the *confidence level*, was preassigned at .95. Then, prior to drawing a sample, we wanted to have $P(L < \mu < U) = .95$, a 95% confidence interval.

Rule

For estimation of the population mean for large samples, the interval of estimation is defined by

$$L = \bar{X} - \left(\frac{Z \cdot \sigma}{\sqrt{n}}\right) \quad \text{and} \quad U = \bar{X} + \left(\frac{Z \cdot \sigma}{\sqrt{n}}\right)$$

Estimation of the Mean—Large Samples

When a specific sample is applied, the statistics L and U take on specific values, the estimates l and u respectively. This notation was used in our first example.

The sample mean has several properties that single it out as "the statistic" for estimating the population mean. We discussed the fact that under random sampling the distribution of sample means is centered at the true value. Then $\overline{X}$ is an unbiased estimator for the population mean. Furthermore, this statistic, among unbiased statistics, has the smallest sampling error. That is, for all sample sizes the values taken by $\overline{X}$ are grouped closer about the true mean than are those for statistics like the median or the mode. So we have the best chance of getting a "good" estimate for the population mean, μ, when using the sample mean, $\overline{X}$.

A second example uses this process for setting a confidence interval, but has a different confidence level.

Example

Since pigs are his major cash product, Farmer Rundell is interested in the average size for litters of baby pigs. If his brood of thirty-four sows average nine baby pigs with a deviation of 1.5, what is the true average size of all litters produced on this farm? Assume a 98% confidence level and consider these observations a random sample for all litters on this farm.

Solution

As before we want an estimate on the true value of a mean, here μ = the mean number of baby pigs per litter on this farm. Sufficient information is given to apply the rule. However, the probability level, .98, requires a new Z-value. For a centered Z-interval this is:

$Z = 2.33$ (see the figure below) so

$$l = 9 - \left(\frac{2.33 \cdot 1.5}{\sqrt{34}}\right) \doteq 9 - .6 = 8.4$$

$$u = 9 + \left(\frac{2.33 \cdot 1.5}{\sqrt{34}}\right) \doteq 9 + .6 = 9.6 \text{ pigs.}$$

Conclusion

Farmer Rundell should anticipate an average of between 8.4 and 9.6 pigs per litter with probability .98. This means that for 98 out of 100 random samples of size $n = 34$, this process should give an interval that includes the true mean. We assume that this sample is one of the good ones with a chance of only 2 in 100 of being wrong. That is, I am 98% confident that this process will give an interval that includes the true value.

But what is the effect of confidence level on the size of the interval? That is, we can explore the relation of the total (confidence) interval length, $u - l$, to the probability level. Here for 98% confidence, the interval length is $u - l = 9.6 - 8.4 = 1.2$ pigs. What change would result if we reduced the probability level to 95%? Using the bell curve this would reduce the Z-interval from ± 2.33 to ± 1.96 and then $u - l$ to $9.5 - 8.5 = 1.0$ pigs (calculations are not shown). This is, of course, a narrower interval for the lower confidence level. Thus the 98% probability statement gives one more confidence about the inclusion of the true value because it covers a longer interval. This is somewhat false security though, because in having a longer interval we have included more values to contend for the true one. This story is not over. The next section concerns what we can do to get improved estimates (meaning narrow intervals) and still have high confidence.

(*Note:* A word about procedure as used both here and for hypothesis testing. For random samples of $n > 30$ we will assume that the value obtained for s will closely approximate that of the true standard deviation, σ. In practice the value of σ is rarely known, and this approach yields satisfactory results.)

Problems

1. A market research study was made in a community of 20,000 families to establish the average expenditure per month for gasoline, oil, and other motor fluids. If in a random sample of 100 families, the average expenditure was $42.13 with a standard deviation of $6.50, estimate the average monthly expenditure for motor fluids? Use a 95% probability level.

2. In their last contract the average increase in biweekly salary for 49 steel workers was $35.00 with a standard deviation of $5.60. Find a 90% confidence interval for the true average biweekly salary increase for all such workers.

3. A random sample of 64 cars passing a checkpoint on a certain highway showed a mean speed of 60.0 mph with standard deviation of 15.0 mph.
   a. Establish a 95% confidence interval for the mean speed of cars at this checkpoint. Interpret the meaning of this interval.
   b. Establish a 99% confidence interval using the same data. Now compare the size, $u - l$, for the two intervals.

4. A psychologist, while performing a stimulus–response experiment, found that a random sample of 49 college students required an average of 6.20 seconds to respond to one stimulus. The deviation in response time for experiments of this type is about 1.00 second. Establish a 94% confidence estimate on the true time required for college students to react to this stimulus.

5. A sociologist selected a random sample of 50 unmarried students in order to estimate the average number of siblings for families with one or more college students in them. The numbers are *other* siblings in the family.

Number of siblings	Frequency
0	25
1	14
2	7
3	4
Total	50

   a. Give point estimates for the mean, μ, and the standard deviation, σ, in number of *other* siblings.
   b. Construct a 95% confidence interval estimate for the mean. Let $\sigma \doteq s$.

2/ Sample Size

The first question the researcher faces in estimation is, "How large should the sample be?" Before point or interval estimates can be made, a reasonable sample size must be determined. Then estimates can be made from the specific sample of n items chosen.

Estimation and Sample Size

The fundamental relation used to determine sample size for estimation of the mean is shown in the rule below.

Rule

Bound of Error = Z-value · standard error of the mean

$$E = Z \cdot \sigma_{\overline{X}}$$

where

E = the bound on error of estimation or, equivalently, the half-width of the interval = $(u - l)/2$
Z = a Z-value for the level of confidence
$\sigma_{\overline{X}}$ = the standard deviation on the sampling distribution for $\overline{X}$, namely $\sigma/\sqrt{n}$.

See the example below.

Example

An estimate is sought for the average amount of money that college students at a large university annually spend in the local economy. How many students should be surveyed?

Before answering this question, we need to develop some tools. In this case we are attempting to estimate a mean. The error of estimation, $|\overline{X} - \mu|$, has bound $E = Z \cdot (\sigma/\sqrt{n})$, i.e., $|\overline{X} - \mu| \leq Z \cdot \sigma/\sqrt{n}$.

Rule

For estimation of the mean with a designated bound of error, E, and for random sampling, $n > 30$,

$$E = \frac{Z\sigma}{\sqrt{n}} \text{ so that}$$

$$n = \left(\frac{Z\sigma}{E}\right)^2.$$

If σ is not known it must be estimated, perhaps by previous experience, e.g., from the sample values found last week, or perhaps by a value obtained by others in a very similar experiment. Thus, in sample size considerations there are four items to consider:

1. n, the sample size
2. Z, the indicator of the probability level
3. σ, the measure of dispersion in the original distribution
4. E, the maximum amount of difference allowable between point estimates and the true value. This is the bound of error.

Knowing or specifying any three allows us to find the fourth. In the example above we seek the sample size, n, after finding or evaluating the other factors.

It is likely that the researcher would be asked to specify values for both the confidence level, a Z-value, and the bound of error, E, in this problem. Suppose these were set at 95% confidence so $Z = 1.96$ and $E = \$50.00$. However, finding the standard deviation of dollars spent by individual students, σ, requires information from the population. A preliminary sample of n_1 (> 30) students might be taken. Thus if $s = \$256.00$ is obtained from the preliminary sample, then

$$n = \left(\frac{Z\sigma}{E}\right)^2 = \left(\frac{1.96 \cdot \$256.00}{\$50.00}\right)^2 \doteq (10.04)^2 \doteq 100.8.$$

Solution

Generally the calculated sample size is not a whole number so we round up to the next larger whole number. This will guarantee that the specified confidence and the maximum error can be met. Here 100.8 would be raised to $n = 101$.

In the last section, it was indicated that we would consider how to get "better" estimates. The next example shows that for fixed confidence (95%) an improved estimate in terms of reduced interval width requires increased sample size.

Suppose that in the previous illustration we wanted less error—say an estimate for the average amount spent by these students with a maximum error of $25.00. We seek an estimate that is twice as precise as the first one, $50.00/2. Assuming the same confidence, 95%, and $\sigma \doteq \$256.00$, then

Example

$$n = \left(\frac{1.96 \cdot \$256.00}{\$25.00}\right)^2 = \left(\frac{1.96 \cdot \$256.00}{(\$50.00/2)}\right)^2 = \left(2\frac{1.96(\$256.00)}{\$50.00}\right)^2$$

$$= 2^2\left(\frac{1.96 \cdot \$256.00}{\$50.00}\right)^2 \doteq 4 \cdot (100.8) = 403.2$$

Rounding up, $n = 404$. (*Note:* Using a different order of procedures may give a slightly different result.)

Because we are asking for a better estimate in the sense of less error, we must pay by taking a larger sample. In fact, cutting the bound of error in half has produced a fourfold increase in sample size. (Recall the A. C. Nielsen example with the picture of the woman in Chapter 1?)

Estimation and Sample Size

In order to compare for a better estimate, observe the effect of requiring a higher probability level while maintaining the original bound of error.

Example

Suppose we want 99% confidence ($Z = 2.58$) and $E = \pm \$50.00$ in the estimate of average amount spent by these university students. That is, compared to the original specifications, we want increased confidence while the allowable error is unchanged. Again this increases the sample size:

$$n = \left(\frac{Z\sigma}{E}\right)^2 = \left(\frac{2.58 \cdot \$256.00}{\$50.00}\right)^2 = (13.21)^2 = 174.5.$$

Rounding up, $n = 175$.

Conclusion

We can say that, generally, better estimates in the sense of increased confidence or decreased bound of error or both can be achieved by taking a larger random sample. This may be expensive, however. For example, it might cost $.50 to interview each student so 101 interviews would cost only $50.50 where 404 interviews would cost $202.00. The latter cost might well be prohibitive for say, a project for a class in marketing research. A balance must be met between cost, confidence, and bound on error of estimation. (Recall the Nielsen example concerning "Why Use a Sample?" in Chapter 1.)

As a postscript to the preceding examples, there is an alternative to increasing the sample size as a means for getting better estimates. This alternative is to use a different, meaning a more complicated, sampling procedure than random sampling. See [4].

The next example illustrates the processes required in a sampling project. Observe that the sample size is determined, then the sample is drawn. Finally, calculations on the sampled units are used to project the mean and related total amounts. The physical procedures of drawing a sample and making the estimate are very similar to those used in the sample inventory example in Chapter 7.

Example

For tax purposes a sample inventory is to be made of the warehouse holdings for a canned foods wholesaler. Generally items are stored in cases of three dozen cans. The warehouse currently contains 23,271 cases. An estimate of the true average price per case, μ, will be projected to an estimate for the total value simply through multiplication by $N = 23,271$. The allowable error for estimating the mean price is $.03 per case. A 95% probability level is chosen.

Solution

The initial step requires determination of proper sample size. Random sampling is used. The sample specifications include $Z = 1.96$ (for 95%

probability) and $E = \pm \$.03$/case. However, the deviation in price per case, σ, is unknown. To estimate σ a preliminary sample of 60 cases is drawn and shows $s = \$1.12$, the deviation in price per case. Sample size can be approximated by

$$n = \left(\frac{Z\sigma}{E}\right)^2 = \left(\frac{1.96 \cdot \$1.12}{\$.03}\right)^2 = (73.17)^2 \doteq 5{,}353.85.^*$$

So, inspection of $n = 5{,}354$ or more cases should give an accurate estimate. "Just to be sure," the company auditors suggest rounding the sample up to an even 5,400 cases, giving a $(5{,}400/23{,}271) \cdot 100 \doteq 23.2\%$ random sample. The sample size is critical since every inventoried case must be inspected. This takes work time, meaning added cost! Next 5,400 sample cases are chosen using invoice numbers with the aid of a random numbers table. Upon inspection, the sample cases show that $\overline{X} = \$17.28$, the estimated average cost per case. From sample specifications this gives ($\$17.28 - \$.03 < \mu < \$17.28 + \$.03$) or ($\$17.25$ to $\$17.31$) per case.

The total inventory value is estimated by

$N \cdot \text{mean} = N\overline{X} = 23{,}271 \cdot \$17.28 = \$402{,}122.88.$

Of the possible intervals (i.e., from different samples) formed through this sampling procedure, 95% will include the true total value. Again we assume this is one of the good ones. The error in the estimate of total value is at most $\$698.13 = 23{,}271 \cdot \$.03$. If the true total is in this interval, then at most it is $\$698.13$ above the point estimate, $\$402{,}122.83$, and not more than $\$698.13$ below this value.

Conclusion

The proverbial question "Which comes first, the horse or the cart?" applies here. In this case, the competitors are sample size and calculation of estimates. In sampling problems, a sample size determination must come first. This is followed by drawing the sample, and then making estimates.

---

*This requirement could be reduced by applying a more exact rule, as in [3], p. 129.

Problems

1. Suppose that the random sample of $n = 101$ suggested in the first example of this section is of students from your college or university. If one such sample showed $\overline{X} = \$1,320.00$ with $s = \$280.00$, what would you estimate as the true mean amount spent by students from your campus in the local economy? Remember the example indicates a 95% probability level.

2. We want to know the average number of kilometers jogged per day by all of the joggers in a certain area. If there are 1,000 regular joggers in this area, what size random sample is required to estimate the mean with 95% confidence and an error of at most 1/4 kilometer? Use $\sigma \doteq 3/4$ kilometer.

3. A normal distribution has $\sigma \doteq 20.0$. How large a sample must one take so that the 99% confidence interval will be not more than 5.0 units in length? Recall length $= u - l = 2E$.

4. A random sample of families is selected to estimate the mean monthly income in a certain area. The standard deviation is approximately $40.00. Use 95% confidence.

   a. For error, E, to be within $10, how large a sample should be selected?
   b. If sufficient funds are available to get a random sample of $n = 100$, determine E, the bound on the error of estimation.

5. A supplier of crushed rock and a road construction firm are seeking an equitable way to measure loads of rock hauled to the job site without having to weigh each one, i.e., gross $-$ net $=$ tare, or load weight. The trucks are all of 20-yard capacity and a single loading operator is used, so conditions are fairly uniform. As a compromise procedure it is agreed to take a random sample of load weights sufficient in number to estimate the true average weight, μ, to within, 0.1 tons. Assume each "yard" of this rock weighs 1.25 tons.

   a. Using a 95% probability level, how many loads should be weighed? Use $\sigma \doteq 0.6$ ton.
   b. If the random sample of the size you suggest in Part a produced $\overline{X} = 25.2$ tons, establish a 95% confidence estimate on μ.
   c. Using information from Part b and the knowledge that a total

of 3,000 loads were hauled for this job, make a point estimate for the total amount of rock (in tons) hauled.

6. In an attempt to determine the wearability of a certain line of tires, a manufacturer decides, primarily because of costs, to road test 20 sets of tires ($20 \cdot 4 = 80$ tires).

   a. What maximum error can be expected in estimating the mean life of all tires in this line? Use 95% confidence and assume $\sigma \doteq 2,000$ miles.

   b. Suppose the maximum error obtained in Part a is to be cut in half while the confidence and standard deviation values remain unchanged. Assuming random sampling, how many sets of tires (a set = four tires) should be tested?

   c. Use the same information as in Part a except now use 94% confidence. Explain the difference between your answers for Part a and Part c.

3/ Estimation and Sample Size in Binomial Experiments

The most common estimates made for the public are estimates of population percentages. These include voter preference polls, college-sponsored polls, and government specifications of such things as prime interest rates, percent of unemployment, the cost of living index, and not least, sports and weather percentages. Although we most commonly see a point or single-valued estimate of percentages, error estimates must be given if the results are to have meaning for statistical projection. That this need is recognized by major polling agencies can be seen in [1].

The basis for estimation of the population proportion from sufficiently large samples is the normal distribution. The point estimator is X/n. That is, for X successes in n trials, the ratio X/n is the relative frequency of success. The central limit theorem yields the result that for samples sufficiently large ($np \geq 5$ and $nq \geq 5$),

$$Z = \frac{(X/n) - p}{\sqrt{\hat{p}(1 - \hat{p})/n}}, \text{ where } \hat{p} = X/n,$$

is approximately $N(0,1)$. From this form, confidence interval estimates on p are derived using the following rule.

Estimation and Sample Size

Rule

For *confidence interval estimates of* p, the population proportion of successes with $np \geq 5$ and $nq \geq 5$ use,

$$L = \frac{X}{n} - \left(Z \cdot \sqrt{\frac{\hat{p}(1-\hat{p})}{n}}\right),$$

$$U = \frac{X}{n} + \left(Z \cdot \sqrt{\frac{\hat{p}(1-\hat{p})}{n}}\right) \text{ with } \hat{p} = X/n.$$

Since we are estimating p, its exact value must not be known. Consequently, the best information available, the estimator $\hat{p} = X/n$, is used in setting the limits. Notice the forms for limits are similar to those for estimating means. The common point estimator is corrected either by addition or by subtraction for a bound on the error of estimation.

The next examples concern estimation of population percentages. In each case the experiment is of the binomial type.

Example

A political candidate has retained a polling agency to determine her voter strength. In its most recent poll, the agency found that (at that time) 225 of a random selection of 400 registered voters would vote for this candidate. What percent of the electorate favored this candidate at the time of this poll?

Solution

Using a 95% confidence level and the last rule, $p =$ the true proportion of voters who prefer this candidate gives

$$\hat{p} = \frac{X}{n} = \frac{225}{400} = \frac{9}{16}, \text{ and}$$

$$l = \frac{9}{16} - \left(1.96 \sqrt{\frac{(9/16)(7/16)}{400}}\right) \doteq .56 - .05 = .51$$

$$u = \frac{9}{16} + \left(1.96 \sqrt{\frac{(9/16)(7/16)}{400}}\right) \doteq .56 + .05 = .61.$$

Conclusion

At this time the candidate likely had a majority of the voter preference. At the 95% confidence level, her estimated voter strength was somewhere between 51% and 61%.

In fact, statistical estimation is widely used as the basis for election projections. It is astounding that, with samples of only 1,500 to 2,500 potential voters, scientific polls can rather accurately predict the outcome of national elections where 70 million or more persons might vote. Major pollsters do, however, use techniques that are more sophisticated than simple random sampling.

Estimation and Sample Size in Binomial Experiments

A second example shows the diversity of application of statistical estimation techniques.

An elementary school principal seeks the true percentage of pupil–day absences in order to determine a standard against which to gauge periods of excessive absenteeism. Her sample is the record of absenteeism for a random selection of 500 pupil–days taken from school records over the past five years. The data gives $X = 30$ absences for the sample of $n = 500$ pupil–days. She chooses a 90% confidence level.

Example

The procedure is as before. Here $p =$ the true proportion of pupil–day absences. Since $\hat{p} = X/n = 30/500 = .06$, and $Z = 1.64$,

Solution

$$l = .06 - \left(1.64 \cdot \sqrt{\frac{.06(1-.06)}{500}}\right) \doteq .06 - \overset{.02}{\cancel{.017}} \doteq .04$$

$$u = .06 + \left(1.64 \cdot \sqrt{\frac{0.6(1-.06)}{500}}\right) \doteq .06 + .02 \doteq .08.$$

The principal can use 4% to 8% as a gauge on the usual percentage of absences. Then, for example, 10% of the students absent on a given day would indicate excessive, possibly unusual, absenteeism.

Conclusion

Suppose in the last example we wanted an estimate with less error. For example, we might want the bound on the error, which was approximately 2%, reduced to 1%. What sample size, under random sampling, would be necessary? This requires new sampling procedures. Sample size in binomial experiments is determined using the fundamental relation:

$$E = Z \cdot \sigma_{\hat{p}} = Z \cdot \sqrt{\frac{\hat{p}(1-\hat{p})}{n}}, \quad \hat{p} = X/n$$

Solving for n is shown in the next rule.

To estimate p in binomial experiments, the sample requirement under random sampling with large samples uses

Rule

$$n = \left(\frac{Z}{E}\right)^2 \hat{p}(1-\hat{p}).$$

This rule is applied to our question.

What sample will be necessary for the principal to estimate the true percentage of absenteeism to within 1% and with 90% confidence?

Example

Estimation and Sample Size

Solution
The probability level is unchanged, but this requires less error than in the preceding example. Logically this sample must exceed the last one, which was $n = 500$. This can serve as one check. Using the rule for sample size gives

$$n = \left[\left(\frac{1.64}{.01}\right)^2 \cdot (.06)(.94)\right] = 1{,}516.93.$$

Conclusion
The principal needs 1,517 (or more) randomly chosen pupil–day records to estimate p to within $\pm 1\%$ and at the 90% probability level.

These two examples illustrate again the fundamental principle that one way to get better sampling results is by taking a larger (random) sample. Of course the best results come from taking a sample of the whole population, a census. However, in most cases this extreme is not economically feasible and so some restrictions are required.

As the next example shows, costs play no small part in the determination of sample size for some projects.

Example
The advertising section for the retailer of a national product wants to know the market share, percentage of total consumer sales, for their particular brand. They have $3,000 for this project and want an estimate to within 2% of their actual market share. Overhead expenses for the project are $500 and it costs $1 per household interview. A confidence of 95% is required. Can their market share be estimated with the required precision for $3,000?

Solution
This question involves two things. First, how much of a sample can the section afford to buy? Since sample funds = total allocation − overhead, or = $3,000 − $500 = $2,500, then at $1.00 per interview, they can afford $n = \$2{,}500.00/\$1.00 = 2{,}500$ interviews. The second requirement is that the bound on error of estimation be $\pm .02$. Since no prior estimate is available for p, we use $\hat{p} = .50$. Can these requirements be met with $n = 2{,}500$, for $\hat{p} = .50$? Using

$$E = Z \cdot \sqrt{\frac{\hat{p}(1-\hat{p})}{n}} \quad \text{for 95\% confidence gives}$$

$$E = 1.96 \cdot \sqrt{\frac{.5 \cdot .5}{2{,}500}} = \left(1.96 \cdot \frac{.5}{50}\right) = .02.$$

Conclusion
The desired precision can just be met. The consumer advertising section should proceed to make its survey. Note that $\hat{p} = .5$ was a guessed value. A better initial value would be the market share for the previous quarter. Of course a previous value is not available if the product is new.

In the last example the value $\hat{p} = .5$ was used yet no justification was given. This value, if used in sample size computations on p, leads to the largest possible (random) sample size required for specified confidence and precision (see [3] for justification of this statement), thus it should be used only as a last resort. One alternative would be to obtain a better initial estimate of p. This might come from a "pilot," or preliminary small-scale survey, or from the scientific results of another closely related survey. Another alternative would be to use more sophisticated sampling techniques. The idea is to maintain design specifications of maximum error bound and specified confidence, yet require a smaller sample. To do so generally takes more planning and more detailed calculations. Less desirable is a third alternative, to relax either the bound on error or the confidence level, and then require a smaller random sample. For more discussion see [4].

Problems

1. In a television quiz show, it was found that a sample of 900 contained 576 people who gave an identical response to a certain stimulus word.
   a. What is the estimated percentage of identical responses?
   b. What is the standard deviation for this estimate? That is, find $\sigma_{\hat{p}}$.
   c. With a probability of .95, what can you say about the error, E, of your estimate in Part a? That is, what is the size of E?
   d. Using the normal approximation, find a 95% confidence interval for the true proportion.

2. A national employment agency, in an attempt to estimate the percentage of college graduates who take a job in the major area in which they were trained, found that 640 in a random sample of 1,000 had done so. Estimate, with 94% confidence, the true percentage who take a job in their specialty.

3. What sample size should be taken in order to estimate p to within 2% with 98% confidence? Assume p is about 0.2.

4. Of 1,000 people treated with a new drug, 200 showed an allergic reaction. With a 90% confidence interval estimate the proportion of the population that would show an allergic reaction to the drug.

5. A high-school teacher wants to estimate the percentage of his students who pass a college boards entrance exam in English. All of the students

take the exam at the end of his course. If, of the 120 students who took his course over the past three years (considered a random sample), 90 passed the boards, estimate the true percentage with 98% confidence.

6. During a production study, a skilled technician is to be observed at randomly chosen points in time during a work day in order to estimate within $\pm.05$ and with 95% confidence the proportion of time that he is productive.

   a. What is the largest sample size required under the foregoing restrictions?
   b. If a study in a similar company resulted in $p = 0.90$, could you use this information to reduce the sample size? To how much?

7. The advertising section for the retailer of a national product seeks the market share, percentage of total consumer sales, for their brand. They have $2,000 for this project and want an estimate of within 2% of their actual market share. Overhead expenses for the project are $500 and it costs $1.00 per household interview. In the last quarter (three-month period) the market share was 25%. A confidence of 95% is required.

   a. Can this proportion be estimated to the desired specifications?
   b. If the cost-determined sample as described above gave $X = 400$, estimate their current market share. Again use 95% confidence.
   c. What sample would be required to estimate the true market share within $\pm 1\%$ (i.e., to 1/2 the error given above)? Assuming a fixed overhead of $500, how much would this sample cost?

4/ Essay Example

This example concerns the Union 76 Fuel Economy tests, developed to measure gasoline consumption of 1975 model cars, that were completed at the Daytona International Speedway in Florida. Determinations were made based on samples of 82 of the most popular 1975 foreign and domestic-make cars. The description that follows not only illustrates a real application of statistical estimation, but also contains several principles of design and control indicative of a well-planned experiment.

In October 1974, Union 76 buying teams went across the country, unannounced, to purchase cars off dealer-showroom floors. No indication

was given as to how the cars would be used. Immediately after purchase the car hoods were sealed and the cars were transported under guard to Daytona Beach. Upon arrival the cars were impounded behind guarded fences on the Speedway grounds. Thereafter each car was assigned a vehicle manager and a professional driver.

The break-in procedure was performed on the Speedway tracks as outlined in each owner's manual. Additional restrictions were made to guarantee uniformity of break-in procedures. For example, all cars were required to maintain break-in mileages within the interval of 2,000 to 2,100 miles. Further, competing cars had similar equipment, such as automatic transmissions, air-conditioning, radial tires, etc. All cars used the same lubricants and unleaded gasoline.

After the break-in period was completed, each car was given a thorough diagnostic check and a tune-up to ensure that it met the manufacturer's specifications. Each car then underwent the Environmental Protection Agency (EPA) emissions test to certify that it met State and Federal emission regulations. The cars were then parked in a guarded and impounded area, locked, with the hoods sealed until the day of testing.

The testing procedure was developed by the Society of Automotive Engineers (SAE). It duplicates three types of driving conditions: stop-and-go driving as encountered in downtown areas of large cities, moderate traffic driving in suburban areas, and high-speed interstate driving. All runs were made on a designated course. For example, the "business cycle" was run on a 2.0 mile course laid out on the Speedway track apron. The cycle consisted of a standing start and finish with seven stops. Average speed was to be 15.6 mph with specified periods of engine idle and a cycle time of 461 seconds. Each car was to run this cycle twice and an average fuel consumption figure obtained.

Test procedures were quite exacting, requiring drivers to accelerate or decelerate at specified rates. Special instrumentation was used on the cars to ensure accuracy and to guarantee that the drivers could in no way significantly affect the mileage results. Starting positions were determined by random drawing and uniformity concerning weather conditions was maintained for all makes in a weight class.

The cars were divided into ten weight classes, with weight defined as "curb" weight (including usual fluids plus 300 pounds). Some of the class ranges were Class 1: 1875–2125 lbs, Class 2: 2126–2375 lbs, etc. All cars in a given weight range were run on the same day. Checks of consistency in mileage readings were made for repeated runs. Some of the official results, point estimates of the economy (average mpg), for these cars appear below. Results are listed by make and model with the engine type (number of cylinders) for each weight class.

Weight Range	Make	Engine type	Urban mpg	Suburban mpg	Interstate mpg
1875–2125 lbs	Honda Civic CVCC (Cal. version)	4	29.6	33.0	31.1
2126–2375 lbs	Datsun B-210 (4-speed)	4	26.5	32.7	31.3
	Datsun B-210 (4-speed, Cal. version)	4	25.5	31.0	32.0
2376–2625 lbs	Toyota Corolla (4-speed)	4	20.2	26.6	27.5
	Toyota Corolla (4-speed, Cal. version)	4	18.5	23.8	26.6
2626–2875 lbs	Chevrolet Vega (3-speed)	4	19.6	25.8	26.5

Note: This test, and subsequently the results, differs from the EPA conducted tests of a similar time period in that the SAE test was an on-the-road test while the EPA test was conducted under lab conditions with cars stationary on a dynamometer test machine.

Statistics on the other make test vehicles are given by the sponsors [5]. Results are also available for tests on the 1976 models, made at Anaheim, California.

REFERENCES

[1] Gallup Opinion Index (Princeton, New Jersey: American Institute of Public Opinion, January 1976).

[2] Thomas A. Harris, M.D., *I'm OK—You're OK* (New York: Harper and Row, 1973).

[3] John A. Ingram, *Introductory Statistics* (Menlo Park, California: Cummings, 1974).

[4] Morris J. Slonim, *Sampling* (New York: Simon and Schuster, 1960).

[5] Union 76 Fuel Economy Tests, 1975 (Palatine, Illinois: Union Oil Company of California, 1975).

Testing Hypotheses (Statistical Decisions)

CHAPTER 9

The second area of inference—testing—has widespread use. In business, government, education, and many other areas, decisions are based on a quantitative appraisal of some population value. That testing affects our lives is evident, for example, in advertising where representative samples are reportedly tested and a determination is made of "a higher average product life" or "a higher effectiveness" or "quicker and longer relief," etc. Other indications that people want test evidence in buying includes current activities in product testing, personal dependence upon "Buyer's Guide" and "Consumer Reports" types of ratings, and wide interest in consumer groups such as the one headed by Ralph Nader.

Life is full of decisions and even the simplest decisions require some planning and consideration of risks. For example, as a student you have the alternatives of either getting up for an early class or of staying at home and getting more rest. If you get up and go to class you might learn something worthwhile. On the other hand, sleeping-in can be enjoyable. You must decide which is the better alternative.

In statistical testing we also consider alternatives and weigh potential positive and negative returns (risks) as a basis for making decisions. The purpose of statistical testing is to develop techniques that afford recognition of the alternatives for action and associated risks. Tests are developed which consider these risks and subsequently lead to the most likely successful path of action (decision). In statistics, however, we are restricted to problems where both the action and the risk are quantifiable.

By eliminating extraneous differences from an experiment, one has a better chance of telling the difference between soft drinks by taste alone.

Testing Hypotheses (Statistical Decisions)

In this chapter we deal with statistical tests on single populations that are based on large samples. Decisions relate to values for (1) a population mean or, in binomial experiments, to (2) the population percentage. The basis for our calculations is the central limit theorem (see Section 7.3). The procedures introduced in this chapter will be used often in later chapters so be sure you know them thoroughly.

1/ Introduction to Testing

Here we will develop techniques for making decisions with the aid of statistics. The procedures in statistical testing are an application of the *scientific method*. These essential steps are followed:

1. Determine the problem or question to be tested.
2. Hypothesize reasonable alternatives to the solution of the problem.
3. Establish (a) procedures for testing the reasonableness of the alternatives and (b) a standard for deciding whether to accept or to reject each.
4. Gather evidence (data) upon which to test the alternatives.
5. Perform the tests and, based on your findings, either make a firm decision or revise the original alternatives and repeat steps 2, 3, and 4.

An example will be used to illustrate the application of the scientific method to statistical testing. At the same time much of the language peculiar to testing will be introduced. Since the problem is stated in all tests that we perform, only steps 2 through 5 are required.

Example

A manufacturer of a sports car wants to make a claim about the good gasoline mileage of its car. In fact they want to state "Our make will average over 30 miles per gallon of gasoline." The question is this: Is the claim of over 30.0 miles per gallon (average) valid? Logical alternatives are either "yes" the mileage claim of over 30.0 miles per gallon is met, or "no" this make car does not average over 30.0 miles per gallon of gasoline. Thus we begin with two *hypotheses,* a *null* hypothesis and an *alternative* hypothesis. In this illustration, the null hypothesis indicates that the mean does not differ *statistically* from 30.0 miles per gallon. This is symbolized H_0: $\mu = 30.0$. The alternative of concern is that the mean for this make car is statistically more than 30.0 miles per gallon. This is symbolized H_A: $\mu > 30.0$. If the manufacturer's claim is correct, representative sample evidence will substantiate the higher mileage.

Definitions

1. An HYPOTHESIS is a statement of belief used as a basis for evaluation of population values or to evaluate the type of probability distribution.

2. The NULL HYPOTHESIS is a statement of no difference or no change, and is symbolized H_0.

3. An ALTERNATIVE HYPOTHESIS is a statement that a difference exists or that a change has taken place and is symbolized H_A.

Several things are noteworthy in the procedures of setting test alternatives. First the null hypothesis is rather a standard or starting point for establishing test procedures. A similar role is played by zero (0) as a starting point for all measure on the scale of real numbers. Thus the null hypothesis will generally be a statement of equality, even though it might not exclude inequalities. The null hypothesis serves primarily to structure the test procedure; it is a test standard.

The alternative hypothesis will be a statement of inequality. In the present test, the manufacturer's claim will be supported if he can substantiate that his car gets more than 30.0 miles per gallon of gasoline. Thus, the alternative hypothesis is what he really wants to test. For this reason it is also called the *test or research hypothesis*.

If other reasonable alternatives exist, they can be tested in turn against the null hypothesis. The standard remains unchanged. It is imperative that for any test the hypotheses be stated so that (1) accepting the null means rejecting the alternative or (2) rejecting the null leads to acceptance of the alternative hypothesis.

Hypotheses should be based on the researcher's knowledge about the question—without leaning on the sample data. It is therefore essential that the null and alternative hypotheses be set *prior to* the gathering of data. That is, should the sample happen to be unrepresentative of its population, the evidence might mislead the researcher to unrealistic test alternatives.

Having set the hypotheses, we next look for a test criterion. Suppose the car manufacturer road-tests a random sample of forty-nine of their make and finds $\overline{X} = 30.4$ miles per gallon with standard deviation of 1.4 miles per gallon. We ask "Is the difference between 30.0 miles per gallon and 30.4 miles per gallon *a significant difference?*"

Testing Hypotheses (Statistical Decisions)

Definition

A SIGNIFICANT DIFFERENCE is one (1) that is too large to be the result of an oddity of the particular sample drawn and (2) that has reasonably high probability that it reflects real differences (change) in characteristics of the population.

A difference small enough to be attributed to chance—the oddity of the particular sample chosen—is termed *a nonsignificant difference.* Our test then will be to decide if 30.0 miles per gallon and 30.4 miles per gallon are significantly different, or if the larger value is just a consequence of this particular sample of 49 cars.

The quest to determine statistical significance leads to the statement of *test errors* and the related *risks* (see Table 1). The two possible errors—types 1 and 2—exist because the test procedures employ samples, imperfect information.

Table 1 / Classifications of Errors in Statistical Testing

Decision based on sample evidence	True Condition	
	H_0 is true	H_0 is false
Reject H_0	Type 1 error	no error
Accept H_0	no error	Type 2 error

In the present test a type 1 error would result from deciding, based on the sample evidence, that this make of car averages more than 30.0 miles per gallon when, if all cars of this make could be tested they actually would show an average of at most 30.0 miles per gallon of gasoline. The risk here is one of faulty advertising, which could result in many dissatisfied customers and a bad dealer image.

A type 2 error would result from deciding that this make car does not average over 30.0 miles per gallon when overall the average is above this value. The risk here is one of forcing the firm to lower its claim on economy and thereby to reduce the desirability-saleability of their product. These are formalized in definitions.

Definitions

1. A TYPE 1 ERROR is rejecting the (null) hypothesis when it is actually true.

2. A TYPE 2 ERROR is accepting the (null) hypothesis when it is actually false.

Although the foregoing discussion adequately describes test risks for this example, it does not afford their quantification. Quantification is essential if the risks are to be considered in statistical testing. The *risks* are quantified as *the probability of making the error,* either of type 1 or type 2.

Introduction to Testing

The structure of the conditional statements for both risks is $P(\text{Decision}/\text{true condition}) = \text{Risk}$. Again, see Table 1.

Definitions

1. The risk of making a Type 1 error $= P(\text{type 1 error}) = P(\text{Reject } H_0/H_0 \text{ is true}) = \alpha$, the α-risk.
2. The risk of making a Type 2 error $= P(\text{type 2 error}) H_0/H_0$ is false$) = \beta$, the β-risk.

Although both risks can be used in setting the test, only the α-risk will be discussed here. With complete information, both risks could be made quite small; but for practical reasons—time and excessive costs—complete information is rarely available. Subsequently, our procedure will be to consider the α-risk more critical. It is fixed at a low value such as 1% or 5%. Then, based on the sample size, a test standard is sought which makes the β-risk as low as possible.

Since $\alpha = P(\text{Reject } H_0/H_0 \text{ is true})$ then $1 - \alpha = P(\text{Accept } H_0/H_0 \text{ is true})$. This suggests a test procedure: For a five-percent α-risk,

$$.05 = P(\text{Reject } H_0/H_0 \text{ is true}) = P(Z < 1.64/\mu = 30.0).$$

The test is on the reasonableness of our assumption that the mean mileage does not exceed 30.0 for this make car. See Figure 1.

Figure 1 illustrates that there is 95% probability that if the true mean is 30.0 miles per gallon, a representative sample from this population of cars will give a Z-score of at most +1.64. Only 5% of the samples will give a Z-score of more than +1.64. Herein, there is only a 5% chance we will get a bad sample and decide erroneously that $\mu > 30.0$ if in fact $\mu = 30.0$ miles per gallon.

Figure 1 / A display of test conditions, one-sided alternative

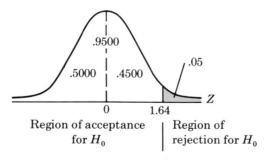

Region of acceptance for H_0 | Region of rejection for H_0

The *central limit theorem* provides the appropriate *test statistic,* the Z. The Z calculated from this sample is,

$$\text{calculated } Z = \frac{\overline{X} - \mu}{\sigma/\sqrt{n}} = \frac{30.4 - 30.0}{1.4/\sqrt{49}} = \frac{0.4}{0.2} = 2.0.$$

Since calculated $Z = 2.0$ is greater than the *critical value* or test value $Z = 1.64$, the sample implies rejection of the null hypothesis; the assumption $\mu = 30.0$ is not supported. This indicates cars of this make average over 30.0 miles per gallon of gasoline. The manufacturer has statistical evidence with which to support his claim.

In summary, the following is offered as a guide for applying *the scientific method for statistical testing* (assuming the problem has been stated).

1. Formulate a null and an alternative hypothesis stated so that accepting (rejecting) the null automatically means rejecting (accepting) the alternative.

2. Based on the alternative hypothesis, the sample size, and the α-risk, establish a test or rule for deciding whether to accept or to reject the null hypothesis. Generally a "does not equal," ($\neq$), alternative hypothesis takes a two-tailed rejection region while a greater than, "$>$," (less than, "$<$") alternative leads to a one-tailed, upper (lower) rejection region. (Identifying these types is a major item in the next section.)

3. A random sample is drawn from which one computes a sample test value. This often requires computing a standard Z-type value.

4. Based on a comparison of the computed value and the critical or test value we decide to either accept the null hypothesis or to reject it. It is desirable to express a conclusion in layman's language.

Rather than accept the null hypothesis many researchers prefer to state that "Based on the given sample we are unable to reject the null hypothesis." This implies two things. First it recognizes the possibility of a type 2 error. Since the alternative hypothesis was not substantiated, this is also an invitation to further testing. That is if we conclude "cannot reject H_0," other viable alternative values can be tested.

Since the α-risk describes a probability measure, it can take any value from 0 to 1, inclusive. But one of the low-risk values, .01, .025, .05, or .10, is commonly used. The choice of α-risk level depends on many factors including, in part, availability of sample and the β-risk level. The α-risk level will be specified in our work.

The test procedure is based on the alternative hypothesis statement. Thus the assertion or claim being tested is quantified as the alternative hypothesis. It expresses some deviation from no change—the null hypothesis. Rejecting the null hypothesis substantiates the claim. Inability to reject the null hypothesis means the sample evidence does not support the researcher's claim.

We have now covered the essentials for statistical testing. The major portion of what follows concerns problems where scientific testing procedures can aid in making decisions.

Problems

1. The workers in a large plant have complained through their union negotiators that they are being underpaid. Both sides (labor and management) agree that the mean wage for all workers in this industry is about $4.75 per hour. State suitable H_0 and H_A making the alternative a statement that supports the workers' claim.

2. What precautions can be taken to reduce the risk of making a type 1 error? A type 2 error?

3. Using $\alpha = .025$ and subsequently the test "Reject H_0 if calculated $Z < -1.96$," complete the decision problem that you started in Problem 1. Suppose a random sample of 49 workers at this plant showed mean wage of $4.54. Assume $\sigma \doteq \$.84$.

4. Based on a sample, the Allied Plastics Corporation wants to test the hypothesis that shipments of plastic pipe meet specifications for compressive strength.
   a. We might refer to the two types of errors in this situation as (1) a seller's risk and (2) a buyer's risk. Describe the error, either type 1 or 2, that would most logically result in a loss for the seller.
   b. What can be done to minimize the seller's risk?
   c. What would you do if you wanted to reduce the probability of a buyer's risk without increasing the seller's risk?

5. A medical researcher is concerned with the effectiveness of a recently developed vaccine.
   a. What null hypothesis is she testing (if she is committing a type 1 error) when she erroneously decides that the vaccine is not effective?
   b. What null hypothesis is she testing (if she is committing a type 2 error) when she says erroneously that the vaccine is effective?

2/ Tests on Means—Large Samples

Tests on means for a single population are subdivided according to the size of the sample used in testing. In this section we consider $n > 30$ which allows use of the central limit theorem and Z-scores. Later on in Chapter 10, where $n \leq 30$ and a population is normally distributed the t-statistic is used. For $n > 30$ it is widely accepted that s^2 gives quite reasonable estimates for σ^2 and comparable results as if σ^2 were known. For random samples of $n \leq 30$, no such claim is made and a new form is necessary.

This section illustrates the testing procedures and interpretations of the results of tests on means. The three types of test alternatives are displayed: (1) "less-than" ($<$), (2) "does not equal" ($\neq$), and (3) "greater-than" ($>$). In the first example, the alternative relation ($<$) is maintained in the "Reject H_0" statement of the test procedure (Step 2). Similarly for all alternative types, the alternative hypothesis determines or structures the test relation. It is appropriate, therefore, to begin each test question by determining a proper alternative relation. This will require careful reading and attention to the question.

Example 1

The faculty at a middle school seeks ways to shorten the school day without, of course, causing any hardship on the learning process. One suggestion is to reduce the present five-minute breaks between classes. In a timed dry run a random sample of 36 students required an average of 4.85 minutes to change classes. The standard deviation in time between classes is about 1/2 minute. Test, using the sample dry-run times, whether the mean time could be reduced for $\alpha = .025$.

Solution

The steps in the solution are those of the scientific method. The first step is the key step because it sets the rejection region and subsequently, the test for $\mu =$ the true mean time, in minutes, required between classes. The test is "lower-tailed" since the problem concerns a "reduced" time.

1. $H_0: \mu = 5.0$ minutes
    $H_A: \mu < 5.0$. This requires a one-tailed (lower) rejection region.

2. For $\alpha = .025$, $n = 36$, the appropriate test value is a Z-score,
    Test: Reject H_0 if calculated $Z < -1.96$
    Cannot reject H_0 if calculated $Z > -1.96$
    Reserve judgment if calculated $Z \doteq -1.96$

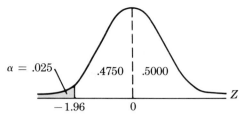

3. Using the dry-run sample times,

$$\text{Calculated } Z = \frac{4.85 - 5.00}{(1/2)/\sqrt{36}} = \frac{-.15}{(1/12)} = -.15 \cdot 12 = -1.80$$

4. Decision: We cannot reject H_0 for $\alpha = .025$ (Since $-1.80 > -1.96$)

These students cannot move between their classes in less than 5.0 minutes. The break between classes should not be reduced.

Conclusion

In Step 2 it was indicated that no decision would be made if the calculated Z was approximately -1.96. Although it is a very coarse rule, the reserve-judgment decision will be used when $|\text{calculated } Z\text{-test } Z| \leq .05$. Generally there is room for at least $5/100$ error due to rounding inaccuracies, etc. When economically feasible, this indecision should be followed by drawing another sample and retesting; the process should be repeated until a firm decision is reached. Hereafter the reserve judgment statement will be deleted from the statement of the test (Step 2). However, it should be applied if the test indicates a close value. The next example uses a two-tailed test.

Bolts used to assemble car engine blocks should average 1.500 inches in length. If a random sample of 900 such bolts, taken from a single machine, gives $\overline{X} = 1.504$ inches with $s = .050$ inches, is the machine within acceptable adjustment? Use $\alpha = .05$.

Example 2

Here $\mu =$ the mean length of bolts made by this machine. As this question illustrates, one must often use logical considerations about the problem to define the test. Logically bolts either too long or too short would not be acceptable—either extreme could cause engine damage. The alternative hypothesis is two-sided.

Solution

Testing Hypotheses (Statistical Decisions)

1. $H_0: \mu = 1.500$
   $H_A: \mu \neq 1.500$

2. $\alpha = .05, n = 900$

   Test: Reject H_0 if calculated $Z < -1.96$ or if calculated $Z > +1.96$.
   Cannot reject H_0 if $-1.96 <$ calculated $Z < +1.96$.

3. Calculated $Z = \dfrac{1.504 - 1.500}{.050/\sqrt{900}} = \dfrac{.004}{.050/30} = 2.40$

4. Decision: Reject H_0, at $\alpha = .05$.

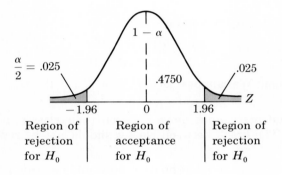

Conclusion

The machine needs adjustment; the mean is not 1.500 inches. The alternative hypothesis does not specify whether the bolts are generally too long or too short. The direction of necessary adjustment is obtained from a confidence interval estimate. For $1 - \alpha = 0.95$, the confidence interval is:

$$1.504 - \left[1.96 \cdot \dfrac{0.05}{\sqrt{900}}\right] < \mu < 1.504 + .003, \text{ or } 1.503 < \mu < 1.507 \text{ inches}$$

Noting that the least value of this interval, 1.503 inches, exceeds the ideal value, 1.500 inches, we conclude that the machine is making bolts that on the average are longer than is desired. Proper adjustment should be made.

In many cases of exploratory research the first alternative is "does not equal." If the null hypothesis is rejected, a more discriminating test, i.e., one-tailed, is then made. New sample data must be used in each succeeding test. Thus, in the last example, another sample might be drawn and the alternative $H_A: \mu > 1.50$ inches tested. Also, in a two-tailed test $1 - \alpha$ is equivalent to the "confidence level" in interval estimation.

A final example employs an upper-tailed rejection region.

Tests on Means—Large Samples

Example 3

A manufacturer of pet foods is considering the issue of bonuses to its distributor salesmen as an incentive for more sales. Forty-nine of the salesmen were randomly chosen to operate on the usual commission plus a bonus for one year. Sales prior to the issue of bonuses averaged $1,550 per distributorship with a standard deviation of $249. During the trial period sales to the same distributors averaged $1,647 (after adjustment for inflation in prices and for the cost of the bonuses). Based on the criterion of statistically increased sales, are the bonuses worthwhile? Use $\alpha = .01$.

Solution

For $\mu =$ the average sales to distributors

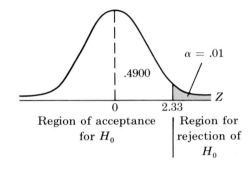

1. $H_0: \mu = \$1,550$
   $H_A: \mu > \$1,550$, one-tailed (upper) test

2. $\alpha = .01$, $n = 49$, so

   Test: Reject H_0 if calculated $Z > 2.33$

3. Calculated $Z = \dfrac{\$1,647 - \$1,550}{\$294/\sqrt{49}} = 2.31$

4. Decision: Reserve judgment as $|2.31 - 2.33| \leq .05$.

Conclusion

This information does not allow a clear decision at the preassigned $\alpha = .01$ risk level. More information should be gathered under the trial system and another test run.

Each example in this section illustrated a different type of alternative hypothesis. The between-class time example illustrates the less-than or lower-tailed alternative test. The does-not-equal alternative is depicted in the bolt example (this is called a two-tailed test because the null hypothesis is rejected if the sample Z-value falls in either tail). The "bonus" sales commission example used an upper-tail or greater-than alternative test.

Testing Hypotheses (Statistical Decisions)

Observe that in each of these three examples the claim of change is in the alternative hypothesis. Also, the alternative hypothesis structures the test procedure. At Step 2 the inequality for rejecting the null, hence for accepting the alternative hypothesis, is the same as that of the alternative hypothesis in Step 1. The conclusion either (1) indicates insufficient evidence to reject the claim or (2) rejects the claim. This structure is important because it is used throughout our testing work.

Problems

1. At the end of an interpersonal group session involving 36 persons, the average tolerance score, as measured on a standardized test, was increased by 12 points. Is the claim justified that this experience will increase the tolerance score by more than 10 points? Use $\alpha = .05$ and $\sigma \doteq 2$. Complete each of the following:

   a. H_0:
      H_A:
   b. Test:
   c. Value of calculated statistic:
   d. Decision:
      Conclusion:

2. Marbles manufactured by a certain company are supposed to weigh 2.00 ounces with a standard deviation of 0.10 ounce. If a sample of size 100 has an average weight of 1.98 ounces, can we say that the average weight is less than 2.00 ounces? Let $\alpha = .10$. Complete each of the following.

   a. H_0:
      H_A:
   b. Rejection region:
   c. Value of calculated statistic:
   d. Decision:
      Conclusion:

3. The mean valuation for single-family homes in a large city is $38,950 with a deviation of $5,200. It is questioned whether the valuation for homes in the University area of this city is statistically the same as the mean valuation. A random sample of 36 homes in the University area showed a mean of $42,460. Are the averages statistically equal? Use $\alpha = .05$ and complete each part, a–d, as in Problems 1 and 2.

4. During the past 33 years a certain university has averaged six first-

place awards in the Drake relays with a deviation of three awards. At $\alpha = .025$ does this indicate this university will *average* over five first-place awards at this competition?

5. An economy-minded family is interested in whether their food costs would be substantially reduced by their raising a garden. Suppose that in a planned test for a random selection of 49 families who grew gardens (1/8 acre each), they spend an average of $300.00 per month for food (including the cost of planting and caring for the garden). Is this expenditure statistically less than expenditures for similar-sized families whose monthly food-bill average, according to national reports, is $320.00 with deviation $50.00? Use $\alpha = .05$.

6. In an elementary school cafeteria the cooks want to prepare, on the average, about the number of dinners that will be served. From their experience they think that about 525 dinners will generally be needed. If a random sample of 36 lunch days gives a mean of 519 with a standard deviation of 24, does the sample evidence agree with the cooks' experience? Use $\alpha = .05$.

3/ Binomial Experiments—Tests on Proportions

Tests on proportions follow the same patterns developed in Section 1 for testing means. The tests are restricted to the case of sufficiently large samples. Again, the test statistic is the Z.

Rule

For binomial experiments with sufficiently large sample, ($np \geq 5$, $nq \geq 5$), the sampling distribution

$$Z \doteq \frac{\frac{X}{n} - p}{\sqrt{\frac{pq}{n}}} = \frac{X - np}{\sqrt{np(1-p)}}$$

is approximately the standard normal, $N(0,1)$.

This Z-form was used in Section 8.3 for estimation of p. In testing, the value for p is that specified under the null hypothesis. For example, for $H_0: p = .4$, the value .4 replaces p throughout the Z-form.

Testing Hypotheses (Statistical Decisions)

Several examples show testing the true proportion of success, p, in binomial experiments. Common applications include tests concerning voter opinion, percentage of defective parts on a production line, and consumer preferences concerning marketable goods. Here is another application.

Example

A company producing door locks has experienced 8% defective items. Each hour a random sample of 100 locks is selected for quality control purposes. If during the 3:00 P.M work hour a sample with 15% defective parts was obtained, should production be stopped to adjust the manufacturing process? Test at $\alpha = .02$.

Solution

Let $p =$ the true percentage of defective door locks being produced. Since we are concerned with protecting against too many defective items the alternative hypothesis is "greater-than."

1. $H_0: p = .08$
   $H_A: p > .08$

2. $\alpha = .02, n = 100$

   Test: Reject H_0 if calculated $Z > 2.05$

3. Calculated $Z \doteq \dfrac{15 - 8}{\sqrt{100 \cdot .08 \cdot .92}} \doteq 2.58$

4. Decision: Reject H_0, $\alpha = .02$

Conclusion

The percentage of defective door locks has increased significantly. Based on this criterion the manufacturing process should be stopped and adjustments made.

The next example illustrates an evaluation on a matter of common interest on most university campuses, student opinions.

Example

In a sample survey conducted at a coeducational university, 400 randomly selected female students were asked if they were in favor of having no restrictions on hours for women. There were 186 women favoring no restrictions. Is there sufficient evidence to indicate that less than a majority of the female students on this campus favor no restrictions on their hours?

Solution

Since the α-risk level is not specified, we are at liberty to make our own choice, say $\alpha = .05$. For $p =$ the percentage of female students at this university who favor no restrictions on their hours.

Binomial Experiments—Tests on Proportions

1. $H_0: p = 1/2$
   $H_A: p < 1/2$

2. For $\alpha = .05$, $n = 400$

   Test: Reject H_0 if calculated $Z < -1.64$

3. Calculated $Z = \dfrac{186 - 400 \cdot (1/2)}{\sqrt{400 \cdot (1/2) \cdot (1/2)}} = -1.40$

4. Decision: We cannot reject H_0 for $\alpha = .05$.

Conclusion

The percentage of female students at this university who favor no restrictions on hours for women is *not* under 50%. Either female opinion is equally divided on this issue or the majority favor no restrictions. Further testing is appropriate.

A 98% confidence estimate is set to get a better feeling for the true proportion of female students who favor no restrictions on hours,

$$\left(\frac{X}{n} - 2.33 \cdot \sqrt{\frac{\hat{p}(1-\hat{p})}{n}}\right) < p < \left(\frac{X}{n} + 2.33 \cdot \sqrt{\frac{\hat{p}(1-\hat{p})}{n}}\right)$$

The confidence interval is based on sample data with $\hat{p} = 186/400 = .465$:

$$\left(\frac{186}{400} - 2.33 \cdot \sqrt{\frac{.465(1-.465)}{400}}\right) < p < (.46 + .06), \quad .40 < p < .52.$$

This appears as a situation where opinion is very nearly evenly split. Since the interval includes values less than .5 as well as some above .5, I conclude that there is not a general consensus for or against this issue.

In the next example the question may seem somewhat indirect. The point is that you should always read the problem carefully to identify the question, hence the test (alternative) hypothesis. In this case the phrase "significantly greater than" is the key to the alternative statement.

Example

The market research section of a soft drink company wants to determine whether teen-agers can identify their cola from other brands just by its taste. An experiment was designed in which two hundred randomly-chosen teen-agers were each given four small glasses of cola, only one of which contained the company's brand. After tasting all four drinks, 64 correctly identified the company's cola. Are the results significantly greater than that expected by chance? Use $\alpha = .025$.

Solution

In deciding what hypothesis to test, we begin by assuming equal chance of selection for each of the four glasses. That is, if an individual

215

cannot actually identify this cola by taste, the chance is 1/4 for his identifying it by guessing. An ability to discern the cola by taste from among the different brands should be indicated by statistically more than 1/4 (something over 25 per 100 tasters) identifying the company brand. Then for p = the true percentage of teen-agers who can identify this cola from the other three by its taste, with $\hat{p} = 64/200 = .32$,

1. $H_0: p = 1/4$
   $H_A: p > 1/4$

   Test: Reject H_0 if calculated $Z > 1.96$ (i.e., $\alpha = .025$)

2. Calculated $Z = \dfrac{(64/200) - (1/4)}{\sqrt{(.25 \cdot .75)/200}} = \dfrac{.07}{.03} = 2.33$

3. Decision: Reject H_0 for $\alpha = .025$

Conclusion

Teen-agers, in general, have some ability in identifying this cola by its taste alone. More than 25% of all teen-agers have the ability to identify this cola in competition with the three other brands, $\alpha = .025$.

Although we have given evaluation of α- and β-risks minimal discussion, these are important considerations. For example one might think it desirable to make α a low value, say .01, in order to reduce the chance of rejecting a correct (null) hypothesis. This is false security for in cases where our conclusion is "accept H_0," a low α-risk causes a higher chance for a false acceptance of H_0; that is, increased β-risk. As long as we can afford only a fixed and not very large sample, like 50 to 100 units, there should be a balance of α- and β-risk levels as protection against both possible wrong conclusions. Remember that as α goes toward zero, the β-risk goes up, generally in a disproportionately quick manner, and vice versa. Consider that in this work we have given α a fairly small value—.10 or .05 or .025, etc.—and have ignored the value of β for the test. First there are ways for determining the probability level of β. See [2]. Secondly it is not always wise to protect the α-risk at the expense of a higher β-risk. In testing medicines (drugs) the reverse is commonly preferred, as for example,

H_0: the drug may be useful (i.e., is harmless or possibly helpful)
H_A: the drug is not useful, (i.e., may in fact harm one's health).

Here the α-risk concerns the possibility of discarding a drug that might be beneficial to human health. Yet the more critical β-risk concerns the possibility of accepting a drug that might harm rather than help people! The potential damage of the latter action is the much greater concern. Here the β-risk must be the one controlled at a low level.

This ends our discussion of applications of the central limit theorem and related forms to single population tests and estimation. Subsequent chapters include hypothesis tests for other cases, including small sample ($n \leq 30$) tests on means for normal distributions, tests for differences in means, and applications to linear forms (regression). The processes shown in Chapters 8 and 9 will be substantially expanded hereafter. The essay by Dunnett in [1] gives an excellent application of the principles of hypothesis testing and also gives an intuitive discussion of the relation between α- and β-risks in the area of drug screening.

Problems

1. During a study of junior high school youth, a random sample of 900 students was given a test on attitude toward littering. The test showed 576 had an attitude *against* littering. Test the hypothesis that 65% of those in junior high have an attitude *against* littering. Use $\alpha = .04$.

2. A random sample of 100 men and 100 women in a city with 10,000 registered voters showed 73 men and 69 women registered as Democrats. Test the hypothesis that 70% of the registered voters in this community are Democrats. Use $\alpha = .075$.

3. A team of doctors claim to have developed a medicine that will, with 80% effectiveness, stop the growth of skin cancer on mice. The medicine was used to treat a random sample of 400 mice that were afflicted with this cancer. The cancerous growths were entirely stopped on 310 mice. Test *against* their claim at $\alpha = .075$.

4. A dairy will change its milk containers from glass to paper if the change is favored by a majority of its customers. In a survey of 200 customers, 110 indicated they were in favor of the change. What should the dairy do? State the statistical hypotheses, then show calculations and critical value. Test at the significance level of .05, and state practical conclusions.

5. In a sample survey conducted at Kent University, 100 randomly selected students were asked if they favored a change from the quarter to the semester system. Fifty-seven answered "yes" and 43 answered "no." Is there sufficient evidence to conclude that a majority of the students were in favor of such a change? State the hypothesis to be tested, test at the .05 level of significance, show calculations, and state your conclusion.

Testing Hypotheses (Statistical Decisions)

6. Adventureland officials are considering a Mother's Day event with free admission for all women aged 18 or older. If for a random selection of 220 admissions, 60 are women aged 18 or older, test the hypothesis that over 25% of Mother's Day admissions would be free. Use $\alpha = .025$. (*Note:* We will assume Mother's Day would be similar to any other day in the proportion of women aged 18 or older in attendance).

4/ Summary for Chapters 7, 8, and 9

Statistical results in inference are based on sampling because it is generally uneconomical and often unnecessary to observe the entire population. With reasonable precautions, quite good results can be obtained from samples. Random sampling is one procedure that yields estimates representative of the population. Moreover probability forms derived from random sampling are available for inference on population means, percentages, and so forth. Random number tables and random number generators (by computer) can be used to select random samples.

The probability distribution for all possible values for a statistic, as a random variable, is called its sampling distribution. The statistic that we use most is the sample mean. For large samples ($n > 30$) probability questions about values of the sample mean are treated using the central limit theorem. This rule has wide use since it says that no matter what form the original distribution, as long as the mean and variance are finite, the sampling distribution of the mean is approximately normal. This allows the use of Z-scores in answering probability questions about values on sample means. The rules hold for $n \leq 30$ as long as X has a normal distribution and σ^2 is known.

The two major subdivisions of statistical inference are (1) estimation, which concerns approximating the values of parameters, and (2) testing in decision-making situations that involve quantifiable risks. Parameters for study include population means and proportions.

Single-valued or point estimators include the common statistics $\overline{X}$ for μ, and X/n for p. Confidence interval estimates on the mean ($n > 30$) and on population proportion ($np \geq 5$, $nq \geq 5$) are based on the central limit theorem. An interval estimate is defined by its limits (l, u). The confidence level indicates the amount of faith that we place in the sampling procedures used to establish the intervals. Common confidence (probability) levels are .90, .95, and .99.

A researcher must first decide upon a sampling technique, then on an appropriate sample size. After these are determined, a sample or samples are drawn and estimates made. The foundation of classical estimation is the technique of random sampling.

Hypothesis testing is an application of the scientific method. The steps in the scientific approach include:

1. Identify the problem.
2. State alternative hypotheses in an attempt to describe true conditions.
3. Establish a test procedure—a method for deciding from sample evidence which alternative is most reasonable.
4. Draw a representative sample and apply the test.
5. Make a firm decision or revise the hypotheses and continue with further tests.

Thus far our tests have concerned the true values for the parameters, μ and p. The testing principles are not, however, restricted to these two cases, but much of what follows depends on the test procedures as defined in Chapter 9.

The "direction" of a test is determined in the alternative hypothesis. A "does-not-equal" alternative sets a two-tailed region for rejection of the null hypothesis. Accordingly, the "less-than" alternative sets a lower-tailed rejection region, while the "greater-than" alternative requires an upper-tailed rejection region.

Two test risks are considered. The α-risk measures the chance for incorrectly rejecting the null hypothesis. The β-risk concerns the chance of accepting a false null hypothesis. A good test requires maintaining low values, near zero, for both risks. But this means taking a substantial sample which may be expensive. Thus a reasonable balance must be attained to moderate both (1) the risks (α and β) and (2) the cost as reflected in the sample size.

REFERENCES

[1] Charles W. Dunnett, "Drug Screening" in *Statistics: A Guide to the Unknown*, edited by Judith M. Tanur et al. (San Francisco, California: Holden-Day Inc., 1972).

[2] John A. Ingram, *Introductory Statistics* (Menlo Park, California: Cummings, 1974).

[3] *The 1975 World Almanac and Book of Facts* (New York: Newspaper Enterprise Corporation, 1975).

Review Problems

1. Assume the mean is 20.6 inches and standard deviation is 2.0 inches for the length of northern pike speared by ice fishermen on Upper Crow Wing Lake.
   a. If the length of these fish is normally distributed, what proportion will be 16.0 inches to 20.0 inches long?
   b. What is the probability that 64 pike speared in this lake (considered a random sample) will have an average length of 21.0 inches or less?

2. In a random sample of 200 voters, 80 were in favor of a candidate and 120 were not. Construct a 95% confidence interval for the true proportion of voters in this population who favor this candidate.

3. Given that $\overline{X} = 20.0, s = 4.0, n = 36$, and with X normally distributed:
   a. compute 95% confidence limits for μ.
   b. decide what sample size would be required to give 1/2 the maximum error, E, indicated in Part a?

4. A machine is set to produce no more than .10 defective parts when properly adjusted. After this machine had been in operation for some time, a sample of 100 pieces was tested. Fifteen defective pieces were observed.
   a. Is there evidence at the 5% level that the machine needs readjustment?
   b. Using your hypotheses, describe the possible error which is controlled at 5% risk.

5. The average life-span for male Americans in 1965 was 66.8 years [3]. If the average had become 67.6 in 1974, does this evidence a significant increase in mean age? Use $\alpha = .05$ and $\sigma_{\overline{x}} \doteq .2$ years.

6. The mean of a certain type of battery is 1,500 days with $\sigma \doteq 90$ days. If 100 batteries are chosen at random, what is the probability that their average life will exceed 1,490 days?

7. Of the letters A, B, C, D, and E, what is the probability that a random sample (without replacement) of:
   a. $n = 2$ would include both letters A and B?
   b. $n = 3$ would include the letters A and B?

8. A student council candidate found that in a random sample of 100 students 54 favored him.
   a. Using the normal distribution, construct a 90% confidence interval on p. Define p.
   b. How large a sample must he take to be sure that his error was no larger than $\pm .04$ with a confidence of 90%?

9. A random sample of households is selected in order to estimate the mean monthly income in a certain area. The standard deviation is approximately $40.00. Use 95% confidence.
   a. For error, E, to be within $10, how large a sample should be selected?
   b. If sufficient funds are available to get a random sample of $n = 100$, how big an error, E, should be expected?

10. A normal distribution has a mean of 26.50 and standard deviation of 1.80. What is the probability that a representative sample of 36 observations will have an average of more than 25.75?

11. Assume the average amount in deposit for the 25,000 depositors at a bank is $425.00, and that the standard deviation is $80.00. What is the probability, in a simple random sample of 400 depositors, that the mean amount in deposit is between $420.00 and $430.00?

12. Paper clips manufactured by a certain company are supposed to weigh 10 grams with $\sigma = 2.0$ grams. If a sample of 100 clips has an average weight of 9 grams, can we say that the average weight is less than 10 grams? Use $\alpha = .025$.

13. A high school teacher will implement an advanced science program if the achievement of her students on a standardized aptitude test is significantly higher than 80 points. For the junior class (assumed a random sample) of 90 students, the average score was 85 with standard deviation $\sqrt{360}$ points. Use $\alpha = .03$ in deciding whether or not she should implement the advanced science program.

14. From experience 5% of certain articles are defective. A new man who has produced 600 of these articles has made 42 defective pieces.
    a. Does this cast doubt on the man's ability to perform the job? Suppose as his foreman you fired him. Make a test to justify your action, $\alpha = .05$.
    b. State the risk involved if the previous finding is "reject H_0." Use everyday practical language.

15. A population of 8,000 students has an average score of 70 on a stan-

dardized test. A random sample of 81 scores is selected. What is the probability that the sample mean will exceed 72 points? Assume scores are distributed with a standard deviation of 12.

16. A random sample of 100 lamps taken from a very large shipment contains 18 with imperfections.
    a. What is the possible size of our error if we estimate this proportion of imperfect lamps as being 0.18? Use a level of confidence of 0.98 and $\hat{p} = 0.18$.
    b. Construct a 98% confidence interval for the actual percent of imperfect lamps in this shipment.

17. A large container of mixed nuts contains peanuts, cashews, almonds, etc., in unknown proportions. If upon inspection a random sample of 100 nuts produces 40 peanuts, can we safely say that among all of the nuts, less than half are peanuts? Design your test so that if exactly half are peanuts, you would make a wrong conclusion 2.5% of the time.

18. How big a sample should you take to estimate the average number of customers per day? The standard deviation is known to be about 20. We want to be 90% sure that our estimate is within five of the true mean.

19. Suppose that you are in charge of a poll to determine how many students favor an exam week. If you wish to estimate the true proportion of students favoring an exam week to within 5% with 90% confidence, how large a sample should you take?

20. A sample of 49 experimental animals is fed a special ration for a two-week period. Their weight gains yield the values $\overline{X} = 92.0$ ounces and $s = 5.0$ ounces.
    a. Establish 95% confidence limits for μ.
    b. How large a sample would you take if you wished $\overline{X}$ to differ from μ by less than one ounce with a probability of .95?

21. From a population of five scores, X: 5, 7, 9, 11, 13, assumed to have a discrete uniform distribution, we take, without replacement, a random sample of $n = 2$. There are $\binom{5}{2} = 10$ possible samples. For the given distribution:
    a. compute the mean and variance for the distribution on $\overline{X}$ (use the table below).
    b. evaluate $P(7 \leq \overline{X} < 9)$, $P(\overline{X} < 5)$, $P(5 < X \leq 10)$, and $P(X < 13)$. (*Note:* The random variable is in some cases $\overline{X}$ and in others X.)

$\overline{X}$	Frequency	$P(\overline{X})$
6	1	.1
7	1	.1
8	2	.2
9	2	.2
10	2	.2
11	1	.1
12	1	.1
Totals	10	1.0

22. A tire recapping shop is attempting to estimate the average defect-free life in miles for their product.
   a. If a random sample of 121 tires showed a mean defect-free life of 12,327 miles with $s = 2,112$ miles, estimate with 95% confidence the true average defect-free miles.
   b. Would the shop be fairly safe in placing a 12,000 mile warranty on their product? Assume normal distribution and make a one-tailed test, $\alpha = .025$.

23. A local driver-training instructor contends that his students average 85 or better on the state licensing exam. An examiner who knows the instructor sometimes stretches the facts surveys records of 49 of his former students. These show $\overline{X} = 81$ with $s = 12.0$. Test against the instructor's claim, $\alpha = .05$.

24. At a university student records kept over a period of many years indicate that 64% pass the entrance examination in English. During the last fall term, 280 of 400 freshmen passed. Is this a significant improvement in the general student aptitude in English? Use $\alpha = .05$.

25. Find a 95% confidence interval for the mean based on the following frequency table. Assume an approximately normal population distribution. First you will need point estimates for μ and σ. Use $\overline{X}$ and s.

Class	Frequency
10 but less than 20	6
20 but less than 30	18
30 but less than 40	24
40 but less than 50	12
Total	60

Inference Using the *t*-Statistic CHAPTER 10

Thus far we have considered inference about mean and percentage values for large samples. We turn now to small sample techniques for tests and estimation about means. Chapter 13 will include small sample inference in binomial experiments.

The development of small sample *t*-tests is attributed to William Sealy Gossett (1876–1937). Much of his research was done while he was working for the Guinness Brewery in Ireland. Because company policy forbade workers to publish research, Gossett presented his findings under the fictitious pen name "Student" [2]. To this day this name remains on much of Gossett's work. His findings were monumental because so many practical inference problems satisfy the condition of normality, but the true variance is unknown and for economical reasons only a small sample is available for its estimation. Apparently this was often the case for experiments in the brewing industry, too. Gossett's findings, which pioneered statistical inference for small samples, remain in wide use.

1/ The *t*-form and Degrees of Freedom

The Student or *t*-distribution is quite similar to the normal. The Studentized form

$$t = \frac{\overline{X} - \mu}{s/\sqrt{n}}$$

is mechanically quite similar to the standard Z-form. In fact there are several similarities that maintain calculations with the *t*-statistic quite

Which spread is butter and which is margarine? How can we be sure we are testing exactly what we want to test? In order for a test to be accurate, we try to block out all differences except for the one that is being tested.

Inference Using the *t*–Statistic

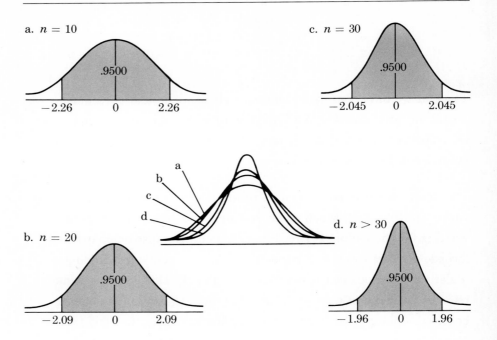

Figure 1 / Plots of the *t*-distribution for alternative degrees of freedom

like those using the Z-form. First, the probability distributions have markedly similar shape, see Figure 1. But this is only reasonable, since the *t*-form assumes the basic variable, X, has a normal distribution.

Thus even for small samples, $n \leq 30$, the distribution for the Studentized random variable takes the bell shape. The distribution centers about zero, but has standard deviation greater than one. Since the distribution is smooth (continuous), tables developed from the probability integral of t (that is, using the calculus) are used for probability considerations. The symmetry of the distribution allows one set of tabular values to be used for locating percentage points above and below the mean.

Unlike the standard normal, the shape of the *t*-distribution is determined by the size of the sample. Contrast Figure 1a through Figure 1d. As the sample size increases, the distribution becomes more peaked. The amount of usable information is described as the degrees of freedom, $df = n - 1$. Logically, as the sample size increases, error in the distribution, $s/\sqrt{n}$, should decrease. Thus in Figure 1 we observe that for fixed probability (95%) as n increases from 10, to 20, to 30 the span of *t*-values decreases. Yet *t*-values remain relatively stable and quite near the common Z-values for $n > 30$. This work takes the approach that for $n > 30$ the *t*-distribution approaches the standard Z-form, then a single table (the Z) can be used for both distributions.*

The t-table (Table II in Appendix B) is constructed somewhat differently from the Z-table (Table I in Appendix B). For the t, percentage points appear in the body of the table, probabilities in the column headings. Since the flatness or peakedness of the t-distribution depends on sample size, this is described in *degrees of freedom,* df; this heads the leftmost column in the Table. For now we will use df $= n - 1$ and consider single sample procedures. A portion of Table II is displayed in Figure 2.

Figure 2/ A segment of the t-table (Table II, Appendix B)

df	0.25	0.10	0.05	0.025	0.01	0.005
⋮						
14	0.6924	1.3450	1.7613	2.1448	2.6245	2.9768
15	0.6912	1.3406	1.7531	2.1315	2.6025	2.9467
⋮						
30	0.6828	1.3104	1.6973	2.0423	2.4573	2.7500
inf	0.6745	1.2816	1.6449	1.9600	2.3263	2.5758

The column headings 0.25, 0.10, . . . , or 0.005 denote the percent of area that is beyond the tabular t-value; then for example, a t-distribution defined by 14 degrees of freedom has 25% of the area to the right of $t = 0.6924$. By symmetry the percentages are for either tail, but since the t-distribution has central value zero, lower-tail t-values are negative. So for df $= 14$, another 25% of the area bounded by the t-distribution is to the left of $t = -0.6924$. The use of Table II is described by several examples.

For one-sided hypothesis tests, the t-value is read at the designated α-risk level. For an upper-tailed test with $\alpha = .05$ and $n = 16$, giving df $= n - 1 = 15$, we read $t_{.05, 15} \doteq 1.75$. Thus 5% of the area falls beyond $+1.75$ for the t-distribution defined by 15 degrees of freedom.

Example

Here the tabular t-values will be rounded to the nearest part of one hundred (two-decimal places). The next example illustrates readings from the t-table for two-tailed tests and confidence interval estimates.

---

*Some texts (e.g., [3]) distinguish t-values for larger degrees of freedom. For example, with 0.025 probability in the upper tail, and for df $= 60$, $t = 2.00$; at the same probability level, but for df $= 120$, exact $t = 1.98$. In this work both are approximated by $t_{\text{inf}} = Z = 1.96$. This is a matter of convenience in requiring fewer tables, and in most cases alters the results very little.

Example

For $n = 16$, giving df = 15, with 95% of the probability in the center of the distribution, the t-percentage points are $t_{.025,15} = \pm 2.13$. That is, $\alpha/2 = .025$ so 2.5% of the area is in either tail and beyond $+2.13$ or -2.13.

We can check the t-percentage points for the two preceding examples by comparison to t_{inf}. Our numbers are reasonable because for single sample inference with $n > 30$, the values t_{inf} are 1.645 and 1.96, respectively. These are equivalent to the percentage points on the Z-scale for an $\alpha = .05$ upper-tailed test and an $\alpha = .05$ two-tailed test, or a 95% confidence interval. See Figure 2.

Degrees of freedom were alluded to in Chapter 3 when determining the divisor for the sample variance, s^2. From here on our work will require evaluation of degrees of freedom requisite to identification of a correct t-, F-, or chi-square tabular percentage point.

The basis for degrees of freedom is tied to sample size. We begin with a general discussion of the meaning of df. Next the general meaning is applied to df in sample variance calculation. Finally a generalization allows our quick identification of df for common studentized t-type statistics. Degrees of freedom are a gauge of the amount of free information available for making decisions. The basic concept is described with relation to the selection of points in 2-space (see Appendix A, p. 363). If asked to choose a point, any point (X, Y), on a rectangular coordinate system (2-space), there are zero (0) restrictions, hence $2 - 0 = 2$ df. We are allowed any value for X and any value for Y. That is, any point in 2-space can be selected. See Figure 3a. Next suppose the point is restricted to those on a line, say to the line $Y = 3 + X$. See Figure 3b. This makes one restriction; that is, taking a value for X then determines another for Y, e.g., when $X = 1$, Y must $= 3 + 1 = 4$. Hence for (X, Y) with the single restriction $Y = 3 + X$ there remains $2 - 1 = 1$ df. Finally for a point in 2-space restricted to satisfy both $Y = 3 + X$ and $Y = 5 - X$, we have imposed two restrictions. But because we have only two dimensions, our freedom is completely removed. Neither the X nor the Y value can be chosen at will. Their values are required to be $(1, 4)$, the point of intersection of the two lines. There are zero degrees of freedom. See Figure 3c.

The idea can be extended from 2-space to n dimensions (n-dimensional space is the case wherein a data set contains n observations). In statistics a common occurrence is n-space with one restriction. Then there remain $n - 1$ degrees of freedom. Consider for $n = 10$ the unspecified values X_1, X_2, ..., X_{10} with the single restriction $X_1 + X_2 + \cdots + X_{10} = 200$; that is, the total of these ten values is 200. Then the arbitrary choice of values for any nine of these Xs fixes the value of the last. There are in fact $n - 1 = 10 - 1 = 9$ degrees of freedom.

Figure 3/ Degrees of freedom (df) in 2-space

a. Unrestricted, 2 df   b. Restricted to $Y = 3 + X$, 1 df   c. Restricted by $Y = 3 + X$ and $Y = 5 - X$, 0 df

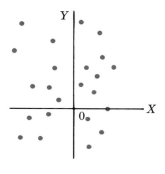
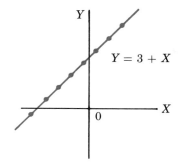
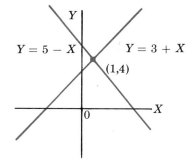

The degrees of freedom concept was used in determinations of sample variance and standard deviation (see Section 2.2). This relates to the fixed sum idea

$$\sum_{}^{10} X = X_1 + X_2 + \cdots + X_{10} = 200, \text{ and consequently } \overline{X} = \frac{\sum_{}^{10} X}{10} = 20.$$

That is, knowing the total and $n \; (= 10)$ means we also know the sample mean. Then

$$\sum_{}^{10} (X - \overline{X}) = \sum_{}^{10} X - 10\overline{X} = 10\overline{X} - 10\overline{X} = 0.$$

Recall $\overline{X}$ as a balance value? Hence if $\sum X = 200$ gives $\overline{X} = 20$, this forces $\sum (X - \overline{X})$ to be 0. Thus for known $\overline{X}$ and any nine specified values of the Xs the tenth value is fixed by $\sum (X - \overline{X}) = 0$. There are $n - 1 = 10 - 1 = 9$ degrees of freedom. Generalizing this for n dimensions, $\sum (X - \overline{X}) = 0$ possesses $n - 1$ degrees of freedom so long as $\overline{X}$ is known. Subsequently $\sum (X - \overline{X})^2 \neq 0$, which is determined directly through $(X - \overline{X})$, also possesses $n - 1$ degrees of freedom. This latter process, based on "sums of squared differences," is the basis for the sample variance and standard deviation forms. Again there are $n - 1$ degrees of freedom.

Now the t-statistic,

$$t = \frac{(\overline{X} - \mu)}{s/\sqrt{n}},$$

depends on the sample variance (or standard deviation) so the amount of usable information in the statistic is given by the "effective sample size,"

Inference Using the t–Statistic

called degrees of freedom or $n - 1$, as outlined in the preceding paragraph. Other t-type statistics in Chapters 10, 11, and 12 are also described by df, but df are *not* always $n - 1$. For a t-type statistic, however, df can always be determined by the following rule.

Rule

For a t-statistic of the form

$$t = \frac{\text{estimator} - \text{parameter}}{s \cdot \text{something}},$$

the degrees of freedom equals the divisor of the expression that defines s.

For example, we have

$$t = \frac{\overline{X} - \mu}{s/\sqrt{n}} \quad \text{where } s = \sqrt{\frac{\sum_{}^{n}(X - \overline{X})^2}{n - 1}},$$

so the degrees of freedom are df $= n - 1$. For a second example of this rule, see p. 253. There the expression that defines the standard deviation is s_p. It (s_p) has divisor $= n_1 + n_2 - 2$, so that the t-statistic has df $= n_1 + n_2 - 2$. You will have occasion to use this in Chapter 11.

The remainder of this chapter concerns inference using the t-statistic. The procedures for estimation and testing are those used in Chapters 8 and 9, but we use the t-statistic and Table II (which requires df) rather than Z-forms.

Problems

1. Assuming a one-tailed (upper) test of hypothesis on the mean, use Table II to find the "test" t-value for

   a. $n = 24, \alpha = .05$ d. $n = 25, \alpha = .01$
   b. $n = 15, \alpha = .05$ e. $n = 25, \alpha = .05$
   c. $n = 5, \alpha = .05$ f. $n = 25, \alpha = .10$

2. Answer Parts a through e of Problem 1 for a two-tailed test of hypothesis, or equivalently for setting confidence interval estimates on μ assuming normal distribution with unknown variance.

3. Answer each of the following questions concerning the use of Table II:

   a. Find the tabular t-value for a lower-tailed test on H_0: $\mu = 10$ based on $n = 22$ and with $\alpha = .05$.

b. A confidence estimate on μ based on $n = 19$ observations uses tabular $t = \pm 2.10$. What is the confidence (probability) level?

c. For a two-tailed t-test, $H_A: \mu \neq -4$, based on $n = 11$ and with $\alpha = .10$, determine tabular t.

d. An upper-tailed test on the mean used $n = 17$ normal observations and test value $t = +2.58$. What is the α-risk level?

4. Discuss similarities and differences in the t- and Z-probability distributions.

5. Write one or two sentences to explain each of the following:
   a. How you would find the 75th percentile using Table II for a t-distribution defined by 17 df.
   b. How you would locate the correct t-value for a two-tailed test with df = 12 and $\alpha = .05$.
   c. Why is Table II so incomplete, e.g., why doesn't the table contain t-values for probability levels .04, .06, .072, etc.?

6. Determine the degrees of freedom for:
   a. points in 2-space restricted to $Y = 1 + X$.
   b. points in 2-space restricted to $Y = X$ and $Y = 1 - X$.
   c. points in 5-space restricted by $\sum X = 100$.
   d. points in n-space restricted by $\sum (X - \overline{X}) = 0$, with $\overline{X} = 10$.

7. Determine degrees of freedom for:
   a. $s^2 = \dfrac{(1-3)^2 + (2-3)^2 + (3-3)^2 + (4-3)^2 + (5-3)^2}{5-1} = 2.5$.
   b. $\sum (X - 12) = 0$ where $\overline{X} = 12$.
   c. points in 2-space restricted by $Y = 3$ and $Y = 4 - X$. (*Note:* $Y = 3 \Leftrightarrow Y = 3 + 0X$).
   d. points in $(n =) 20$ space; that is, suppose no restrictions. Determine df.

2/ Estimation and Sample Size Considerations

Use of a t-statistic presupposes that the basic random variable, X, is normally distributed. Then for unknown variance and samples less than thirty, the sampling distribution for sample means is the t. Procedures for estimation are those outlined in Chapter 8, but of course now t-replace Z-values.

Inference Using the *t*–Statistic

Confidence intervals are set according to the rule:

Rule

For a $(1 - \alpha) \cdot 100\%$ confidence interval on the mean, where $n \leq 30$ from a normal distribution with unknown variance use:

$P(L < \mu < U) = 1 - \alpha$ with

$$L = \overline{X} - \left(t_{\alpha/2, n-1} \cdot \frac{s}{\sqrt{n}}\right)$$

$$U = \overline{X} + \left(t_{\alpha/2, n-1} \cdot \frac{s}{\sqrt{n}}\right)$$

This rule is illustrated through several examples. Check the *t*-values to be sure you can use Table II.

Example

Suppose a random sample of 25 high school basketball players have mean weight 150.00 pounds with $s = 22.5$ pounds. Construct a 95% confidence estimate for the true mean weight. Assume weights for the population of all high school basketball players to be normally distributed.

Solution

Since $n = 25$ is a small sample, the true variance is not known, and the parent distribution is normal, the *t*-statistic is appropriate. So $t_{.025, 24} = 2.06$ gives

$150.0 - [2.06 \cdot (22.5/\sqrt{25})] < \mu < 150.0 + 9.3$ or

140.7 pounds $< \mu <$ 159.3 pounds.

Conclusion

The true mean weight for the players in this league is estimated to be between 140.7 pounds and 159.3 pounds. Notice that the α-level is split giving $[\alpha/2 \cdot 100] = 2.5\%$ probability in either tail.

Evaluation of the *t*-statistic requires calculation of sample values for $\overline{X}$ and s. Of course our evaluations in Chapters 8 and 9 utilized these statistics, too. But we generally did not display the data, nor require calculations for samples which were commonly of fifty or more observations. However, someone must evaluate a mean and a standard deviation for *t*- or *Z*-statistics. This is something we can practice (with reasonable effort) for smaller samples. The next example illustrates the added steps. You may want to review the procedures from Chapter 2.

Example

The following scores were achieved by a contestant in diving competition (assume 6.0 is a perfect score): 5.6, 5.4, 5.7, 5.5, 5.7, 5.6, 5.7. Assuming a normal distribution of scores, we seek a 90% confidence interval estimate on the true mean score for this individual.

Estimation and Sample Size Considerations

The small-sample and normal distribution requirements again lead to the t-statistic procedure. But first we need values for $\overline{X}$ and s:

Solution

X	$(X - 5.6)$	$(X - 5.6)^2$
5.4	$-.2$	.04
5.5	$-.1$	.01
5.6	0	0
5.6	0	0
5.7	.1	.01
5.7	.1	.01
5.7	.1	.01
Totals 39.2	0	.08

Giving $\overline{X} = \dfrac{\sum_{}^{7} X}{7} = \dfrac{39.2}{7} = 5.6$

$s = \sqrt{\dfrac{\sum (X - 5.6)^2}{7 - 1}} = \sqrt{\dfrac{.080}{6}} = .12$

(Remember that page 20 shows, for $n = 7$, $s \approx \text{range}/2 = (5.7 - 5.4)/2 = .15$ as a coarse check). Then for $t_{.05,6} = 1.94$,

$l = 5.6 - \left(1.94 \dfrac{\sqrt{\tfrac{.08}{6}}}{\sqrt{7}}\right) = 5.6 - \left(1.94 \cdot \sqrt{\dfrac{.08}{6 \cdot 7}}\right)$

$l = \quad\quad\quad\quad\quad \doteq 5.6 - 0.1 = 5.5$

$u = \quad\quad\quad\quad\quad \doteq 5.6 + 0.1 = 5.7$

That is, $5.5 < \mu < 5.7$.

The mean rating on the performance for this individual is estimated between 5.5 and 5.7 on a 6-point scale. (*Note:* Simplification in arithmetic, plus added accuracy in the answer, were achieved by leaving s as a square root until combining as $\sqrt{s^2/n}$. Using the mathematical equivalence between the "quotient of roots" and "the root of a quotient" has allowed us to take one square root instead of two!)

Conclusion

A third example compares numerically the results of estimation under large samples (with Z-form) and those with small samples (t-form).

A new drug given to sixteen patients with high blood pressure lowered their blood pressure an average of eleven points with a standard devia-

Example

Inference Using the *t*–Statistic

tion of six points. Assuming that reduction in blood pressure is a normal random variable, set a 99% confidence estimate on the true mean reduction in blood pressure.

Solution

Again the *t*-statistic is used, here with $t_{.005,15} = 2.95$. This yields

$$11 - [2.95 \cdot (6/\sqrt{16})] < \mu < 11 + 4.42 \quad \text{or} \quad 6.58 < \mu < 15.42.$$

Conclusion

The true mean reduction in blood pressure affected by this drug is estimated to be 6.58 units to 15.42 units.

Estimates based on the *t*-statistic have larger error than those based on the Z-statistic, but one should expect more error when less information is used. In the last example, if we had obtained approximately the same mean, $\overline{X} = 11$, and standard deviation, $s = 6$, from a random sample of $n = 64$ patients, the error for a 99% confidence estimate would be $E = [2.58 \cdot (6/\sqrt{64})] = 1.94$. This is a reduction to less than one-half the previous error of 4.42 units. Yet, realize the price we must pay for the better estimate; smaller error requires a larger sample, and usually substantially increased cost. This we may not be able to afford.

We have observed that the spread of the *t*-distribution is a function of the sample size; *t*-scores are based on degrees of freedom, here $n - 1$. This dependence on sample size limits our ability in certain computations.

Rule

Bound of Error = *t*-value · Standard Error

$$E = t_{n-1} \cdot \frac{s}{\sqrt{n}}$$

The sample size is used to specify both the *t*-value and the standard error. Consequently, for small sample inference on means, only (1) the error of estimation and (2) the standard deviation are readily estimated through the error relation. Finding sample size, *n,* would require knowing a *t*-value. But the *t*-score depends on *n*. Thus a *direct* estimate of sample size cannot be made from the error relationship. This is a severe limitation on the *t*-statistic.

The following illustrate sample size considerations for small sample inference on means.

Example

A car dealer wants to estimate the average age of cars traded to his firm. A preliminary sample gives $s = 1.75$ years. What error of estimation can he expect from a random sample of $n = 10$ cars if he wants 90% confidence?

Solution

Using $E = t_{.05,9} \cdot (s/\sqrt{n}) = 1.83 \cdot (1.75/\sqrt{10}) \doteq 1.0$, the dealer could expect to be in error in estimating the mean age of trade-ins by at most one year. He has 90% confidence in the procedure used to size this error.

The next example, although unlikely as a practical application, shows determination of a value for the standard deviation from the confidence interval.

Example

From a sample of 16 taken from a normal distribution, the 98% confidence limits for the mean are $\overline{X} \pm 10.4$. Estimate the standard deviation of the population.

Solution

Solving for s in the error of estimation equation gives

$$s = \frac{E \cdot \sqrt{n}}{t} \quad \text{so} \quad s = \frac{10.4 \cdot \sqrt{16}}{2.60} = 16.$$

That is, $\sigma \doteq s = 16$.

Having looked briefly at the form for estimation of the mean using small samples, we turn to testing. The forms and procedures are very much like those of Chapter 9.

Problems

Assume that the basic variable X has a normal distribution with *unknown* standard deviation for each of the following:

1. Records of 24 auto thefts in a certain city indicate that the mean age for those responsible is 20 years with a deviation of 2.6 years. Use this information, $\alpha = .05$, to estimate the average age for auto thiefs in this city.

2. For $\overline{X} = 25.0, s = 5.0, n = 19$, and with X normally distributed, compute 95% confidence limits for μ. Interpret your result.

3. How big a sample should one take to estimate the average number of emergency calls per week to a hospital? The standard deviation is known to be about 20. We want to be 90% sure that our estimate is within five of the true mean. Assume n will exceed 30.

4. In a study of the effectiveness of a reducing diet, the following weight-gain figures were obtained from a random sample of ten men in the age group 50 to 60. Note that "$-$" indicates a loss of weight. Given measures $-3, -7, -1, 0, +3, +1, -1, -6, -10,$ and -6 pounds and using $\alpha = .10$,

estimate the true mean loss effected by this diet. Assume normal distribution of weight loss.

5. For a random sample of 22 observations from a normal population we find $\overline{X} = 29.35$, and $s = 17$.
   a. Find a 98% confidence interval for μ.
   b. What sample size would be required to reduce the error of estimation obtained in Part a by one-half? (*Hint: n* will be > 30.)

6. A 95% confidence interval for the average amount of money spent by State University students during a week was $32.00 ± $2.13, that is, between $29.87 and $34.13. If a random sample of 16 students was surveyed, what is the standard deviation of amounts spent? Assume a normal distribution.

7. Explain how procedures would change in finding the 95% confidence interval in Problem 6 for a random sample of 36. Would you expect the interval based on $n = 36$ to be wider or narrower than one based on $n = 16$? Explain.

8. Suppose scores of 97, 93, 99, 100, 103, 101, 99, 94, 90, and 94 on a social adjustment test. Using $\alpha = .05$, estimate the true mean score for this test. Assume normal distribution of scores.

3/ Tests on Means for Small Samples

The principles of "Introduction to Testing Procedures" in Chapter 9 remain the foundation for tests on the mean with unknown variance. Comparable small sample test procedures require that the basic random variable have a normal distribution. Here "small samples" will mean thirty or less. Several examples follow.

Example

The commanding officer of a paratrooper unit knows that the average tension between the parachute snap ring and the automatic release on the airplane should be at least 5.0 pounds. He randomly inspects 25 parachutes and finds the tension averages 4.7 pounds with a deviation of 0.5 pounds. Is the average tension less than 5.0 pounds? We use $\alpha = .025$.

Solution

The officer seeks a course of action which, if wrong, might lead to injury to some of his men. The parameter is μ = the mean tension between the snap ring and its release. Thus

1. $H_0: \mu = 5.0$ pounds
   $H_A: \mu < 5.0$

2. Test: Reject H_0 if calculated $t < -2.06$ ($t_{.025,24}$)

3. Calculated $t = \dfrac{\overline{X} - \mu}{s/\sqrt{n}} = \dfrac{4.7 - 5.0}{.5/\sqrt{25}} = -3.0$

4. Decision: Reject H_0 for $\alpha = .025$.

This sample indicates the average tension is less than 5.0 pounds. If the tension is not or cannot be increased the paratroopers should be advised of this potential problem.

Conclusion

Again, we observe the earlier test procedures in action; only the test statistic is changed. The next example also demonstrates calculation of the sample mean and standard deviation.

The following weights are for a random sample of twelve-ounce packages of bacon. Assuming a normal distribution of weights, test the hypothesis that an average of 12.0 ounces of weight is being maintained. Use $\alpha = .05$, given weights 11.8, 11.8, 11.9, 11.9, 12.1, 11.9, 12.0, 12.1, 12.1, 12.1, 12.2, 12.3, 12.3, 12.4, and 12.4 ounces.

Example

The solution follows the same steps for applying the scientific method, but with one additional step. Since the raw data is available, we first compute values for $\overline{X}$ and s. Preliminary calculations give:

Solution

$$\overline{X} = \frac{11.8 + 11.8 + \cdots + 12.4}{15} = \frac{181.3}{15} \doteq 12.1 \text{ ounces}$$

$$s^2 = \frac{(11.8 - 12.1)^2 + (11.8 - 12.1)^2 + \cdots + (12.4 - 12.1)^2}{15 - 1}$$

$$= \frac{.58}{14} \doteq .04 \text{ ounces}^2$$

$s = 0.2$ ounces

$H_0: \mu = 12.0$ ounces

$H_A: \mu \neq 12.0$ ounces

Test: Reject H_0 if $|\text{calculated } t| > 2.14$ ($t_{.025,14}$)

Calculated $t = \dfrac{\overline{X} - \mu}{s/\sqrt{n}} = \dfrac{12.1 - 12.0}{0.2/\sqrt{15}} = 1.94$

We cannot reject H_0. From this evidence, with $\alpha = .05$, we conclude that the average weight is essentially 12.0 ounces.

Conclusion

Inference Using the *t*–Statistic

The next example poses an interesting question. The claim is not atypical of some commercial advertising. But is the claim substantiated by the data?

Example

Crown Manufacturers claim it's new cars use a minimal amount, at most one quart, of oil in the first year of operation. From a sample of 16 new Crown cars, the mean oil consumption was 1.4 quarts with $s = 0.4$ quarts. As a potential Crown owner, would you agree with the company statement? Let $\alpha = .05$ and assume a normal distribution of oil consumption.

Solution

The test is one-tailed.

$H_0: \mu = 1.0$ quart
$H_A: \mu > 1.0$ where $\mu =$ mean amount (quarts) of oil used in the first year for Crown cars

Test: Reject H_0 if calculated $t > 1.75$ ($t_{.05,15}$)

Calculated $t = \dfrac{1.4 - 1.0}{.4/\sqrt{16}} = 4.00$.

Conclusion

Reject H_0. The claim is not substantiated. The statistical evidence indicates that Crown cars use more than one quart of oil in the first year of operation, $\alpha = .05$.

The following example, in addition to displaying a *t*-test, illustrates an equivalent way of computing the sample standard deviation. Bear in mind that this is just an alternative; the earlier procedure would work, too.

Example

Following are downtimes in hours for a computer during nine consecutive months: 18, 5, 9, 10, 13, 7, 2, 11, and 6. Assuming that downtimes are normally distributed, test the claim that this computer installation has been nonoperable an average of less than 10 hours per month. Use $\alpha = .025$.

For $\mu =$ the true mean downtime per month

$H_0: \mu = 10$ hours per month
$H_A: \mu < 10$

Test: Reject H_0 if calculated $t < -2.31$ ($t_{.025,8}$) so

$$\overline{X} = \frac{18 + 5 + \cdots + 6}{9} = \frac{81}{9} = 9$$

$$s^2 = \frac{n \sum (X)^2 - \left(\sum X\right)^2}{n(n-1)} = \frac{9(18^2 + 5^2 + \cdots + 6^2) - (81)^2}{9 \cdot 8}$$

$$s^2 = \frac{1620}{72} = 22.5$$

$$\text{Calculated } t = \frac{9 - 10}{\sqrt{22.5}/\sqrt{9}} = \frac{-1}{\sqrt{2.5}} \doteq -.63.$$

We are unable to reject H_0. The mean downtime is not less than 10 hours per month.

The computing form for s^2 in this example gives results equivalent to the definition form used earlier. This *machine procedure* is easier if you have access to a computer-calculator. Remember, either form will do. For more on this see the workbook or [3].

Although the β-risk has been ignored in these examples, careful consideration should be given its value in t-tests. One should realize the value of additional information, that is, a larger sample. Something is lost when we test with smaller samples; namely, there is an increase in the risk (β-) of accepting a false null hypothesis. A standard research procedure following an "accept H_0" decision should be computation of the β-risk for reasonable alternative values of the parameter. If, for example, an "accept H_0" decision also produces a β-risk of over 50% for the most reasonable alternative, then the decision is most likely wrong! Of course a safer procedure—one with lower β-risk—results from using larger samples. It can be shown that the tests we use have the smallest β-risk possible for given α and n.

In conclusion, testing on the population mean is much the same with large samples or with small samples coming from a normal distribution. Either a Z- or t-statistic is used, respectively. In any case, every effort should be made to obtain a sample sufficiently large to maintain reasonable decision risks.

Problems

In each of the following problems assume the basic random variable X has a normal distribution, but an unknown standard deviation.

1. Given scores $X_1 = 7$, $X_2 = 3$, $X_3 = 4$, $X_4 = 5$, $X_5 = 3$, and $X_6 = 5$, compute $\overline{X}$, s, and test $H_0: \mu = 5.0$ against $H_A: \mu < 5.0$. Use $\alpha = .05$.

2. In a learning experiment, it took nine three-year old youngsters an average of 5.2 attempts to pronounce correctly the word "supercali-

fragilisticexpialidocious." The standard deviation in number of attempts was 2.1. Test the claim that it takes three-year old persons an average of over five trials to say this word correctly for the first time. Use $\alpha = .05$.

3. A manufacturer of ski-lift slings claims that his product has a mean breaking strength of 4,000 pounds. A test of the breaking strengths of ten of these slings showed mean strength of 3,910 pounds and standard deviation of 90 pounds. Test the manufacturer's claim using $\alpha = .01$. Make a one-tailed test.

4. Mrs. C. has complained about the amount of time Mr. C. spends for his evening nightcap at the corner bar. He says he averages less than 50 minutes for each trip. Without his knowing it, she has kept records of his last eight evening trips: 39, 35, 42, 50, 51, 40, 42, and 53 minutes. Using this data, which gives $s = 6.5$ minutes, can the husband justify an average of under 50 minutes? Use $\alpha = .025$.

5. In the test of a gasoline additive, a group of carefully engineered cars were run at a testing site under rigorously controlled conditions. The number of miles obtained on a single gallon of gasoline were 15, 12, 13, 16, 17, 11, 14, 15, 13, and 14. Hundreds of prior trials under similar conditions with the same gasoline minus the additive yielded an average of 12.62 miles per gallon. Using $\alpha = .05$, can one conclude that the additive has improved gasoline mileage? Describe the possible type 2 error.

6. An electronic calculator dealer claims that his make possesses an average defect-free life of three years or more. A random selection of three sales produced failure times in years: 2.4, 2.9, and 2.8, giving $\overline{X} = 2.7$ and $s \doteq .26$ years.

    a. Does this evidence contradict the distributor's claim? Use $\alpha = .05$ and assume normal distribution of lifetimes.

    b. Does your conclusion agree with intuition? If not, explain what further testing you might do before making a final conclusion.

4/ Paired Differences Tests

The procedures of the last section, slightly modified, are used for inference where the data is differences on paired observations. This is a two-sample experiment, but because of the dependence or expected similarity of responses for pairs, calculations are made on a single set of numbers—the differences for pairs. For example, comparisons of achievement for twins would likely be quite similar. Even if each is given a quite different treatment or conditioning, there is too much commonality for their responses to be considered independent. Similarily, in before–after experiments the two responses for an individual must be related. Other cases where paired differences tests are used include comparisons for siblings, animals matched with littermates, matched plots in agriculture, test–retesting in education, etc. It is noteworthy that many works present this technique with other two-sample procedures. Yet the computing procedures are nearly those of the single sample *t*-test of Section 3; hence, I have included it with one-sample procedures.

One modification of the preceding section requires finding differences of scores for each set of paired observations. Differences are denoted $X_d = X_A - X_B$. Then these differences rather than the original values are the basis for calculations. For example, in before–after experiments $X_d = $ response after treatment minus response before treatment. The following rule describes the *t*-statistic appropriate for inference.

Rule

For small samples coming from normal distributions the sampling distribution of mean differences for paired responses, $\overline{X}_d$, is described by the *t*-statistic,

$$t_{n-1} = \frac{\overline{X}_d - \mu_d}{s_d/\sqrt{n}}, \quad s_d^2 = \frac{\sum_{}^{n}(X_d - \overline{X}_d)^2}{n-1}$$

with $\overline{X}_d = $ the mean of differences on paired responses
$n = $ the number of pairs
$X_d = $ the difference of paired responses.

For dependent samples a major source of differences is "between" pairs; the matching, pairing, or repeated measures on individuals results in small differences "within" pairs. In the rule, the "between" differences form the basis for the standard deviation.

The following example, illustrative of a before–after experiment, demonstrates both testing and interval estimation procedures.

Example

A reading teacher wants to estimate the increase in comprehension for students who have completed her course. She selects a random sample of ten students. The students were given a pretest, and an equivalent test was administered at the end of the course. We will also test the hypothesis that the average score has increased by *over* 30 points. Assume that the distribution is normal.

Inference Using the *t*–Statistic

Increase in Reading Test Scores

Student	Post-test	Pretest	Scores X_d	$(X_d - \overline{X}_d)$	$(X_d - \overline{X}_d)^2$
1	51	15	36	2	4
2	63	21	42	8	64
3	55	20	35	1	1
4	68	36	32	−2	4
5	48	12	36	2	4
6	38	9	29	−5	25
7	54	17	37	3	9
8	73	42	31	−3	9
9	49	26	23	−11	121
10	57	18	39	5	25
Totals			340	0	266
Mean		$\overline{X}_d = 34$			

Solution

We apply the paired differences test with μ_d = the mean increase in test score. Let's use $\alpha = .05$.

$H_0: \mu_d = 30$ points
$H_A: \mu_d > 30$ points

Test: Reject H_0 if calculated $t > 1.83$ ($t_{.05,9}$)

$\overline{X}_d = 34$, $s_d^2 = \dfrac{266}{10-1} \doteq 29.56$

Calculated $t = \dfrac{34 - 30}{\sqrt{29.56/10}} \doteq \dfrac{4}{1.72} \doteq 2.33$

Conclusion

Reject H_0. This suggests an average increase of more than 30 points for post-course over precourse scores, $\alpha = .05$.

Additional information can be gained by setting a confidence estimate on the true mean increase. Thus for 95% confidence,

$(L, U) = [\overline{X}_d \pm (t_{.025} \cdot s_d/\sqrt{n})]$ so

$(l, u) = 34 \pm (2.26 \cdot 1.72) = 34 \pm 3.89$

The true mean improvement is estimated to be within 30.11 and 37.89 points.

A second example uses matched swatches of cloth to block out differences for pairs. Some fabrics are inherently brighter than others. Recog-

Paired Differences Tests

Example

A researcher would like to decide if one laundry soap is better than another with respect to brightness of the wash. He takes swatches of twelve different types of cloth, each uniformly soiled, tears each piece in half, and by random choice washes one part in detergent A and the other in detergent B. He then measures the brightness of the resulting washes with a special meter. The resulting paired observations appear below. Is there a significant difference in the brightness induced by the two detergents for $\alpha = .05$?

Solution

We assume that the sample differences come from a normal distribution. Then

$H_0: \mu_d = 0$
$H_A: \mu_d \neq 0$

Test: Reject H_0 if |calculated t| > $2.20\,(t_{.025,11})$

The data, with differences and t-calculation, is

Swatch	1	2	3	4	5	6	7	8	9	10	11	12
X_A	5	4	4	1	6	3	4	5	5	3	6	4
X_B	1	3	4	2	4	3	2	3	6	5	3	2
$X_d = X_A - X_B$:	4	1	0	−1	2	0	2	2	−1	−2	3	2

(detergent)

Giving $\overline{X}_d = \dfrac{\sum X_d}{12} = 12/12 = 1$, $s_d^2 = 48/(12-1) = 48/11$

Calculated $t = \dfrac{1 - 0}{\sqrt{(48/11)}/\sqrt{12}} = \dfrac{1}{\sqrt{48/(11 \cdot 12)}} = 1.0/.60 = 1.67$

Decision: we cannot reject H_0.

Conclusion

This data evidences no statistical difference in brightness for the fabrics washed in these two detergents, $\alpha = .05$.

As stated earlier we should always recognize the possibility of a type 2 error when the conclusion is to "accept H_0," especially for small samples.

Inference Using the t–Statistic

The paired-differences test is one of a class of statistical designs that use grouping techniques to improve test precision. The paired procedure is appropriate for two-sample tests on the mean when the responses are from normal distributions, but are *not* independent.*

In this chapter we have stressed some relations between the Z- and t-distributions. In the remaining chapters we will view two related probability distributions for (1) the chi-square (χ^2) statistic and (2) that of the F-statistic.

Problems

1. The following figures were obtained from a random sample of ten women in the age group 40 to 50 for an eight-week trial on a new diet. Using $\alpha = .01$, test to determine if the diet produces a substantial loss of weight. Assume normal distribution of weight losses and $H_0: \mu_d = 0$.

	Weight	
Subject	Before	After
1	145	142
2	156	149
3	138	137
4	149	150
5	152	155
6	161	160
7	128	128
8	153	147
9	174	164
10	142	136

---

*In fact, the original distributions need not be normal as long as the distribution of the difference values is normal. However, normal parent distributions guarantee a normal distribution for the X_d.

2. Suppose in an owner maintenance-cost analysis for cars of one make, that maintenance costs were kept on the cars during their first two years. Assuming normal distributions test the hypothesis, at $\alpha = .05$, that costs go up during the second year. Also establish a 95% confidence interval for the true mean difference. The numbers in the table are (average) weekly maintenance costs.

	Age	
Car	One year	Two years
1	$4.72	$4.67
2	4.81	5.12
3	3.90	5.01
4	5.16	4.93
5	4.56	4.98
6	4.13	4.82
7	4.55	4.56
8	4.38	5.28

3. A systems specialist has studied the work flow of clerks, all doing the same processing job. He has designed a new work-flow layout for a clerical work station and wants to compare average production for the new method with the older method. Nine clerks are available for the test. After ample familiarization with the new station each clerk is assigned a common task. The order of stations is randomized.

	Production Rate (units processed)	
Clerk	Old	New
1	66	71
2	76	90
3	65	72
4	73	80
5	84	90
6	82	91
7	72	68
8	71	76
9	74	80

Assuming the necessary assumptions are met, test the hypothesis that the new station speeds up the work rate. Use $\alpha = .025$.

Inference Using the *t*–Statistic

4. In a service test of traffic paints, two paints were tested at 12 different locations. The locations can be considered pairs. The measures of visibility were taken after a comparable period of exposure to weather and traffic. Assume that each measure is normally distributed, and test the hypothesis of equal visibility for the two paints. Use $\alpha = .05$.

Location	Paint 1	Paint 2	Location	Paint 1	Paint 2
1	7	8	7	6	8
2	9	10	8	5	5
3	8	8	9	4	5
4	6	5	10	6	9
5	5	3	11	3	5
6	7	9	12	4	7

5. The second-grade teachers at Bryant Elementary School have decided to test their children for reading improvement over the span of the school year.

   a. Assuming necessary test conditions are met, test at $\alpha = .025$ that the average increase in reading score is over 30 points. The results are for a random selection of 16 children. Scores are 100 points maximum.

   b. Estimate, with 95% confidence, the true average increase in reading scores for second graders at this school.

Child	Beginning of year	End of year	Child	Beginning of year	End of year
1	55	89	9	66	98
2	62	86	10	57	88
3	41	76	11	66	100
4	70	93	12	61	97
5	36	74	13	72	98
6	59	92	14	50	88
7	56	85	15	68	94
8	39	69	16	43	86

5/ Essay Example

Following are excerpts from *The Design of Experiments* by R. A. Fisher [1]* first published in 1935. The discussion includes some of Fisher's comments concerning the book "The effects of cross- and self-fertilization in the vegetable kingdom," by Charles Darwin (1876). Therein Darwin describes some of the considerations which guided him in the planning of his experiments and in data presentation. Reference is also made to Francis Galton (1822–1911) who apparently communicated with Darwin on some statistical matters.

> The object of the experiment is to determine whether the difference in origin between inbred and cross-bred plants influences their growth rate, as measured by height at a given date. . . .
>
> The disturbing causes which introduce discrepancies in the means of measurements of similar material are found to produce quantitative effects which conform satisfactorily to a theoretical distribution known as the normal law of frequency of error. It is this circumstance that makes it appropriate to choose, as the null hypothesis to be tested, one for which an exact statistical criterion is available, namely that the two groups of measurements are samples drawn from the same normal population. . . .
>
> We must now see how the adoption of the method of pairing determines the details of the arithmetical procedure, so as to lead to an unequivocal interpretation. The pairing procedure, as indeed was its purpose, has equalised any differences in soil conditions, illumination, air currents, etc., in which the several pairs of individuals may differ. Such differences having been eliminated from the experimental comparisons, and contributing nothing to the real errors of our experiment, must, for this reason, be eliminated likewise from our estimate of error, upon which we are to judge what differences between the means are compatible with the null hypothesis, and what differences are so great as to be incompatible with it. We are therefore not concerned with the differences in height among plants of like origin, but only with differences in height between members of the same pair, and with the discrepancies among these differences observed in different pairs. . . .

---

*Reprinted with permission of Macmillan Publishing Co., Inc. Copyright © 1971, The University of Adelaide.

Inference Using the *t*—Statistic

... The mathematical distribution for our present problem was discovered by "Student" in 1908, and depends only upon the number of independent comparisons (or the number of degrees of freedom) available for calculating the estimate of error. With 15 observed differences, we have among them

Table 3 Differences in eighths of an inch between cross- and self-fertilized plants of the same pair

49	23	56
−67	28	24
8	41	75
16	14	60
6	29	−48

14 independent discrepancies, and our degrees of freedom are 14. The available tables of the distribution of t show that for 14 degrees of freedom the value 2.145 is exceeded by chance, either in the positive or negative direction in exactly 5 percent of random trials. The observed value of t, 2.148, thus just exceeds the 5 percent point, and the experimental result may be judged significant, though barely so.

18. Fallacious Use of Statistics

We may now see that Darwin's judgment was perfectly sound, in judging that it was of importance to learn how far the averages were trustworthy, and that this could be done by a statistical examination of the tables of measurements of individual plants, though not of their averages. . . .

It may be noted also that Galton's skepticism of the value of the probable error, deduced from only 15 pairs of observations, though, as it turned out, somewhat excessive, was undoubtedly right in principle. The standard error (of which the probable error is only a conventional fraction) can only be estimated with considerable uncertainty from so small a sample, and, prior to "Student's" solution of the problem, it was by no means clear to what extent this uncertainty would invalidate the test of significance. From "Student's" work it is now known that the cause for anxiety was not so great as it might have seemed. Had the standard error been known with certainty, or derived from an effectively infinite number of observations, the 5 percent value of t would have been 1.960. When our estimate is based upon 15 differences, the 5 percent value, as we have seen is 2.145, or less than 10 percent greater. Even using the inexact theory available at the time, a calculation of the probability error would have provided a valuable guide to the interpretation of the results.

In a historical perspective this work by Darwin and Galton leaned heavily on the intuitive. Yet they realized that proper comparison required matching like plants as a means of removing extraneous and unwanted variations. It was "Student," however, who later described the probability distribution appropriate for this test. Most of the language in Fisher's discussion is current; yet some terms, like "probable error" (50% confidence), have lost favor in modern times.

This example displays some of the processes in the development of scientific knowledge. Intuition and reason can lead to systematic procedures, but this often takes time and the contributions of many people.

REFERENCES

[1] R. A. Fisher, *The Design of Experiments,* 8th Edition (New York: Hafner Publishing Company, Inc., 1971).

[2] E. S. Pearson and John Wishart, editors, *Students' Collected Papers* (Cambridge: The University Press, 1958).

[3] Robert G. D. Steel and James H. Torrie, *Principles and Procedures of Statistics* (New York: McGraw-Hill, 1960).

Tests on Means—Analysis of Variance

CHAPTER 11

In business it is commonly desirable to compare one's achievement against that for competitors. In advertising statements are made such as "Our brand lasts longer than that of major competitors." Certainly an important factor in consumer purchasing is the stamp or guarantee for an average defect-free life of X months, especially if competing brands carry no such claim.

In education the relative achievement for several groups is measured by comparing mean scores. For example, a teacher might want to compare the work of his students in two sections of the same course. One approach would use a comparison of the difference of mean achievement on a common standardized test.

The preceding are just a few cases wherein groups might be compared through some measure of average performance. In this chapter we consider statistical procedures for inference on two or more groups. In addition to the earlier standard normal and studentized distribution forms, procedures include the F-statistic with some of its applications.

1/ Comparing Means for Two Samples

The foundation for inference concerning the means for independent samples from two populations is based on the linear form $\overline{X}_1 - \overline{X}_2$. With known variances and sufficiently large samples, the probability rule is based on a Z-statistic.

By using the analysis of variance, we can measure if this year's crop of vegetables is as good as last year's crop and the one for the year before, etc. In other words, we can tell what the differences are among two or more groups of something and what the differences are for items within a group. Then a comparison of the "among" and "within" measurements can provide a test.

Tests on Means—Analysis of Variance

Rule

For two distributions with unknown means but with *known* variances and independent samples of sizes n_1 and n_2, the linear combination $(\overline{X}_1 - \overline{X}_2)$ is distributed with mean $(\mu_1 - \mu_2)$ and variance

$$\sigma^2_{\overline{X}_1 - \overline{X}_2} = \frac{\sigma_1^2}{n_1} + \frac{\sigma_2^2}{n_2}.$$

Applications of this rule use the standard normal form,

$$Z = \frac{(\overline{X}_1 - \overline{X}_2) - (\mu_1 - \mu_2)}{\sigma_{\overline{X}_1 - \overline{X}_2}}, \quad \text{with } \sigma_{\overline{X}_1 - \overline{X}_2} = \sqrt{\sigma^2_{\overline{X}_1 - \overline{X}_2}}$$

which is approximately $N(0, 1)$, when $n_1 > 30$ and $n_2 > 30$. The form $\sigma_{\overline{X}_1 - \overline{X}_2}$ is the standard deviation for the difference in means. Observe in the rule that the variance of the difference is the sum of the variance in a first population and the variance in the second population. This also requires independence to assure no (0) common variance (covariance).

In education, comparison is frequently made of averages for two or more groups. Consider one example.

Example

Two methods for teaching logic are being compared. For a random sample of $n_2 = 64$ students taught by the traditional method, the mean score on a standardized test on logical thinking was 68.8 with standard deviation 5.2. For a random sample of 80 students who studied the Socratic method, the mean score on the same test was 70.5 with standard deviation 5.6 points. We test the hypothesis that the Socratic training results in a statistically higher mean at $\alpha = .025$.

Solution

This is a test on $\mu_1 - \mu_2$ = true difference in mean performance (1 = Socratic method, 2 = traditional method). Then

1. $H_0: \mu_1 = \mu_2$ (or $\mu_1 - \mu_2 = 0$)
   $H_A: \mu_1 > \mu_2$ (or $\mu_1 - \mu_2 > 0$)

2. For $\alpha = .025$

   Test: Reject H_0 if calculated $Z > 1.96$

3. Then $\sigma_{\overline{X}_1 - \overline{X}_2} = \sqrt{\dfrac{(5.6)^2}{80} + \dfrac{(5.2)^2}{64}} = \sqrt{.3920 + .4225}$

   $= \sqrt{.8145} \doteq .90$

   giving $Z = \dfrac{(\overline{X}_1 - \overline{X}_2) - (\mu_1 - \mu_2)}{\sigma_{\overline{X}_1 - \overline{X}_2}} = \dfrac{(70.50 - 68.80) - 0}{.90}$

   $= \dfrac{1.70}{.90} = 1.89$

4. Decision: We cannot reject H_0.

The evidence does not support a higher performance for those studying under the Socratic method, $\alpha = .025$.

Conclusion

Several alternative forms exist for null and alternative hypothesis statements, e.g., $H_0: \mu_1 - \mu_2 = 0$ may be preferred over $H_0: \mu_1 = \mu_2$ since the former indicates the value zero (0) is to be used for the true mean difference at Step 3. In fact, differences other than zero can be hypothesized. For example, $H_0: \mu_1 - \mu_2 = +2$ might be preferred over $H_0: \mu_1 = \mu_2 + 2$. The first expression has an advantage in that it indicates the hypothesized value (here $+2$) to be substituted into the Z-form. For example, we might hypothesize that average miles per gallon, mpg, for a car of Make 1 exceeds that for Make 2 by $+ 2$ mpg.

A second complexity is in computing a value for the divisor, $\sigma_{\bar{X}_1 - \bar{X}_2}$. My experience is that trying to take shortcuts in this calculation can lead to gross arithmetical errors. Looking back at the solution to the last example, a correct approach is first squaring each numerator value, then dividing (in two places), then adding the resulting quotients, and finally taking a square root. This gives the value for the divisor in the Z-form.

We are not always so fortunate to have large samples, nor do we always know population variances; yet lacking these conditions does not void all inference. For independent samples from two normal distributions, valid inference on differences in means can be derived from a t-statistic.

For independent samples, each of size less than or equal to 30, and coming from normal distributions which have a *common, but unknown variance*, σ^2, the sampling distribution for the difference of means is described in

Rule

$$t = \frac{(\bar{X}_1 - \bar{X}_2) - (\mu_1 - \mu_2)}{\sqrt{(s_p^2/n_1) + (s_p^2/n_2)}} = \frac{(\bar{X}_1 - \bar{X}_2) - (\mu_1 - \mu_2)}{s_p\sqrt{(1/n_1) + (1/n_2)}}$$

$s_p^2 = $ a pooled estimator for the common variance, σ^2

$$= \frac{(n_1 - 1)s_1^2 + (n_2 - 1)s_2^2}{n_1 + n_2 - 2}$$

The name "pooled variance" for s_p^2 indicates estimating the common variance by pooling or combining the measures of variability from both samples. It is an "average" of variances with total variability (numerator) divided by the effective total degrees of freedom, df* $= (n_1 - 1) + (n_2 - 1) = n_1 + n_2 - 2$. That is, a first sample contains an "effective" sample size

---

*Recall the discussion of df given in Section 10.1.

$n_1 - 1$, while for a second sample the useful information is $n_2 - 1$. This is combined in a weighted and pooled variance to estimate σ^2. For the case when $n_1 = n_2$, $s_p^2 = (s_1^2 + s_2^2)/2$.

A first example may be of interest to any person aspiring to become a millionaire.

Example

A sociologist studying self-made millionaires found the following for a random sample of nine such individuals in each of two countries. Sample mean-age and variances are given (*Note:* This is hypothetical data):

$\overline{X}_1 = 40.33$ years $\qquad \overline{X}_2 = 36.54$ years
$s_1^2 = 25.25 \qquad\qquad s_2^2 = 34.61$

where μ_i = The (true) average age at which the individual becomes a millionaire, i = Country 1 or 2. We will assume both age distributions to be normal and with a common, but unknown variance. Is there sufficient evidence to conclude a significant difference in mean age of attainment for self-made millionaires in these two countries? Let $\alpha = .05$.

Solution

For μ_1 = mean age at which those individuals in Country 1 become self-made millionaires, etc.

$H_0: \mu_1 - \mu_2 = 0$
$H_A: \mu_1 - \mu_2 \neq 0$

Test: Reject H_0 if |calculated t| > 2.12 ($t_{.025, 16}$)

For $s_p^2 = \dfrac{(9-1)25.25 + (9-1)34.61}{9+9-2} = 29.93$, then

calculated $t = \dfrac{(40.33 - 36.54) - 0}{\sqrt{(29.93/9) + (29.93/9)}} = \dfrac{3.79}{2.58} = 1.47$

Decision: We cannot reject H_0

Conclusion

From this evidence and for $\alpha = .05$, the sociologist must conclude no significant difference for the mean age of actualization for self-made millionaires in these two countries.

For a more pointed indication of the difference in mean ages in the preceding, we can establish a 95% confidence estimate on the true difference.

Comparing Means for Two Samples

For a $100(1-\alpha)\%$ confidence estimate on the difference in means for small independent samples from normal distributions with common, but unknown variance, use

Rule

$$(\overline{X}_1 - \overline{X}_2) \pm \left(t_{\alpha/2, n_1+n_2-2} \cdot \sqrt{\frac{s_p^2}{n_1} + \frac{s_p^2}{n_2}} \right)$$

This is applied to the data for self-made millionaires.

The preceding rule with

Example

$$\overline{X}_1 - \overline{X}_2 = 3.79, \sqrt{\frac{s_p^2}{n_1} + \frac{s_p^2}{n_2}} = 2.58, \text{ and } t_{.025,16} = 2.12$$

gives

$l = 3.79 - (2.12 \cdot 2.58) \doteq -1.68$
$u = 3.79 + 5.47 \doteq 9.26$,

giving $-1.68 < \mu_1 - \mu_2 < 9.26$ years

The average age for those attaining the status of self-made millionaire is as much as 1.68 years younger in Country 1, to as much as 9.26 years older than for those in Country 2.

Conclusion

A second example using the t-form adds the requirement of computing values for s_1^2 and s_2^2. Also this example illustrates the case wherein the null hypothesis indicates nonzero difference.

A used car dealer is interested in the relative age for trade-ins of two makes—A and B. Having had the dealership for some years, his experience makes him believe that, generally, Make A cars traded-in are over one year older than Make B cars. Test his belief using $\alpha = .025$ and the data in the table.

Example

Age at Trade-in

Car Make							
A				B			
3	3	4	5	2	3	3	4
5	5	6	6	4	4	4	4
7	8	8		4	4	4	5
				5	6		

Solution

For μ_A = mean age of Make A cars at trade-in

$H_0: \mu_A - \mu_B = 1$
$H_A: \mu_A - \mu_B > 1 \quad (\mu_A > \mu_B + 1)$

Test: Reject H_0 if calculated $t > 2.07\,(t_{23,.025})$

Summary statistics (calculations are requested in Problem 3)

$$\overline{X}_A = 5.45$$
$$\overline{X}_B = 4.00$$
$$s_p^2 \doteq 1.85$$

$$\text{Calculated } t = \frac{(5.45 - 4.00) - 1}{\sqrt{(1.85/11) + (1.85/14)}} = \frac{0.45}{\sqrt{.3003}} \doteq 0.8.$$

Conclusion

We cannot reject H_0; Make A car owners generally trade to this dealer at most one year later than do Make B car owners. Again, because of the small samples you should be aware of the possibly high β-risk, and subsequently of the chance that this decision is wrong.

In this, as well the first example of the section, we used an upper-tailed test. This was a matter of convenience; with proper adjustments each could be treated equally well as a lower-tailed test. The conclusions would be unaltered. One precaution concerning the one-tailed test; the difference of sample means must be used in an order consistent with the test hypothesis. Using $\overline{X}_B - \overline{X}_A$ with the preceding hypotheses would give a grossly incorrect calculated t-value.

Inference concerning the difference of means for dependent samples from normal distributions was discussed in Section 4 of Chapter 10. See [1] for inference where the variances cannot be considered equal, but the samples are independently drawn from two normal distributions.

Problems

In each of the following assume the variables being compared have normal probability distributions.

1. Two friends who are teaching American history at separate high schools want to compare achievement for their students. Scores on a standardized test provided the following values:

$\overline{X}_1 = 76.2 \quad s_1^2 = 18.6$
$\overline{X}_2 = 83.1 \quad s_2^2 = 38.2$

for $n_1 = 18$ and $n_2 = 12$ students, respectively. Is there a significant difference in achievement on this measure for students in these two classes, $\alpha = .05$? Assume common variances.

2. A chemist discovered a drug called S.O.S. which he claims will give people longer relief from colds than will other drugs. To test this claim, a sample of thirty-six people with colds were given the drug and subsequently found relief for an average of 8.6 hours with a standard deviation of 3.0 hours. Thirty-six others took a different cold medicine for which the mean relief was 7.2 hours with a standard deviation of 4.0 hours. Is the chemist's claim justified at $\alpha = .05$? At $\alpha = .025$?

3. Perform the calculations suggested in the used-car dealer example on p. 255. That is,

   a. Compute s_A^2 and s_B^2 then,
   b. Use these to check s_p^2. Check the other calculations in that solution.

4. A training director in a large hospital claims that orderlies who take her course perform better on the job than do those not receiving this training. Of 72 more recently-hired orderlies, 36 were randomly selected to receive the training. The other 36 received no special job training. Six months later, on-the-job evaluations produced the following performance statistics. For the trained group $\overline{X}_1 = 84.63$, $s_1 = 3.5$. For the orderlies without this training $\overline{X}_2 = 81.45$ and $s_2 = 4.0$. Test the director's claim using $\alpha = .05$.

5. Use this summary information for independent samples from two normal distributions to test $H_0: \mu_1 = \mu_2$ against a two-tailed alternate. Use $\alpha = .10$.

   $n_1 = 10 \quad \overline{X}_1 = 16 \quad s_1 = 7$
   $n_2 = 12 \quad \overline{X}_2 = 13 \quad s_2 = 5$

   Assume population variances are statistically equal.

6. Following are summary statistics concerning the time (in minutes) required by two separate fire squads to put out a series of simulated fires.

Squad	Mean	Standard deviation	n
1	23.0	4.2	15
2	15.0	3.7	17

Test the claim that Squad 1 takes an average of over five minutes longer to do the job. Assume that all essential test conditions are met. Use $\alpha = .10$.

7. The work in a lawn-spraying service is fairly simple and requires a high percentage of semi- and unskilled laborers. Should the workers be hired (1) on a hit-or-miss basis such as a quick interview or (2) on the basis of the results of a job-test? Both methods were tried for several months and a record kept of the productivity in lawns sprayed per day. A random selection of workers gave the results displayed.

Productivity (Number of Lawns Per Day)

Group 1 Hit-or-miss selection	Group 2 Selection by test
4 8 3	6 8 9
5 7 5	8 6 7
5 4 6	5

Using $\alpha = .025$, test the hypothesis that the average production is higher for those selected by the job test. Assume normal distributions and use $\sum(X_1 - \overline{X}_1)^2 = 19.56$ and $(n_2 - 1)s_2^2 = \sum(X_2 - \overline{X}_2)^2 = 12.00$.

8. In a learning experiment designed to test the time required to solve a complicated puzzle, thirty-five individuals using direction set 1 required an average of $\overline{X}_1 = 57$ minutes with $s_1^2 = 64$ squared minutes. Thirty-two individuals each using direction set 2 took $\overline{X}_2 = 54$ minutes with $s_2^2 = 25$ squared minutes. Assuming independence of trials and using a 5% α-risk, test whether mean times required to solve the puzzle are significantly different.

2/ Relation between the t- and F-Statistics

We have explored several techniques for comparing means of samples from two distributions. Yet limiting comparisons to two groups is rather restrictive; e.g., a manufacturer may be required to compare their product with that of all competitors, and generally that's more than two. Although the t-test (or Z for larger samples) will allow comparison against other means one at a time, the F-statistic provides a more economical approach.

Relation between the t– and F–Statistics

This form allows a single test on the equality for a large number of means. Related tests are available if we care to single out *individual* comparisons.

The F-ratio statistic is described by

$$F = \frac{U/\mathrm{df}_1}{V/\mathrm{df}_2},$$

where U and V are generalizations for statistics derived from the sample data. One form is used to compare two population variances. For independent samples of n_1 and n_2 from normal distributions. This F-statistic is:

$$\frac{s_1^2}{s_2^2} = \frac{\sum (X_1 - \overline{X}_1)^2/(n_1 - 1)}{\sum (X_2 - \overline{X}_2)^2/(n_2 - 1)}$$

with $\mathrm{df}_1 = n_1 - 1$ and $\mathrm{df}_2 = n_2 - 1$. This form is used in tests on the equality of two population variances.

In a test of equal variances for independent samples ($n_1 = 21$, $n_2 = 15$) from two normal distributions, $s_1^2 = 25$ and $s_2^2 = 10$,

Example

$H_0: \sigma_1^2 = \sigma_2^2$ (or $\sigma_1^2/\sigma_2^2 = 1$)

$H_A: \sigma_1^2 > \sigma_2^2$ (or $\sigma_1^2/\sigma_2^2 > 1$)

For $\alpha = .05$ with $n_1 = 21$ and $n_2 = 15$ we use Table IV

$\mathrm{df}_1 = n_1 - 1 = 20$, $\mathrm{df}_2 = 14$

Test: Reject H_0 if calculated $F > 2.39$ ($F_{.05, 20, 14}$)

Calculated $F = 25/10 = 2.5$

Decision: Reject H_0.

These two normal distributions do not have equal variances, $\sigma_1^2 > \sigma_2^2$.

Conclusion

This or another test for *homogeneity of variance* is appropriately used prior to a Z- or t-test on the equality of means. See [4].

Our main interest is the form of the F-ratio statistic as used in tests on means. The curves in Figure 1 generally have a skewed, or nonsymmetrical appearance with the general shape depending on the sample size. Several characteristics are observed. First, values on the F-scale (horizontal axis) are zero or positive; none are negative. Each curve is described by its degrees of freedom, which in turn is defined by the sample sizes. As the samples get larger, and accordingly degrees of freedom increase, the F-curve becomes less asymmetric. If the numerator df (the first number) is one, then as the denominator df (the second number) is increased, the F-distribution becomes more nearly bell-shaped.

Figure 1 / The F probability distribution for various degrees of freedom

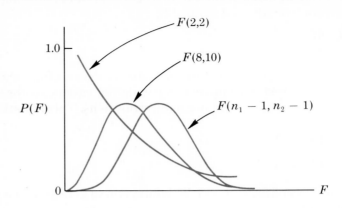

The use of the F-table requires two sets of degrees of freedom. In the preceding there were two estimators for σ^2, s_1^2 and s_2^2. From the rule for df on p. 230, the numerator contains $n_1 - 1 = \text{df}_1$, for a sample of n_1, while the denominator contains $\text{df}_2 = n_2 - 1$, for an independent sample of n_2. See Figure 1. Since the exact form of the F-distribution depends on degrees of freedom, alternative values for numerator, df_1, and denominator, df_2, must be specified. That is, numerator and denominator df are required in order to read Table IV. F-values are presented for 1%, 2.5%, 5%, and 10% areas in the upper tail. Check Table IV for the values given in the following examples.

Example 1   The value of tabular F, based on samples of $n_1 = 13$ and $n_2 = 16$, for $\alpha = .01$ is 3.67; that is, $P(F_{12,15} > 3.67) = .01$.

Example 2   For the same sample sizes, so $\text{df}_1 = 12$ and $\text{df}_2 = 15$, but for $\alpha = .025$, the F-value is $F_{.025,12,15} = 2.96$.

Example 3   Testing $H_0: \sigma_1^2 = \sigma_2^2$ against $H_A: \sigma_1^2 < \sigma_2^2$ for $n_1 = 10$ and $n_2 = 17$ can be done by interchanging the roles of σ_1^2 and σ_2^2; that is, by testing $H_A: \sigma_2^2 > \sigma_1^2$. Then $F = (s_2^2/s_1^2)$ with numerator $\text{df} = n_2 - 1$, denominator $\text{df} = n_1 - 1$ and test statistic $F_{.05,16,9} \doteq 3.01$ (see Table IV). The exact F-percentage value for $\text{df}_1 = 16$ is not in Table IV, so the closest entry with $\text{df}_1 = 15$, $\text{df}_2 = 9$ is used.

See [2] if you want more complete F-tables.

The F- and t-distributions are related in the following way: $t^2 = F$ if there is one df in the numerator. Since t^2 is nonnegative, the F-distribution upper-tail percentage points include *twice* the area of comparable t-values.

For example, $t_{18} = 2.10$ cuts 2.5% of the area from the upper-tail while $(t^2_{.025,18}) = F_{.05,1,18} \doteq 4.41$ cuts 5% of the area from the upper-tail. This relation is used to explore the one degree of freedom (numerator) F-values and the relation of t- and F-statistics for tests on equality for two means. See Table 1.

Table 1 / Selected t_{df_2} and F_{1,df_2} Tabular Values

2.5% t-values, two-tailed percentage points		5% F-values, right-tail percentage points	
df_2	t_{df_2}	df_2	F_{1,df_2}
1	12.706	1	161.45
2	4.303	2	18.513
3	3.182	3	10.128
⋮	⋮	⋮	⋮
30	2.042	30	4.171
⋮	⋮	⋮	⋮
∞	1.96	∞	3.842

Comparison of entries in Table 1 reveals that the t-value with df_2, when squared, yields an F-value with one df in the numerator and corresponding df_2 in the denominator. For example, $t^2_{.025,30} = (2.042)^2 = 4.17 = F_{.05,1,30}$. Thus a two-tailed t-test at $\alpha = .05$ (that is, with $\alpha/2 = .025$ risk in each tail) is comparable to a one-tailed F_{1,df_2} test at $\alpha = .05$. F-tests on equality of means will employ a one-tailed, upper, rejection region. The F-, like the t-distribution is based on the assumption that the original random variables have a normal distribution. In hypothesis tests a two-tailed t-test with "df_2" degrees of freedom is comparable to a one-tailed F-test having one and "df_2" degrees of freedom, both being α-level tests.

The remainder of this chapter concerns tests on means which use the F-ratio statistic. Because calculations become involved, a systematic approach called the *analysis of variance,* is used for computing F-values.

Problems

1. Find or approximate each of the following using Table IV:
   a. $F_{.025,13,8}$
   b. $F_{.01,13,8}$
   c. $F_{.01,8,13}$
   d. $F_{.05,19,19}$
   e. $F_{.05,100,120}$
   f. $F_{.01,1,1}$

2. Check your understanding of the use of F-tables by evaluating the following:
   a. For $P(F > F_{.05,3,2}) = .05$, find $F_{.05,3,2}$.
   b. If $t_{.025,10} = 2.23$, determine $F_{.05,1,10}$ (use $t^2_{df_2} = F_{1,df_2}$).
   c. Determine $F_{.05,1,\infty}$.
   d. Does $F_{.05,1,5} = F_{.05,5,1}$?
   e. If $F_{.01,1,10} = 10.044$, does $t_{.005,10} = \pm 3.17$?
   f. Does $F_{.05,4,7} = F_{.01,4,7}$?

3. From Table IV for $\alpha = .05$, observe $P(F_{1,1} > 161.45) = .05$, while $P(F_{10,12} > 2.7534) = .05$, and $P(F_{20,18} > 2.196) = .05$. Noting that as sample sizes increase, upper-tail 5% points are less positive, explain how sample size affects the positioning of the F-curve.

4. For the test $H_0: \sigma_1^2 = \sigma_2^2$ against $H_A: \sigma_1^2 > \sigma_2^2$, the calculated $F = 4.17$. Given independent samples of $n_1 = 5$ and $n_2 = 12$, at which of the levels, .01, .025, .05, or .10, would H_0 be rejected?

5. The F-ratio statistic is $F = (s_1^2/s_2^2)$. Why is it never negative?

3/ The Analysis of Variance Procedure

Common statistical analyses require comparison of more than two groups. Consider, for example, a psychology experiment concerning the effect of shock on the time required to lace a shoe. Each of the 18 participants is randomly assigned to one of three experimental conditions.

Group 1: the control group, no shock

Group 2: medium level shock
Group 3: high level shock.

Responses are the times, in seconds, required to lace a three-eyelet shoe. We will assume that the necessary assumptions are met including independent samples from normal distributions with equal variances.

The hypothesis of no statistical difference in mean times is tested against the alternative that one or more groups show a statistically different average. Symbolically, $H_0: \mu_1 = \mu_2 = \mu_3$, and H_A: at least one μ is not equal to the others. Rejecting the null hypothesis would indicate one or possibly two groups had faster reaction times. Other tests could be used to identify specific differences. On the other hand if we are unable to reject equality, then we must conclude that the intensity of shock does not appreciably affect this activity for those measured. Assuming equal variances, if the null hypothesis is true, i.e., if $\mu_1 = \mu_2 = \mu_3$, then these scores will be considered three independent samples from the same normal distribution. With known mean and variance, the distribution of normal scores is completely defined.

We explore the explicit form of the F-ratio test appropriate to this experiment. The test incorporates two independent estimators of the common but unknown variance, σ^2, in the form

$$F = \frac{\text{``Among'' estimator of } \sigma^2}{\text{``Within'' estimator of } \sigma^2} = \frac{S_1^2}{S_2^2}.$$

The first or numerator estimator for σ^2 measures the amount of variation *among* the k group means by comparison of the individual means to the grand mean.

For $S_1^2 =$ the "among" groups mean square   Rule

$$S_1^2 = \frac{\text{Among sum of squared differences}}{\text{Among degrees of freedom}} = \frac{\sum\limits^{k}[n(\overline{X}_i - \overline{X})^2]}{k - 1}$$

where

$k =$ the number of groups
$n =$ the (same) number in each group
$\overline{X}_i =$ the mean for any group, $i = 1, 2, 3, \ldots, k$
$\overline{X} =$ the mean for all observations, the grand mean.

This estimator compares each group mean to the grand mean, in essence a comparison "among" the k means. To determine numerator df, we observe that there are k numbers (means) used in the comparison. Since correction is made in each calculation to the approximated grand mean, we lose one df. There remain $k - 1$ degrees of freedom in this computation.

Tests on Means—Analysis of Variance

The second or denominator estimator for σ^2 describes the amount of variation "within" groups. Calculations are *within* each group, with comparison for individuals to their respective means.

Rule

For $S_2^2 =$ the "within" groups mean square

$$S_2^2 = \frac{\text{Within sum of squared differences}}{\text{Within degrees of freedom}} = \frac{\sum\limits^{k}}{k}\left(\frac{\sum\limits^{n}(X - \overline{X}_i)^2}{n-1}\right)$$

where

$n =$ the *common* number of individuals in each group
$X =$ the scores for individuals within groups.

This form is analogous to the "pooled variance" estimator described on p. 253. For k groups with equal numbers in each group,

$$S_2^2 = \frac{s_1^2 + s_2^2 + \cdots + s_k^2}{k} = \frac{\sum\limits^{k}(s_i^2)}{k}.$$

Since each term, s_i^2, has divisor $n - 1$, then the divisor, $df_2 = (n - 1) + (n - 1) + \cdots + (n - 1) = k(n - 1)$. The F-standard becomes $F_{k-1, k(n-1)} = S_1^2/S_2^2$. That is, for tests on the equality of means from k normal distributions, use an F-ratio statistic with $k - 1$ degrees of freedom in the numerator and $[k \cdot (n - 1)]$ degrees of freedom in the denominator.

We now have the tools for a test on equality of means. If the (null) hypothesis of equal means is true, then numerator and denominator

Table 2/ Display of Data with Equal Numbers of Observations
(Includes Totals and Means for a One-way Analysis)

	Groups			
	1	2	3	
	5	8	11	
	7	9	13	
	8	10	14	
	8	11	14	
	10	12	17	
	10	13	18	
Totals:	48	63	87	Grand total = 198
Means:	$8 = \overline{X}_1$	$10.5 = \overline{X}_2$	$14.5 = \overline{X}_3$	Grand mean = $\overline{X} = 11$
$\sum(X - \overline{X})^2$:	18.00	17.50	33.50	

The Analysis of Variance Procedure

estimators are unbiased, meaning each has average value σ^2, and we should expect $F = S_1^2/S_2^2 \doteq 1$. That is, if the sample produces a calculated F which is near 1, the assumption of "equal means" is reasonable. On the other hand, if the sample means (here $\overline{X}_1, \overline{X}_2$ and $\overline{X}_3$) differ widely, the estimate S_1^2 will be large. Yet S_2^2 will be small so long as the values within each group are relatively internally constant. The resulting F-ratio would be large and we should expect to reject the assumption of equal means.

Essential calculations appear in Table 2. This is part of an analysis of variance for the shock–response experiment. Again, the numbers are times, in seconds, required for different persons to lace a three-eyelet shoe.

For Sums of Squares, SS, calculations using Table 2

$$\text{Among SS} = \sum n(\overline{X}_i - \overline{X})^2 = 6(8-11)^2 + 6(10.5-11)^2 + 6(14.5-11)^2 = 129.00$$

$$\text{Within SS} = \sum\sum(X - \overline{X}_i)^2 = \{[(5-8)^2 + (7-8)^2 + \cdots + (10-8)^2] + [(8-10.5)^2 + (9-10.5)^2 + \cdots + (13-10.5)^2] + [(11-14.5)^2 + (13-14.5)^2 + \cdots + (18-14.5)^2]\} = \{18.00 + 17.50 + 33.50\} = 69.00$$

The following table, an analysis of variance, is used in executing the F-ratio test. It provides a systematic way of performing the test. Take $\alpha = .05$.

Table 3/ The Analysis of Variance and F-test for the Shock–Response Experiment

Source of variation	Degrees of freedom	Sum of squared differences, SS	Mean squares	F
Among	$3 - 1 = 2$	129.00	$S_1^2 = \frac{129}{2} = 64.50$	$\dfrac{S_1^2}{S_2^2} = 14.02$
Within	$3(6-1) = 15$	69.00	$S_2^2 = \frac{69}{15} = 4.60$	
Total	17	198.00		

From Table IV the test requires a calculated $F > 3.68$ ($F_{.05,2,15}$) for rejection of the null hypothesis. Since calculated $F = 14.02$, we reject the hypothesis of equal means. One or more of the groups react more quickly than others. A word of caution: our conclusion indicates only that one or more differences exist. Procedures for isolating all differences are described in [4] as *multiple comparisons*.

That degrees of freedom and sums of squares are separately additive allows checks on the calculations recorded in Table 3. That is,

Total SS = Among SS + Within SS or
Within SS = Total SS − Among SS

Tests on Means—Analysis of Variance

In the example,

$$\text{Total SS} = \sum^{k}\sum^{n}[(X - \overline{X})^2]$$
$$= [(5 - 11)^2] + [(7 - 11)^2] + \cdots + [(18 - 11)^2] = 198.00.$$

This additive relation is commonly exploited by directly computing Total SS and Among SS from the observations. Within SS, being a more awkward calculation, is then obtained by subtraction. That is, Within SS = 198.0 − 129.0 = 69.0. See Table 3. Direct calculation of all three values gives a check, but this is usually not necessary. Generally only Total SS and Among SS are computed directly. In a similar fashion degrees of freedom are additive. That is,

Total df = Among df + Within df
$(kn - 1) = (k - 1) + k(n - 1)$
or
Within df = Total df − Among df.
$k(n - 1) = (kn - 1) - (k - 1)$

In each case the degrees of freedom counts the number of terms in the summation less one for each mean used as a standard for difference calculations. Again the additive property can be used to check the values.

The next illustration is presented without the original data, but assumes all previous calculations are correct.

Example

The principal at Grace Lake Elementary School wants to determine if kindergarten absences relate to the session the child attends. Youngsters were randomly assigned to a session—either early, 8–11; midday, 10–1; or afternoon, 12–3—at the beginning of the school year. A selection of 24 pupils from each session gave the following number of days absent. We assume absences in each class follow a normal distribution, that the assumption of equal variances is reasonable, and that others have found Total SS = 1,104.90. The test is for equal numbers of absences in the three sessions at $\alpha = .05$.

	Early, E	Midday, M	Afternoon, A	
Total	82	112	43	Grand total = 237
Mean	3.42	4.67	1.79	Grand mean = 3.29

Solution

The test is for equality of means with, for example,

μ_E = true average number of pupil absences per day for the early session.

Then $H_0: \mu_E = \mu_M = \mu_A$
H_A: at least one $\mu \neq$ others
Test: Reject H_0 if calculated $F > 3.15 \, [F_{.05, 2, 3(23)}]$
(the nearest tabled value—$F_{2, 60}$—was used).
Calculations:
Among SS $= [24(3.42 - 3.29)^2 + 24(4.67 - 3.29)^2$
$\qquad + 24(1.79 - 3.29)^2] = 100.11$

Remembering that Total SS $= 1{,}104.90$ is given, we exploit the additive property for sums of squares to complete the analysis of variance, and then perform the F-test.

Source of variation	Degrees of freedom	Sum of squares	Mean squares	F
Among	2	100.11	50.06	3.44
Within	$3(24 - 1) = 69$	$1{,}104.90 - 100.11 = 1{,}004.79$	14.56	
Total	$3 \cdot 24 - 1 = 71$	$1{,}104.90$		

Decision: Reject H_0, $\alpha = .05$.

Absence rates are not the same for the three sessions, $\alpha = .05$. Observation of sample means should give an intuitive feeling about individual differences.

You may have noticed that the groups or treatments in this last example are levels of a qualitative variable. That is, we named or qualified the groups, but numerical values were not assigned to "early," "midday," and "afternoon." The groups may, in fact, be quantitative; their (name) values do not enter the computations.

The next chapter, on regression, is related to analysis of variance, but differs in that the so-called treatment variable is quantitatively evaluated.

$$SS_B = \sum_{i=1}^{k} (\overline{X}_i - \overline{X})^2$$

$$SS_T = \sum_{i=1}^{k} \sum_{j=1}^{n} (X_j - \overline{X})^2$$

Problems

1. Using $\alpha = .01$ and the information below, complete a test on the $H_0: \mu_1 = \mu_2 = \mu_3 = \mu_4$ and H_A: at least one of the $\mu_i \neq$ the others. Assume normal distribution for groups.

Tests on Means—Analysis of Variance

	Group			
	1	2	3	4
	13	7	15	11
	15	10	18	10
	20	13	21	15

Source of variation	df	Sum of squares	Mean squares	F
Among groups				
Within groups		76		
Total				

2. Complete the test for equality of means using the following information and $\alpha = .05$.

	Group		
	1	2	3
	30	35	28
	25	42	28
	31	38	30
	38	39	24
	36	31	25
Means	32	37	27

Source	df	Sum of squares
Among groups		
Within groups	12	
Total		450

State conclusions assuming that the groups are exam scores for random samples of students from three sections of a high-school biology course. Assume normal distributions.

3. An experiment with five groups and 12 observations per group showed:

$F = S_1^2/S_2^2 = 44.1/17$.

Assume a test on the equality of means at $\alpha = .05$. What can you conclude? For $\alpha = .025$ what can you conclude? Why should the α-risk level be set before any statistical calculations are made?

4. A psychoanalyst is experimenting with three treatments for patients who show marked withdrawal symptoms. The treatments are: control, concentrated visitation with the analyst; Condition 1, complete rest; Condition 2, physical release of tension (i.e., pounding with a sledgehammer on a junk car for 15 minutes each day). Treatments were randomly assigned to patients. Response is measured by scores on the NORM test. Low scores indicate high withdrawal. Use the procedures of this section and $\alpha = .025$ to test for the best treatment.

	Treatment				
Control	16	19	18	15	17
Condition 1	9	7	8	5	6
Condition 2	12	6	10	7	15

5. Ten subjects are randomly assigned to a control group and ten others to an experimental group. The control group was asked to make immediate recall of an event. The experimental group was asked to respond several hours later. The response is a measure of recall time in seconds. Assume these samples are from normal distributions.

	Control	Experimental
Mean	13	15
s^2	3.2	4.6

a. Test $H_0: \mu_C = \mu_E$ using the t-test for independent samples with equal, but unknown variances. Use $\alpha = .05$.

b. Use the F-ratio test developed in this section to answer part a. Compare your results. Does calculated $t^2_{df_2}$ = calculated F_{1,df_2} for this experiment?

6. The table lists scores on three placement exams for each of five college freshmen. Calculate F by a one-way analysis of variance and test at $\alpha = .05$ whether there is a significant difference in mean scores among at least two of these persons. Assume these are samples from normal distributions.

Anne	Brenda	Dale	Lee	Shelly
171	161	139	119	172
148	156	126	125	182
176	184	152	155	204

4/ Tests on Means—Unequal Numbers

Although it is convenient to have equal numbers, it is most common to observe unequal numbers in the groups. In practice irregularities just seem to occur; when testing pupils one or more youngsters might be absent on the test day; in the psychological treatment of drug addicts, some might be incoherent or otherwise unable to participate. What then can be done to get meaningful results from planned experiments which result in unequal numbers?

A first principle in treating statistical data requires that every observation produced under design standards be included in the analysis. There is no place in research for work based on willfully distorted data. Do not throw out valid observations just to get equal numbers. Most experiments, even those with unequal numbers, will adapt to a realistic analysis. The following experiment contains unequal numbers in the groups.

Example

A test concerning attitude toward war was given to a random sample of single men aged 20 to 30. Following are scores for three groups: (1) combat veterans, (2) noncombat veterans, and (3) nonveterans. A high score on a scale of 400 points indicates an unfavorable attitude toward war. The results appear in Table 4. The information affords a one-way analysis on service experience. The test is $H_0: \mu_1 = \mu_2 = \mu_3$, where, for example, μ_1 = mean score on the attitude test for those in Group 1, etc., against H_A: at least one $\mu \neq$ others. Then:

Total SS = [$(396 - 382.28)^2 + (397 - 382.28)^2 + (398 - 382.28)^2$
$+ \cdots + (368 - 382.28)^2$] = 2,877.61
Groups SS = [$7(395.29 - 382.28)^2 + 5(382.60 - 382.28)^2$
$+ 6(366.83 - 382.28)^2$] = 2,617.55
Within groups SS = Total SS − Groups SS = 260.06.

The analysis is the same as in the last section except that we must adjust for the unequal numbers, here $n_1 = 7, n_2 = 5, n_3 = 6$. In general the following adjustments are necessary for calculations with unequal numbers.

1. The sum of squares for groups computation becomes:

$$\sum_{}^{k} [n_i(\overline{X}_i - \overline{X})^2].$$

2. The partition of total degrees of freedom is:

Total df = Among (Groups) df + Within df

$$\left(\sum_{i}^{k} n_i\right) - 1 = (k - 1) + \sum_{i}^{k}(n_i - 1).$$

Accordingly, the test for equality of means, generalized to H_0: $\mu_1 = \cdots = \mu_k$, follows the procedures of the last section.

Table 4 / Scores on an Attitude-Toward-War Test

	Group 1 Combat veterans	Group 2 Noncombat veterans	Group 3 Nonveterans	
	396	383	368	
	397	380	369	
	398	387	366	
	400	385	366	
	398	378	364	
	395		368	
	383			
Totals	2767	1913	2201	Total = 6881
Means	395.29	382.60	366.83	$\bar{X} = 382.28$

Solution

Here Within groups df $= \sum(n_i - 1) = (7 - 1) + (5 - 1) + (6 - 1) = 15$. The analysis of variance is given in the table. Let $\alpha = .01$.

Source	Degrees of freedom	Sum of squares	Mean square	F
Groups (service experience)	2	2617.55	1308.78	75.5
Within groups	15	260.06	17.34	
Total	17	2877.61		$F_{.01,2,15} = 6.36$

We conclude mean scores are not the same for these three groups. The calculated $F \doteq 75.5$ may seem large. However close observation of Table 4 indicates that within experience groups the readings are quite consistent. This has induced a quite low *Within* group mean square, subsequently a large F-ratio.

As a second example with unequal numbers consider the following experiment wherein a pharmaceutical company was interested in two

Tests on Means—Analysis of Variance

growth stimulants for beef cattle. The stimulants both were liquid solutions given by periodic hypodermic injections. These solutions were (1) iron, and (2) a compound rich in vitamin A. Let's call these Treatment 1 and Treatment 2, respectively. A control group, Treatment 3, included only cattle that received neither stimulant. The experimental animals were feedlot beef cattle in several different locations.

The initial groups were of equal size, $n = 20$ but some animals died, resulting in unequal numbers. The observations were weight gains with comparison for equal mean gains, in pounds. See Table 5.

Table 5/ Weight Gain in Pounds for Beef Cattle

	Treatment						
	1		2		3		
	101	158	125	114	126	141	
	115	114	169	165	117	152	
	116	115	158	166	135	149	
	129	117	178	144	130	78	
	102	166	168	130	144	99	
	120	104	127	140	122	129	
	150	110	153	140	142	122	
	156	100	110	141	143	104	
	100	111	126	152	146		
	114	148	130		134		
Totals	2,446		2,736		2,313		
Mean	(1) 122.30		(2) 144.0		(3) 128.50		$131.49 = \overline{X}$
n_i	20		19		18		

Using the numbers in Table 5 and the procedures for unequal numbers,

$$\text{Total SS} = [(101 - 131.49)^2 + (115 - 131.49)^2 + \cdots + (104 - 131.49)^2] = 26{,}880.25$$

$$\text{Treatment SS} = [20(122.30 - 131.49)^2 + 19(144.0 - 131.49)^2 + 18(128.50 - 131.49)^2 = 4{,}823.55.$$

Total df is, as before, the total number of observations less one. Total df $= (20 + 19 + 18) - 1 = 57 - 1 = 56$. Treatments (groups) df remains $k - 1 = 3 - 1 = 2$. The F-ratio test is presented in the following analysis of variance format. For these three groups, the mean gains were not the same. By observation of means, in Table 5, Treatment 2 has induced a "significantly" larger mean gain than Treatment 1. Identification of other differences would require further testing. Most importantly we have illustrated reasonable procedures for testing the equality of means with unequal numbers in the groups.

Source	Degrees of freedom	Sum of squares	Mean square	F
Treatments	2	4,823.55	2,411.78	5.90
Within	54	Diff. = 22,056.70	408.46	
Total	56	26,880.25	$F_{.05,2,54} \doteq 3.15$	

The procedures of this chapter include the most basic *experimental plans* or *designs*. This area is most fundamental to more advanced statistics. More extensive treatment is given in [3], in [4], and in [1].

Problems

1. In an analysis of variance comparing four treatments, the following data were recorded: Total SS = 265. Assuming necessary conditions are met and using $\alpha = .01$, test for a significant difference between groups. Observe that this experiment contains unequal numbers. Use $\overline{X} = [\sum(n_i \overline{X}_i)/\sum n_i]$.

Group	Sample Size	Mean
A	4	50.0
B	4	45.3
C	3	41.7
D	4	47.3

2. Complete the following analysis of variance table on the next page and test $H_0: \mu_1 = \mu_2 = \mu_3$ at $\alpha = .05$:

Given:	Treatment	n_i	Mean
	1	6	104.4
	2	5	146.8
	3	7	155.4

Tests on Means—Analysis of Variance

Source	df	SS	Mean Square	F
Treatments	2	9209.1		
Within	15	11433.51		
Total	17	20,642.61		

3. A socioeconomic study was made comparing IQ's for five-year olds at a day-care center. Generally each youngster's home is economically (1) normal, (2) subnormal, or (3) welfare. A random selection of youngsters showed the following IQ's. Using $\alpha = .025$, test for significant differences in IQ between the youngsters in these groups.

	Groups	
1	2	3
112	105	86
125	98	73
121	103	90
112		80
110		

Σ 580 306 329

4. A sales manager feels that there is a substantial difference in total business performed depending on the sales region. This is a real concern because salesmen get a fixed percentage of sales. Given the following random selection of monthly intake for three regions, what can you conclude? Use $\alpha = .05$ and Total SS = 136.50. Assume normal distributions.

	Sales by region ($1,000)		
	1	2	3
	$25.95	$24.65	$20.14
	20.93	28.37	30.61
	21.80	28.27	21.17
	25.66	27.25	21.46
Totals	$94.34	$108.54	$93.38

5. In a stimulus–response experiment, psychologists measured the time required by gerbils to run a maze. Gerbils were randomly assigned to one of three paths. All were the same, having two hallways at each intersection. The first path (Control) had a single reward—food—at the end of the maze; the second (Experimental 1) had a reward at every

intersection, and the third (Experimental 2) had a reward at every other intersection. Times required to run the maze are rounded to the nearest whole minute. Using $\alpha = .05$, test for a significant difference in time to run the maze. Assume the necessary test assumptions are met.

	Group	
Control	Experimental 1	Experimental 2
4	7	7
7	3	6
8	6	7
7	5	8
	4	7

5/ Summary for Chapters 10 and 11

The standard normal (Z) and the t-distributions have numerous similarities. Both are bell-shaped and center around a mean value of zero. The Z-form is appropriate, where $n > 30$, for large sample inference about means. Use of the t-distribution presupposes that the basic random variable has a normal distribution, but that the standard deviation is unknown. Then for inference on the mean for small samples, we use the t-statistic. A basic difference is that the spread of the t-distribution depends on the sample size; t-scores depend on "n."

Estimation forms based on the two statistics are quite similar, but the t-form, based on small samples, yields longer confidence intervals. With increased sample size the t-distribution approaches the Z-, so larger samples give smaller confidence intervals. Because t-scores are a function of the sample size, calculations on n cannot be made directly from the t-distribution. This is a real limitation.

In statistical testing the reseacher can make the α-risk arbitrarily low, but if the sample is fairly small, the β-risk may be excessive. Then if the sample evidence leads to accepting the null hypothesis, there is a high probability that the conclusion is wrong. One should avoid unnecessarily small samples in hypothesis testing.

Inference on the difference of two means with *dependent* samples (see Section 10.4) requires a sampling distribution different from that with independent samples (see Section 11.1). The risk of not recognizing a dependence and of inappropriately using the independent samples procedure is excessive error of estimation or, in testing, too frequent acceptance of the null hypothesis.

The F-ratio statistic is used in tests on the equality of means. The testing requires random samples from two or more normal distributions having equal variances. When only two population means are compared, the t-test for independent samples is equivalent and can be used. The analysis of variance is a systematic tool for the orderly development of the F-ratio test.

The results of planned experiments can be evaluated in many cases where there are unequal numbers of observations. Moreover, for the researcher who wants to go beyond just knowing that differences exist, procedures are available for identifying which groups are different. See the References for more advanced developments in these areas.

REFERENCES

[1] William C. Cochran and Gertrude M. Cox, *Experimental Design,* second edition (New York: John Wiley & Sons, 1957).

[2] R. A. Fisher and F. Yates, *Statistical Tables for Biological, Agricultural and Medical Research,* 6th edition (Edinburgh: Oliver and Boyd Limited, 1970). Published in U.S.A. by Hafner Publishing Co., Darien, Conn.

[3] John A. Ingram, *Introductory Statistics* (Menlo Park, California: Cummings, 1974).

[4] Robert G. D. Steel and James H. Torrie, *Principles and Procedures of Statistics* (New York: McGraw-Hill, 1960).

Review Problems

1. Check your understanding of the use of Z, t, and F distribution tables by finding:
   a. $P(Z > 2.04)$
   b. $t_{.05,13}$
   c. $P(-t_{.025,8} < t < +t_{.025,8})$
   d. $F_{.01,5,5}$
   e. $P(F > F_{.05,10,12})$
   f. $F_{.025,3,\infty}$

2. Find the following:
   a. If $t_{.025,18} = 2.1$, find $F_{.05,1,18}$.
   b. If $F_{.05,1,\infty} \doteq 3.84$, find $t_{.025,\infty}$.
   c. Determine $P(-2.61 < Z < -1.61)$.
   d. Find $F_{.01,31,32}$.

3. An analysis of variance on three groups is begun below.

Source	Degrees of freedom	Sum of squares	Mean squares	F
Treatments		198		
Within				
Total		392		

Complete the table and test $H_0: \mu_1 = \mu_2 = \mu_3$ assuming equal sample sizes, $n_1 = n_2 = n_3 = 9$. Use $\alpha = .05$.

4. If all of the possible samples of size 24 are drawn from a normal distribution with mean 20.0 and standard deviation 4.0, within what range will the middle 90% of the sample means fall? That is, set up the appropriate 90% confidence interval. What is the maximum error, E, for your estimate?

5. Following are ages for random samples of master's degree graduates in political science from two large universities. Assume normal distributions of ages.

University A:	25.0	23.5	24.6	23.7	23.1	24.1	25.4
University B:	26.7	27.5	23.5	27.4	23.8	26.3	25.1
	24.4	25.7					

a. Using $s_A^2 = 0.70$ and $s_B^2 = 2.24$, test the hypothesis of equal variances at $\alpha = .05$.

b. For $\alpha = .05$, test $H_0: \mu_A = \mu_B$ using the t-test for differences in means, independent samples.

c. Answer Part b by using an F-ratio test with unequal numbers and Total SS $= 29.82$.

6. Consider the following results from two different treatments in the processing of rubber. The measurements are for elastic strength.

	Treatment	
	A	B
Units in sample	64	100
Sample mean	1.74	1.81
Standard deviation	0.5	0.70

a. Test $H_0: \sigma_1^2 = \sigma_2^2$ against $H_A: \sigma_1^2 \neq \sigma_2^2$ with $\alpha = .05$.

b. Do the two treatments produce different products with respect to elastic strength? Test $H_0: \mu_A = \mu_B$ with $\alpha = .05$. (*Note:* You should answer this question only if your conclusion to Part a was "do not reject H_0.")

7. In testing $H_0: \sigma_1^2 = \sigma_2^2$ against $H_A: \sigma_2^2 > \sigma_1^2$, independent samples from two normal distributions of size $n_1 = 12$ and $n_2 = 16$ gave $s_1^2 = 12$ and $s_2^2 = 53$, respectively. Perform the test, stating conclusions for $\alpha = .025$.

8. Explain what procedural changes would be required for Problem 3 preceding if $n_1 = 9, n_2 = 10$, and $n_3 = 8$? Do *not* attempt an analysis.

9. The following was observed in a one-way analysis of variance for testing the equality of means:

Source	Degrees of freedom	Sum of squares	Mean squares	F
Groups	2			
Within	12	1,244.3		
Total	14	1,397.0		

The group means, *each* based on five sample observations, were:

Group	1	2	3	Total
Mean	26.4	32.3	36.4	31.7

Somehow this data is inconsistent. Explain?

10. Eight sets of identical twins ages two to three (pairs $A, B, C, \ldots$), were selected at random from a population. One child, chosen at random from each pair became a member of the control group. These children were allowed the usual occasions to see numbers and counting, e.g., counting story books, watching Electric Company on TV, and parent-sibling discussions of numbers. The other twin, in the experimental group, was given formal instruction in counting. Number comprehension scores were obtained at the end of the experiment. Assume normal distributions.

a. Does this indicate that two- and three-year olds can learn numbers by formal instruction? Use $\alpha = .10$.

b. Establish a 95% confidence interval on the true difference in means for pairs.

Pair	Experimental	Control	Pair	Experimental	Control
A	55	51	E	67	61
B	85	75	F	58	53
C	63	60	G	77	77
D	79	75	H	81	74

11. Following are two ratings by a judge on the performance of nine figure skaters. The highest possible rating is 6.5. Using $\alpha = .05$, test the claim that, in general, the first rating is higher. Assume normal distribution.

Contestant	Rating 1	Rating 2	Contestant	Rating 1	Rating 2
1	5.0	3.8	6	5.1	5.4
2	4.6	4.2	7	4.3	4.2
3	6.1	6.2	8	5.4	4.7
4	3.8	3.2	9	4.6	4.3
5	6.1	5.4			

12. For the following scores chosen randomly from a normal distribution, compute $\overline{X}$ and s values. Test $H_0: \mu = 5.0$ against $H_A: \mu < 5.0$ at the .10 α-risk level. Scores are: 5, 0, 4, 7, 3, 4, 1, 3, 3, 3, 3, 2, 4, 7, 6, 2, 1, and 5.

13. Experience shows that a fixed dose of a certain drug causes an average increase in the pulse rate of 10.0 beats. A group of nine patients given the same dose of this drug showed increases of 16, 15, 14, 10, 8, 12, 13, 20, and 9 beats per minute. Assume normal distribution and a standard deviation of 4.0 beats per minute.
    a. Is there evidence that this group reacts differently in response to the drug? Use $\alpha = .05$.
    b. Establish a 95% confidence interval estimate for μ based on this sample data.

14. Thirty students were tested before *and* after taking a math course. The average increase in their scores was 8 points with $s = 16$ points. Using a significance level of .01, test the hypothesis that the students have improved from the first test to the second test. Under what circumstances would a type 2 error occur? Assume normal distribution.

15. Random samples of three brands of gasoline gave the following performance scores. Brands were randomly assigned to cars and the other conditions of the experiment were maintained as uniformly as possible. Assume normal distributions.

    a. Test for equal performance, $\alpha = .05$.

Brand	Mileage to nearest mile	Total
A	27 30 25 26	108
B	21 20 22 21	84
C	27 31 30 32	120

    b. Set a 95% confidence interval for the mileage for Brand C (only). Use $s_C = 2.16$.

16. The mean age for a random sample of 15 professional hockey players is 23.0 years with deviation 1.5 years. An equal number of professional football players showed a mean age of 25.2 with standard deviation 2.6 years. Can we conclude that the mean age for professional football players exceeds that of hockey players for these two populations? Use $\alpha = .025$ and assume normal distributions.

17. The annual starting salaries for a random sample of 50 elementary education majors has a mean of $\overline{X}_1 = \$9{,}675$ with standard deviation $s_1 = \$400$. The comparable values for 50 newly graduated secondary education majors were $\overline{X}_2 = \$9{,}500$ and standard deviation $s_2 = \$300$. Is it reasonable to say that newly trained elementary education majors are being paid significantly more than newly trained secondary education majors? Use $\alpha = .05$. Assume normal distributions.

18. The mean height of 50 male students who showed above average participation in college athletics was 70.2 inches with standard deviation 2.5 inches. An equal number of other male students gave a mean of 69.5 inches with deviation 2.8 inches.

    a. Test $H_0: \sigma_1^2 = \sigma_2^2$ using $\alpha = .05$ and a does-not-equal alternative hypothesis.

    b. Test the hypothesis that male students with above average participation in college athletics are taller than other male students. Again use $\alpha = .05$ and assume independent samples from normal distributions.

19. Samples of the hourly wages paid to city employees in two cities showed:

i	$\overline{X}_i$	$\sum (X - \overline{X}_i)^2$	n_i
1	$3.40	$.75	8
2	$2.76	$.70	9

Assuming the necessary assumptions are met, test the hypothesis that $\mu_1 > \mu_2 + \$.50$, where μ_1 = mean wage paid to employees in City 1, etc. Use $\alpha = .025$. State in words the possible type 1 error. [*Note:* $(n_i - 1)s_i^2 = \sum (X - \overline{X}_i)^2$.]

20. Driving test scores were compared for a random selection of Kansas youth, all age 16. Group 1 had no formal training; Group 2 had one term of driver education; and Group 3 had one year of driver education. Test for equal achievement using $\alpha = .025$ and total SS = 14,820.0. This was the first time each person had taken the driver's exam. Again, assume normal distributions. Use 77.5 as the overall mean.

i	Group	Mean score	n_i
1	No formal training	72.4	16
2	One term of driver education	76.8	15
3	One year of driver education	84.2	14

Regression, Paired Observations, and Linear Models

CHAPTER 12

In regression work our first concern is to study the descriptive relation between two or more variables. For example, most of us who own cars realize that their initial cost is, among other things, related to their age. Cost and age for cars are commonly related in "blue-book" valuations. For the high school graduate aspiring to a college degree, the unanswered question is "Will I succeed in college?" For him entrance examination marks are indicators that relate to his chances for achievement. A commonality of these examples is the desirability of describing one unknown through knowledge of its relation to another variable. A purpose of regression is to make quantitative descriptions that relate two or more variables.

1/ Simple Linear Regression

A synonym for *regression* is the word "relation." We discuss relations that can be described using mathematics. Examples of mathematical relations include $Y = 2 + 1X$, and $Y = 4 + 3X - X^2$, and $Y = 2^X$. Data plots appear in Figure 1. The expressions described in Figures 1a and 1b are polynomial curves in the sense that the X-terms are X, X^2, etc. The third relation contains the quantity X as an exponent so the description in Figure 1c is exponential. There are many distinct curves in mathematics. Each describes a relation between two quantities, which we label X and Y. The *independent* or X variable is assigned certain values. The values taken by Y, the *response* or *dependent variable,* depends on (1) the value taken by X and (2) the

Have you ever returned a tire under warranty? By using statistical methods like linear regression, we should be able to determine accurately how much life remains in the used tire. This might, for example, be used for refunds under the warranty for this brand of tire.

Figure 1 / Various math relations for paired data

a. Linear

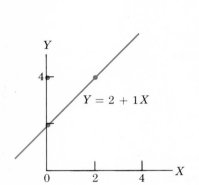

$Y = 2 + 1X$

b. Quadratic

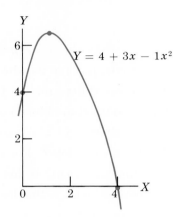

$Y = 4 + 3x - 1x^2$

c. Exponential

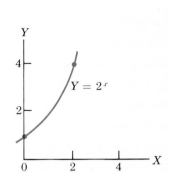

$Y = 2^x$

specific relation to X. Our work will concern only the simplest relations, those that graph as a line, but our developments can be extended to other types.

The line describes one of the simplest mathematical relationships. Moreover linear relations, or approximate linear relations, are common in experience. For these reasons our primary concern will be with the simple linear regression model. The linear (math) model in Figure 1a can be described by two numbers: 2—the value of the Y or vertical scale intercept; and 1—the slope of the line. The positive coefficient in $+1$ indicates a positive slope; that is, as X-values increase so do those for Y. (For more discussion see Rectangular Coordinates in Appendix A, p. 363.)

The diagrams in Figure 1 display cases wherein all the points fall on the curve. The associated relations are mathematical, that is, without error. Unfortunately such exact patterns are rarely achieved with real data. This is demonstrated by example.

Example

Being interested in the price of used cars of one make, an individual obtained the following data concerning age, X, in years, and price, Y, in thousands of dollars, from a single sales dealership.

X, age (years)	1	2	2	3	4	5	6	8	9	10
Y, price ($1,000)	3.8	3.4	3.3	3.0	2.5	2.0	2.2	1.0	1.0	0.8

Since the age is easily determined this person wants to describe price through knowledge of the car's age. But what is a reasonable relation between these two variables?

A first approach to this problem is to plot the data. The plot will generally indicate some discernable pattern and then an intuitive appraisal is made for a reasonable relation. By applying appropriate methods the relation might even be described by statistics. The age–price data plot comes first.

Solution

Figure 2/ A plot of price against age for a sample of used cars of one make

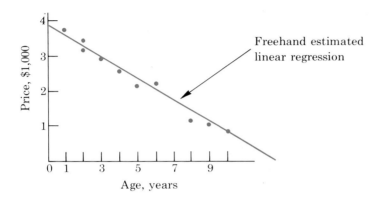

By observation it is not practical to attempt a mathematical *linear* model; these paired observations do not fall in a straight line. But by the "freehand" estimate they are not, in total, far off from a line. The freehand estimate is made by eye and without any calculations. It is subject to individual evaluation and so serves only as a first approximation to the true relation. The idea is to place a line that lies close to *all* of the points.

One approach for a quantitative description is to allow for *error* in the linear relation. The result is a probabilistic model which includes an error term. The assumed model is $Y = B_0 + B_1 X + E$. The B_0 denotes a Y-intercept, while B_1 is the slope of the line, and the E indicate errors or deviations from linearity.

In statistical modeling we accommodate the fact that scores are sample readings. We cannot expect a fixed pattern to recur without error for every possible sample. That is, were we to draw eight more cars some of their prices would be different from ones in this sample. Consequently we attempt a general pattern subject to error. Hopefully the error component will (1) be generally small and (2) have a statistically well-defined pattern that can be described as a random variable having a probability distribution.

Next we use the paired observations to approximate the slope and the Y-intercept, thereby identifying a linear relation between price and age. In the process a generalization is developed for finding slope and intercept values.

Problems

1. Construct a graph for the line described by $Y = \frac{1}{2}X + 1$.

2. Find slope and Y-intercept values for the following linear description. Write this as a linear relation. You may want to see Appendix A, page 365, for the rule for computing slope.

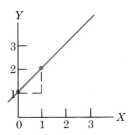

3. The following scores are a random sample from a much larger class. Plot the data on rectangular coordinates (as in problem 2), then make a visual freehand plot of the *linear* relation. Approximate Y-intercept and slope.

Entrance score, X:	93	59	85	78	60	90	72	73	60	85
English grade, Y:	95	64	80	87	64	88	74	74	80	84

4. For the pairs of values below, answer the same questions as for problem 3.

X:	−3	−2	−1	0	1	2	3
Y:	1	2	3	5	8	11	12

Is the slope positive or negative? Explain.

5. For the data plots below:
   a. which, A, B, or C, have a reasonable *linear* relation with negative slope?
   b. which, by letter, *appear* to hold a fairly good *linear* statistical relation?

c. which, for data set C, is a more likely relation: $Y = a + bX + e$ or $Y = c + dX + fX^2 + e$. Why? Here $a, b, c, d,$ and f are generalizations for some unspecified numbers. Recall Figure 1.

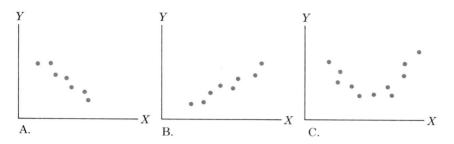

2/ The Regression Equation by Least Squares

It is unreasonable to assume that the observations in Figure 2 for age–price of used cars are linear without error. However, it may be reasonable to assume that the average or expected price is a linear function of a car's age. We therefore assume that a linear expression, $\mu_{Y/X} = B_0 + B_1 X$, denotes the true mean value of Y for a specified value of X. B_0 represents the Y-intercept, and B_1 denotes the slope of the population regression line.

Our present task is to discover how the coefficients B_0 and B_1 can be approximated. Since we have only a sample, we can at best get estimates. So an estimated regression, $\hat{Y} = b_0 + b_1 X$, is used to describe the line. We use the least squares estimation procedure to obtain slope and intercept values. This assumes a sample model $Y = b_0 + b_1 X + e$.

THE LEAST SQUARES CRITERION Estimators for B_0 and B_1 in a linear model are obtained by minimizing the sum of squares for errors,

$$\sum (e)^2 = \sum (Y - \hat{Y})^2,$$

with respect to b_0 and b_1. The results include least squares estimators (statistics) (1) b_0 for estimation of the true intercept, B_0, and (2) b_1 for estimation of the true slope, B_1.

We pass a line close to the sample points. Closeness is described by the vertical distances from the sample points to the regression line. The least

Criterion

squares procedure guarantees that these distances, the errors, will be as small as is possible. The minimization can be made by calculus [2] or with algebra [3]. Figure 3 illustrates the error distances that we hope to make small. Minimization implies two normal equations.*

Figure 3/ Deviations (e) for observed (Y) against regression ($\hat{Y}$) values

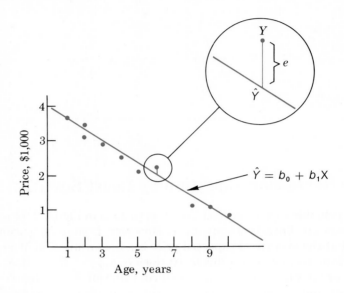

1. $\sum Y = nb_0 + b_1 \sum X$

2. $\sum (XY) = b_0 \sum X + b_1 \sum (X^2)$

These forms yield estimators for slope and Y-intercept values. The results are summarized in the following rules:

Rule

The least squares estimators for linear regression coefficients are:

Y-intercept: $b_0 = \overline{Y} - b_1 \overline{X}$

$$\text{slope: } b_1 = \frac{n \sum (XY) - \sum (X) \cdot \sum (Y)}{n \sum (X^2) - \left(\sum X\right)^2}$$

---

*The name "normal equation" is unrelated to the normal probability distribution, but is used in many statistics books.

The Regression Equation by Least Squares

By these rules evaluation of the intercept estimator, b_0, requires a value for slope, b_1. Thus the latter value must be determined first.

These forms are used on the age–price data from the preceding section.

We want an estimated linear regression for used car prices, Y, on age, X. The data with the essential sums appear in Table 1.

Example

Table 1 / Calculations for Least Squares Estimation of Linear Regression Coefficients—Age on Price for Used Cars

Age, X	Price, Y	XY	X^2	Y^2
1	3.8			
2	3.4			
2	3.3			
3	3.0			
4	2.5			
5	2.0			
6	2.2			
8	1.0			
9	1.0			
10	0.8			
Totals 50	23.0	$\sum(XY) = 84.4$	$\sum(X^2) = 340$	$\sum(Y^2) = 63.62$
Means $\overline{X} = 5$	$\overline{Y} = 2.3$			

Substituting the totals into the correct forms gives:

$$b_1 = \frac{10 \cdot (84.4) - (50)(23)}{10 \cdot (340) - (50)^2} = \frac{-306}{900} = -.34 \text{ and}$$
$$b_0 = 2.3 - (-.34)5 = 4.0.$$

The estimated regression equation is

$\hat{Y} = 4.0 - .34X$ where
X = age in years
$\hat{Y}$ = estimated average price (in \$1,000) for used cars of this make and given age sold at this dealership.

The regression constant is $b_0 = 4.0$. This means that for the graph of the line $\hat{Y} = 4.0 - .34X$, when $X = 0$, the Y-intercept is 4.0. This can be of help in graphing the line. But since cars have a dramatic decrease in price for the first year and since no new cars were in the sample, 4.0 (\$4,000) may not be a good estimate for true average price when $X = 0$. In fact the line gives reasonable estimates for the X values included (one through ten years), but is likely unrealistic for $X < 1$ or $X > 10$ years. We can use $b_0 = 4.0$, the Y-intercept, to identify one point $(0, 4.0)$ on the locus. With one more point, we can plot the graph. The second point is determined from $b_0 = \overline{Y} - b_1 \overline{X}$. Since $\overline{X}$ and $\overline{Y}$ are used to determine the regression constant, the line ($\hat{Y}$) is forced through this point: $\overline{Y} = b_0 + b_1 \overline{X} = 4.0 - .34(5) = 2.30$. A second point is $(\overline{X}, \overline{Y}) = (5, 2.30)$.

The coefficient b_1 (the slope estimate) $= -.34$ indicates that for each additional year of age, the price for this make at this lot *decreases* on the average by $340. Then for example, I would *expect* to pay $340 less for a two-year old car than for a one-year old car of this make at this lot. The negative slope is consistent with my experience that "older cars usually sell for less."

A coarse check on the reasonableness of the regression coefficients is made by substituting one or more of the observed ages from Table 1 into the regression equation. The resulting $\hat{Y}$-values can be checked against the observed price(s). For example,

$X = 2$ gives $\hat{Y} = 4.0 - .34(2) = 3.32$; observed $Y = 3.3, 3.4$ (two values).

Comparison of $(Y - \hat{Y})$ values indicates the estimated average prices are reasonably close to observed prices. Some of the values are compared below:

X	Y	$\hat{Y}$	$(Y - \hat{Y})$
2	$3,400	$3,320	$ 80
	3,300	3,320	-20
4	2,500	2,640	-140
6	2,200	1,960	240
8	1,000	1,280	-280

The regression coefficients are reasonable since projected values are close to the sample observations. Also the theory assures us that among all *lines* that might describe this data, the least squares line will have least overall distance from the points. The least squares line can be placed on the graph by identifying any two points on its locus. These can be (1) the Y-intercept value, and (2) the point $(\overline{X}, \overline{Y})$. The regression line and data points appear in Figure 4. The vertical distances or errors are visible for all sample points. For a good fit the regression line will fall above some of the observed points and below others with somewhat random displacement of points and generally with small errors between the line and observed values. Also the freehand estimate and the more exacting least squares line should be reasonably close.

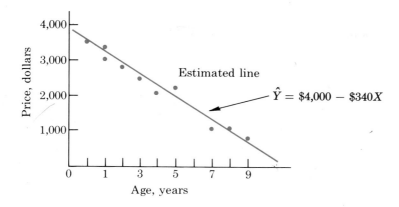

Figure 4/ A linear regression for price on age of used cars

The next section gives some indication of just how well the linear model describes this data. This is evaluated in a coefficient of correlation. A second procedure, used to test the assumption of "a meaningful regression," is discussed in Section 5.

Problems

1. The table contains values for two variables—X and Y.
    a. Plot the data. Does a linear regression appear reasonable? Make a freehand plot of a regression line.
    b. Establish the estimated regression of Y and X. Use least squares results.

X:	−2	−1	0	1	2
Y:	0	1	0	1	3

2. Given a sample of seven pairs of values:

X:	4	5	6	7	8	9	10
Y:	4	6	8	12	18	24	26

a. Find the equation for a regression of Y on X (assumed to be linear) given that

$$\sum(X^2) = 371, \quad \sum(XY) = 798, \quad \overline{X} = 7, \quad \overline{Y} = 14.$$

b. Does the point with coordinates $(7, 14)$ fall on your regression line? Why or why not?

c. What is the estimated value of Y given $X = 5.5$?

3. Use the data from problem 3 in Section 1 to find estimated regression coefficients b_0, b_1, in $\hat{Y} = b_0 + b_1 X$. Use $\sum X = 755, \sum Y = 790, \sum(XY) = 60{,}627, \sum(X^2) = 58{,}477, \sum(Y^2) = 63{,}338$. Compare these values with your first estimates.

4. a. Use the information below to develop an estimated regression equation, $\hat{Y} = b_0 + b_1 X$, by evaluating b_0 and b_1.

X:	−3	−2	−1	0	1	2	3
Y:	1	2	3	5	8	11	12

b. Plot the data, and make a freehand regression line. Now impose the least squares regression line on your figure. Compare to Problem 4 in Section 1.

5. An industrial psychologist has initiated the following plan aimed at reducing the percentage of absenteeism for workers in a large manufacturing plant. Every workday (5 days per week) each worker who arrives at work on time is given a playing card from a standard deck of cards. On Fridays that worker in each section with the best poker hand wins a bonus of $15. Suppose after five weeks the results were as follows:

Week, X	1	2	3	4	5
Percent absent, $Y(\%)$	7	6.5	7	6.5	6

a. Plot the data. Does there appear a reasonably strong linear relation?

b. Using the following summary statistics, compute estimates for the regression coefficients and plot your estimated regression.

$$\sum(X^2) = 55 \qquad \sum(X - \overline{X})^2 = 10.0$$
$$\sum(Y^2) = 218.5 \qquad \sum(Y - \overline{Y})^2 = 0.7$$
$$\sum(XY) = 97 \qquad \left[\sum(XY) - \frac{\sum X \sum Y}{n}\right] = -2.0$$

c. Use your regression equation to estimate percent absent for the fifth week. Compare the estimated and observed values.

d. Does the approach seem to have reduced absenteeism? Consider the sign and size of the slope for your regression.

6. The solution of the normal equations is essentially a problem of the simultaneous solution of linear equations from algebra. Solve the following for b_0 and b_1:

$$b_0 + 2b_1 = 5$$
$$3b_0 + 4b_1 = 6.$$

3/ Pearson's Coefficient of Correlation

We have described an assumed linear relation first by graphing, then by estimation of slope and Y-intercept values. But to what extent is a linear relation meaningful? One attempt to answer this question uses the coefficient of correlation defined in about 1903 by Karl Pearson. It is used as a measure of the strength of the linear regression. The population symbol is ρ, but most commonly we have only a sample and so can compute only the estimator r.

The sample coefficient of *linear* correlation between X and Y is: *Rule*

$$r = \frac{n \sum (XY) - \sum X \cdot \sum Y}{\sqrt{\left[n \sum (X^2) - \left(\sum X\right)^2\right] \cdot \left[n \sum (Y^2) - \left(\sum Y\right)^2\right]}}$$

To illustrate Pearson's form, consider again the price–age data for used cars.

The *linear* correlation between age and price for used cars of the given make at one dealership is approximated by the sample value, *Example*

$$r = \frac{10 \cdot (84.4) - 50(23)}{\sqrt{[10 \cdot (340) - (50)^2] \cdot [10(63.62) - (23)^2]}}$$

$$= \frac{-306}{\sqrt{900 \cdot 107.2}} \doteq -.985$$

(*Computation note:* Look back to page 289. We have already calculated two parts of r in the evaluation for b_1; that is, we already have the values -306 and 900. There is no need to recompute these unless you want to check your arithmetic.)

But how do we interpret this coefficient? First we define bounds for values that can be taken by r (also by ρ). These appear in Figure 5. Figures 5a and 5b illustrate cases of a perfect linear relation. In Figure 5a an increase of one unit in X corresponds to an increase by b_1 units in the response, Y. The slope is positive, the fit is perfect. In diagram 5b each increase of one unit in X is accompanied by a decrease of b_1 units in Y. The slope is negative, yet again the fit is perfect. In Figure 5c the appearance is much that of a shotgun splatter. There is no apparent linear pattern. The best fit is the line $\hat{Y} = \overline{Y}$. The mean response, $\hat{Y}$, does not depend on X in a linear fashion. There is essentially zero *linear* correlation. Again, in Figure 5d, $r \doteq 0$. However $r \doteq 0$ does *not* mean "no relation" between X and Y. In fact, a quadratic regression looks quite reasonable for the data in Figure 5d.

Figure 5/ Bounds on the linear correlation coefficient

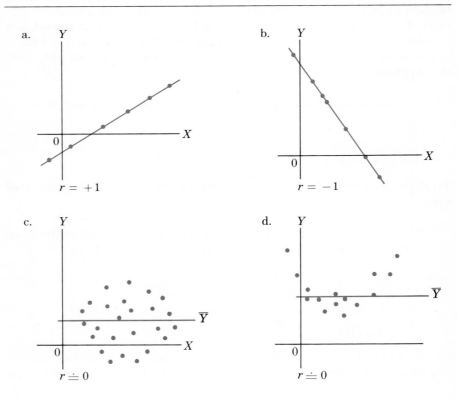

The extreme values $(+1, -1, 0)$ are rarely achieved with real data. They do, however, specify bounds on the size of the coefficient of correlation.

> The coefficient of correlation for sample data takes values between -1 and $+1$, inclusive; that is, $-1 \leq r \leq +1$. Similarly the population coefficient of correlation, ρ, is bounded, $-1 \leq \rho \leq +1$.

Rule

A coefficient $r = -.985$ appears quite strong since it is near -1. However, for objective comparison of the reasonableness of the "linear" assumption, we use the coefficient of determination, $100r^2$.

> The COEFFICIENT OF DETERMINATION describes the percent of variation in observed Y-values that is explained by the linear regression on X.

Definition

$$100 \cdot r^2 = \frac{\text{explained variation} \cdot 100}{\text{total variation}}$$

Rule

The coefficient of determination is most meaningful when two or more coefficients are compared.

> Consider two predictor variables, say X_1 = wingspan and X_2 = age, with Y = body weight for migrating geese. For $r_1 = .4$ and $r_2 = .8$, the relative descriptive strength is:

Example

$r_1^2 \cdot 100 = (.4)^2 \cdot 100 = 16\%$ while
$r_2^2 \cdot 100 = (.8)^2 \cdot 100 = 64\%$.

> Thus $r_2 = .8 \,(= 2r_1)$ is four times as strong as $r_1 = .4$. Hence a linear description on age accounts for 64% of the observed differences in body weight. The wingspan describes only 16%, or 1/4 the amount of variation.

The sign of the correlation coefficient indicates the direction of slope for the regression line. The strength of regression is expressed by the coefficient of determination. Thus for example, $r = -0.9$ shows a stronger linear relation than does $r = +0.71$ even though the lines slope in opposite directions.

In studies using real data, strong correlations commonly range in $.5 \leq |r| \leq .8$. That is, in *some* experiments $100 \cdot r^2 = 100 \cdot (\pm .5)^2 = 25\%$ explained differences is a reasonable criterion to assume linearity. One should investigate related studies to interpret the relative strength of a regression description.

At the beginning of this section we asked "To what extent is an assumed linear relation meaningful?" This has been answered using statistics. Sometimes one may ask a second question, "If chance quantities are (linearly) related, is there any reason that they should be?" Usually this second question, inferring causation, cannot be answered. Consider a set of basketball records [5]. A cursory look indicates a strong (linear) relation between a player's total points and fouls committed. This is strengthened by a data plot and (eye) estimated regression. Both the table and the graph indicate that more fouls generally accompany high total points. The relation is positive; the (linear) regression has a positive slope. But is this reasonable? If a player repeatedly fouls, he would commonly foul-out of games and thereby lose his chance to score many points. Surely few coaches would encourage fouling as a means of increasing his players' points! The fact that a player commits more fouls is reasonably *not the cause* for him scoring more points. But perhaps both acts, more fouls and more points, are due to some players playing more time than others. Or possibly both are due to some players being more aggressive than others.

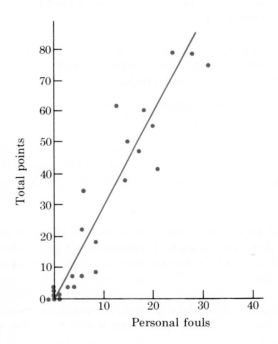

Pearson's Coefficient of Correlation

Basketball Scoring and Fouls

Player	Total points	Personal fouls
1	0	0
2	0	0
3	0	1
4	0	1
5	1	0
6	2	0
7	2	1
8	2	3
9	2	4
10	3	0
11	6	6
12	7	3
13	9	9
14	18	9
15	21	7
16	35	5
17	37	16
18	42	24
19	46	20
20	48	17
21	57	22
22	59	18
23	60	15
24	75	24
25	75	31

The number of fouls and points scored have a high positive (linear) correlation. Still we have no right to assume one is the cause of the other. The cause or causes remain unknown. The point of this is a word of caution. Correlation is a source for description but *not* for causation. A high (linear) coefficient, near $+1$ or -1 *does not* mean cause and effect. Quite often a high correlation exists because two variables are both highly related to a third but common variable.

The difficulty in justifying causation is well illustrated by events leading to the Surgeon General's Report concerning the relation of smoking to lung cancer [6]. See the essay example in Section 12.6.

Problems

1. For

X	2	6	4	7
Y	−15	41	13	55

    , $r = 1$

    while for

X	5	20	25
Y	42	12	2

    , $r = -1$.

    Graph both of these and explain the meaning of each r-value.

2. These data were obtained in a correlation study:

    $n = 10, \overline{X} = 4, \overline{Y} = 7, \sum X^2 = 260, \sum Y^2 = 890, \sum(XY) = 180.$

    Compute the coefficient of correlation. Interpret the result using $100 \cdot r^2$.

3. The following data indicate yields in pounds of potatoes and amounts of fertilizer in pounds used by six individuals in competition. The objective was to see how many pounds of "spuds" one could grow from one seed potato.
    a. Plot the data. Does a strong correlation appear to exist?
    b. Compute r. Describe $100r^2$.

Yield, Y (in pounds)	2	4	6	6	4	2
Fertilizer, X (in pounds)	0.5	1	2.5	4.5	4.5	6

4. In two related research studies one person got $r_1 = -.65$, while the second found $r_2 = +.58$. Which and by how much is the stronger linear relation?

5. A study of New White detergent surveyed 15 test markets to determine the relation of two variables to sales. The first variable, X_1, was expenditures for television advertising measured in thousands of dollars.

The second, X_2, was the unit price of the product. Sales, Y, was measured as the proportion or share of the market held by New White. Given:

$[n_1 \sum(X_1 Y) - \sum(X_1)\sum Y] = 11.18 \qquad [n\sum(Y^2) - (\sum Y)^2] = .88$
$[n_2 \sum(X_2 Y) - \sum(X_2)\sum Y] = -.21 \qquad [n\sum(X_1^2) - (\sum X_1)^2] = 300$
$\hspace{6cm} [n\sum(X_2^2) - (\sum X_2)^2] = .12.$

Compute the correlation of Y with X_1, and the correlation of Y with X_2. Which variable, X_1 or X_2, gives the stronger description of sales?

6. Use information from Problem 1 in Section 2 to find a value for r. Interpret by using $100r^2$.

7. Use information and calculations from Problem 4 in Section 2 to find a value for r. See the Computation note on page 293.

8. Suppose someone ran a correlation analysis between age and score (100 point scale) in a college course. If he or she found $r = +1.00$, is this sufficient indication to conclude that older persons *will* get higher grades in this course? Please explain a *yes* or a *no* answer.

4/ Other Linear Models—Time Series

In business much of the action for future activity is based on statistical evidence. Businessmen must often base decisions relating to supply and demand, prices, wages, etc., on patterns or trends observed from past experience. A common assumption is that in the near future business will follow patterns similar to those of the present and of the recent past. One basis for such projections is from a *time series*.

> A TIME SERIES is a sequence of observations recorded at successive intervals of time.

Definition

Common examples of time series include daily temperature records, New York stock "averages," and the sales charts (quarterly, etc.) used by retail firms.

The *classical theory* of time series envisions fluctuations in time data, (i.e., variation) attributable to either (1) a secular or long-term trend, (2) seasonal variations, (3) cyclical variation (e.g., business cycles), or

(4) irregular variations. Our concern is in describing the long-term regular patterns called *trend*. For a study of the other variations see [2].

To describe a trend, or the regular movement of a series over a long period of time, we use smooth (not jagged) curves. The simplest case is the linear trend. The several methods for identifying trends include (1) freehand construction, (2) moving averages, and (3) least squares. This work considers the first and third methods, the second is described in [2].

Example

Suppose the following data are a record of sales by a single firm for the past eight years. A trend is sought whereby projections can be made for sales in the next year. Both the freehand and least squares approaches are explored.

Table 2/ Data for Sales by Year (in $100,000)

Year:	1968	1969	1970	1971	1972	1973	1974	1975
Sales:	6	5	9	8	10	11	13	10

Solution

As with other regression problems we begin with a data plot and a freehand estimate of a meaningful relation. From the freehand estimate, the series appears to follow a linear trend moderately but not extremely well. Visual inspection indicates that a quadratic trend equation might give better description, but we will restrict calculations to the linear form.

Figure 6/ Linear trend for a time series problem

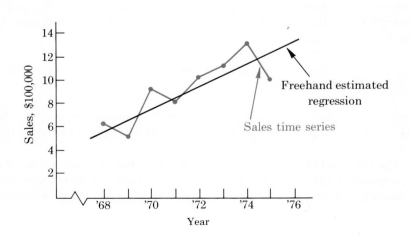

The least squares analysis for linear trend begins with coding the years. In cases where n, the sample size, is an even number, we use one-half of the basic time interval as the unit. Here sales records are by years (annual) with $n = 8$, so the unit is six months. Thus the span of one year covers two units. That is,

Years:	1968	1969	1970	1971	1972	1973	1974	1975
Years coded, X_c:	−7	−5	−3	−1	1	3	5	7

The data is centered at the time span between 1971–1972. The advantage of this center is that $\sum X_c = 0$. This grossly simplifies calculations and so is the usual procedure for time series or other equally spaced data.

The least squares linear trend (regression) equations are the same as before; $\sum X_c = 0$ simplifies the calculations.

Year, X	X_c (coded)	Sales, Y	X_c^2	Y^2	$(X_c Y)$
1968	−7	6			
1969	−5	5			
1970	−3	9			
1971	−1	8			
1972	1	10			
1973	3	11			
1974	5	13			
1975	7	10			
Totals	0	72	168	696	76
Means	0	9			

Then estimating the linear trend model by $\hat{Y} = b_0 + b_1 X_c$,

$$b_1 = \frac{8 \cdot (76) - 0(72)}{8 \cdot (168) - (0)^2} = \frac{8(76)}{8(168)} = .45$$

$b_0 = 9 - (.45) \cdot 0 = 9.$

$\hat{Y} = 9 + .45 X_c$ with
$\hat{Y} = $ estimated sales in \$100,000
$X_c = $ years, coded with origin 1971–1972 and six-month time increments.

The plot of this line (not shown) is quite close to the freehand description given in Figure 6. Projecting for 1976, which is two (six-month) units above 1975, makes $X_c = 7 + 2 = 9$. This gives

$\hat{Y} = 9 + .45(9) = 13.05$ (\$100,000).

Since each Y-unit represents $100,000, estimated sales are $1,305,000. A word of caution, however, this projection is beyond the time interval of the observations so we are assuming that factors quite similar to those in 1968–1975 will exist in 1976. If this is not reasonable, e.g., because of changing economic conditions or new competition, etc., then the projection can be quite wrong!

As with other linear regressions, the strength of the linear trend in the preceding can be described through a correlation value. The earlier procedures apply.

Rule

For equally spaced data, coded so that the X_c-values sum to zero,

$$r = \frac{n \sum X_c Y - \sum X_c \sum Y}{\sqrt{\left[n \sum X_c^2 - \left(\sum X_c\right)^2\right]\left[n \sum Y^2 - \left(\sum Y\right)^2\right]}}$$

$$= \frac{n \sum X_c Y - 0}{\sqrt{\left[n \sum X_c^2 - 0\right]\left[n \sum Y^2 - \left(\sum Y\right)^2\right]}}$$

The interpretation of strength of the linear trend uses the coefficient of determination, $100 r^2$. For the sales data given earlier

$$r = \frac{8 \cdot 76}{\sqrt{[8 \cdot 168] \cdot [8 \cdot 696 - 72^2]}} = .85,$$

$$r^2 = .85^2 \doteq .72$$

$$100 r^2 \doteq 100 \cdot .72 \text{ or } 72\%.$$

That is, about 72% of the variability in sales is accounted for by the linear relation on time.

Simplified computations on equally spaced data result from coding to allow $\sum X_c = 0$. For n time periods with n odd, the coding uses the whole period for units as illustrated in the following example.

Example

A national retail firm wants to relate disposable personal income (D.P.I.) over time. Their main interest is to get estimates of the disposable income for the next few years. What trend is suggested by the data? We use this to project the D.P.I. for year six given:

Year:	1	2	3	4	5
D.P.I. ($1,000,000):	3.2	3.4	3.5	3.6	3.8

Data is given for $n = 5$, an odd number of periods, so the data are centered with Year 3 as origin and with annual increments. The coding is as follows:

Year	X_c	D.P.I., Y	$X_c Y$	X_c^2	Y^2
1	−2	3.2			
2	−1	3.4			
3	0	3.5			
4	1	3.6			
5	2	3.8			
Totals	0	17.5	1.4	10	61.45
Means	0	3.5			

Using the least squares procedure,

$$b_1 = \frac{5(1.4) - 0(17.5)}{5(10) - 0^2} = .14$$

$b_0 = \bar{Y} - b_1 \bar{X}_c = 3.5 - .14(0) = 3.5$

$\hat{Y} = 3.5 + .14 X_c$ where

$\hat{Y}$ = estimated disposable personal income in millions of dollars

X_c = years coded to Year 3 = 0 and with annual increments.

Then for Year six $X_c = 3$, so $\hat{Y} = 3.5 + .14(3) = \$3.92$ million. The data indicates a strong linear relation. This claim could be reinforced by a data plot and by calculation of the coefficient of determination.

Equal spacing procedures illustrate only one way that coding can be used to simplify regression calculations. Another that can be used for any regression problem is discussed in [4]. Coding is especially helpful if you must calculate by hand.

Problems

1. Determine the least squares linear regression equation for corporate sales (in $1 million) over time for the following data. Use your trend equation to estimate sales in the next period, the sixth.

Regression, Paired Observations, and Linear Models

Period:	1	2	3	4	5
Sales:	4.7	4.7	4.8	5.1	5.2

2. The following numbers indicate production, in millions of units, for the past six years.

Year:	1	2	3	4	5	6
Production:	15	14	18	20	14	24

   a. Plot the data as a time series. Freehand sketch a linear trend. Is the slope positive or negative; that is, is production generally increasing or decreasing with time?

   b. With proper coding, develop the least squares trend line. Be sure to label your trend equation, specifying origin and X-unit increments.

3. Code the weeks, using time-data format, and then find the least squares (regression) trend line for the manufacturing data given. Use this to estimate the percent defective for week six. Now compare your answer with that of Problem 5 in Section 2 (they should agree). Explain why your projection for week 6 is questionable. (*Hint:* Can we expect the percent defective to decrease indefinitely?)

Week, X:	1	2	3	4	5
Percent defective, Y:	7	6.5	7	6.5	6

4. Suppose the following shipments of steel (in millions of tons) by United Steel Corporation for an 11 year period: 5.9, 5.8, 5.9, 5.9, 5.4, 4.7, 6.1, 6.4, 5.1, 6.4, and 6.1.

   a. Determine the least squares trend line.

   b. Use the least squares trend line to compare projections for Years 12 and 13 for steel shipments against the actual values, 6.0 and 7.1, respectively. Did your linear trend persist?

5. A machine is inspected for the number of defective articles it produces. The data seems to indicate this number relates somewhat to machine speed (in revolutions per second).

Speed (rps), X:	10	11	12	13
Defectives per minute, Y:	4	8	9	11

Code, using the origin between 11 and 12 and with one-half-unit increments. Determine the linear regression equation. Why is it unreasonable to move the speed to either extreme, e.g., 0 rps or 50 rps, to optimize production?

6. Plot the data for D.P.I. and years given on p. 303 and then calculate the coefficient of determination. Compare this with the value obtained for sales and years, p. 301.

5/ Inference in Regression

We have described an *assumed* linear regression. Calculation of slope and intercept estimates can be simplified if X-values are equally spaced, which is common for time (series) data. Since our discussion of correlation coefficients remains subject to individual interpretation, we have as yet no objective method for testing the reasonableness of the assumption of a linear relation. Now an objective procedure is introduced.

Sound regression inference requires several assumptions which concern the distribution of points around a true regression line. As with inference on means, comparisons are to a central value. But now we are trying to describe not just a single value, but a whole set of points that collectively form a line. The possibilities for the (sample) estimated line are many. Figure 7 displays a few. Consider any value X_0, an arbitrary choice from one sample. Let the associated regression value be $\hat{Y}_0$ (see Figure 7). The

Figure 7/ Possible regression lines

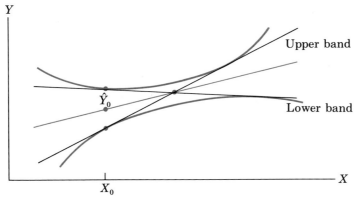

"upper" and "lower" limits for probable values of $\hat{Y}$ are specified in a manner similar to the confidence limits of previous chapters. The collection of these upper and lower points considered over all X-values determine a *confidence band*. Several properties are generalized:

1. For fixed X-value, the possible Y-values will vary about a true mean with a normal probability pattern. The distribution is normal for each observed X-value.
2. The true mean values, i.e., going from one X to the next, fall in a straight line. This line is the true regression.
3. The variability among possible Y-values for each X-value depends on a constant variance, σ_e^2. I will call this the *error variance*.

Although these assumptions are all important, we will concentrate on the last. A constant error variance is illustrated in Figure 8. Regression inference is based on the estimator for a common error variance. Although we will not bring in computational forms, these, along with a rationalization, appear

Figure 8/ The assumptions for linear regression

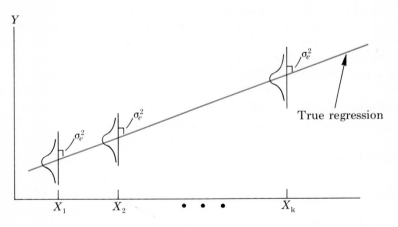

in [4]. The important thing is that, because we have specified a probability distribution, it is possible to determine upper and lower confidence values for the true regression line. These are the upper and lower bands shown in Figure 7.

This work will address only inference about the true slope for an (assumed) linear regression. That is, as concerns the lines in Figure 7, we will make inferences about the true slope, but will make no attempt to identify the (Y-) intercept. We treat two (related) problems: (1) tests and

Inference in Regression

(2) estimation of the true slope. Both are derived from the following t-statistic.

For inference about the slope for an assumed linear regression we use

$$t_{n-2} = \frac{b_1 - B_1}{s_{b_1}}$$

Rule

where

b_1 = the sample estimate for the slope
B_1 = the true slope
s_{b_1} = the standard deviation for b_1 values.

I must warn you that this t-form appears deceptively simple. Its evaluation requires calculation of an estimate for error variance, σ_e^2, and more. Since it is basic, this work will always provide s_{b_1} values.*

First let's look at a test on slope. To test $H_0: B_1 = 0$ against $H_A: B_1 \neq 0$ is a test for a meaningful linear regression. Observe that the alternative hypothesis is "does not equal" so that one-half of the α-risk is taken from each tail.

In estimating the relation of price to age of used cars (See Section 2), we have assumed there is a "meaningful" linear relation. Let's test that assumption at $\alpha = .05$, again using the sample values from Section 2.

Example

We have already found that $b_1 = -.34$. So for $s_{b_1} = .02$,

Solution

$H_0: B_1 = 0$ and $H_A: B_1 \neq 0$

Test: Reject H_0 if $|\text{calculated } t| > 2.45 \, (t_{.025, 8-2})$

Calculated $t = \dfrac{-.34 - 0}{.02} = -17.$

Reject H_0. The test substantiates the assumption of a meaningful linear relation of price on age for this make of car, ages one–ten years, sold at this dealership. Our values have predicative meaning.

Conclusion

A word of caution. The conclusion that we have a meaningful linear regression is not a wholesale ticket to use a linear model. The model can be used

---

*The general form is $s_{b_1}^2 = \dfrac{\Sigma(Y - \hat{Y})^2/(n-2)}{\Sigma(X - \overline{X})^2} = \dfrac{s_e^2}{\Sigma(X - \overline{X})^2}.$

Thus $df = n - 2$ because two parameters are estimated in computing s_e^2. Recall the rule for df on page 230.

for predictions, but the test does *not* rule out the possibility for even better projections. Consider Figure 9. Both 9a and 9b indicate data sets for which there is a strong linear relation between X and Y. Probably these would both lead to a test conclusion $B_1 \neq 0$. In Figure 9a a linear relation is likely "best"; yet Figure 9b depicts data for which a linear regression can probably be improved upon by using a higher-order model, here a quadratic model. That is, a linear regression would be meaningful, but is likely not the best (most descriptive) relation. Better projections would probably come by using a quadratic regression. In both 9a and 9b the correlation would be nonzero.

Figures 9c and 9d give data plots indicative of zero slope ($B_1 = 0$). In both cases a relation between Y and X is other than linear. For 9c a horizontal line $\hat{Y}$ = a constant (the Y sample mean) describes the data. In 9d a higher-order polynomial, $\hat{Y} = A_0 + A_1 X + A_2 X^2$, is appropriate. Also recall Figures 5c and 5d of Section 3; they are the same. $\rho \doteq 0$ implies that a linear regression is not meaningful.

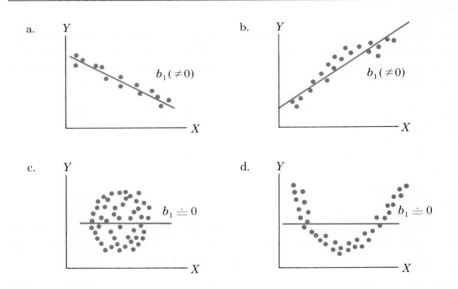

Figure 9/ Sample points and alternative values for b_1

The point to remember is that both the slope estimator, b_1, and the sample (linear) correlation coefficient, r, will give consistent information. Low values—toward zero—indicate the regression is other than linear. Values far away from zero indicate that a linear regression is meaningful. However, the latter case does not rule out a better relation in a higher-order polynomial model. In any case, a data plot and critical observation of the data can suggest what model to consider next.

Estimation of the true slope, B_1, follows the logic for estimation of means. Observe that the statistic takes the t-form of the last rule.

Let's estimate the true mean change in price from the used car data of Section 2. Again we need to assume normality and constant error variance. The distribution of $[(b_1 - B_1)/s_{b_1}]$ is t_{n-2}.   *Example*

For a 95% confidence estimate on B_1,   *Solution*

$$l = b_1 - t_{n-2} \cdot s_{b_1} = -.34 - (2.45 \cdot .02) = -34 - .05 = -.39$$
$$u = b_1 + t_{n-2} \cdot s_{b_1} = -.34 + (2.45 \cdot .02) = -.34 + .05 = -.29.$$

For each additional year of age, the estimate average decrease in price for cars of this make, sold at this lot, is $290 to $390.   *Conclusion*

Recall the comment by the Nielsen Agency in Chapter 1 concerning the error in their estimation of percent of viewer audiences. It is reasonable to expect a researcher to indicate the accuracy of his or her estimates. The preceding gives an indication of the error of estimation for b_1-values in this experiment. Similarly, sound research will indicate the error in prediction of $\hat{Y}$-values. For example, in projecting that a two-year old car (of this make and sold by this dealer) has average price $\hat{Y} = 4.00 - .34(2) = 3.32$, or $3,320, we should indicate the error in this point estimate. However, to do so requires more involved forms. So again see [4]. Generally the bounds of error for estimation of $\hat{Y}$-values are the upper and lower (confidence) bands shown in Figure 7.

These techniques are in no sense limited to the used car data and can, for example, be applied to time series and other equally spaced data. To illustrate let's extend the example on sales over years first given in Section 4.

Earlier we found $\hat{Y} = 9 + .45X_c$ with   *Example*

$\hat{Y}$ = estimated sales (in $100,000)
X_c = years, coded with origin 1971 $\frac{1}{2}$ and with six months increments.

The data appears on p. 301. Using $s_{b_1} = .12$ and assuming normal deviations and constant variance, we might want to test for a "meaningful" linear sales trend. Suppose $\alpha = .10$. We will also estimate the change in sales per six-month period and place bounds on the error of estimation.

The test is for zero slope: *Solution*

H_0: any sales trend is other than linear ($B_1 = 0$)
H_A: a linear sales trend is meaningful ($B_1 \neq 0$)
Test: reject H_0 if $|\text{calculated } t| > 1.94\,(t_{.05,6})$

$$\text{Calculated } t = \frac{b_1 - B_1}{s_{b_1}} = \frac{.45 - 0}{.12} = 3.75.$$

309

Conclusion
Reject H_0, so a linear sales trend is meaningful, $\alpha = .10$.
The estimation procedures follow:

$$l = b_1 - (t_{.05} \cdot s_{b_1}) = .45 - (1.94 \cdot .12) = .45 - .23 = .22.$$
$$u = b_1 + (t_{.05} \cdot s_{b_1}) = .45 + (1.94 \cdot .12) = .45 + .23 = .78$$

Conclusion
For the period 1968–1975 sales have increased by an average estimated at $22,000 to $78,000 per six-month period, or $44,000 to $156,000 per year.

We have only scratched the surface in regression inference. Those who want to do regression–correlation studies should get a deeper understanding of inference; see reference [2] or [4].

Problems

1. For each of the data plots, A and B, indicate which options from among a–e that you would try. One or more options should be appropriate for each data plot.

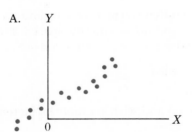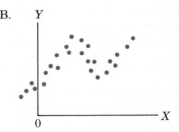

   a. Test the hypothesis $H_0: B_1 = 0$ against $H_A: B_1 \neq 0$.
   b. Test for a higher order polynomial regression.
   c. Compute r as a (point) estimate of the linear correlation.
   d. Gather more data before attempting any analyses.
   e. Compute an estimate for the true slope.

2. Indicate which options from among a–e in Problem 1 that you would try for data plots A, B, and C in this problem.

Problems

A.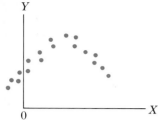

B.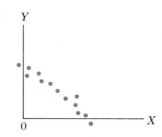

C.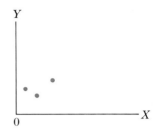

3. a. Plot the data given in the table. Does a linear regression appear meaningful?

   b. This data is the same as that for Problem 1 in Section 2. Use the values for b_0, b_1 found there and $s_{b_1} = .28$ to test $H_0: B_1 = 0$ against $H_A: B_1 \neq 0$. Use $\alpha = .05$ and consider assumptions for inference are met. State your conclusions. (*Note: r* was required in Problem 6 of Section 3.)

X:	−2	−1	0	1	2
Y:	0	1	0	1	3

4. The table shows sales, Y, for each of eight salespersons. Years of sales experience are also recorded for each:

X (years):	6	3	2	4	1	6	3	5
Y ($1,000):	8	5	2	3	3	9		6

   a. Given $\bar{Y} = 5$, find the value missing from the table.

   b. Estimate B_1, the mean increase in sales for each additional year of experience. Use $s_{b_1} = .27$ and 95% confidence. Give a point estimate, then place bounds of error on your estimate give (l, u).

5. For the equally spaced data in the table, compute a 95% confidence interval for estimation of B_1. Use $\hat{Y} = 6 + 2X$ and $s_{b_1} = .17$.

X:	−3	−2	−1	0	1	2	3
Y:	1	2	3	5	8	11	12

(*Note:* this data was also considered in Problem 4 of Section 2 and in Problem 7 of Section 3.)

6. Suppose you are the production manager in a firm where productivity for each worker can be described as the number of units of output, Y. For $\hat{Y} = 400 + 2X$, X is years of experience in this trade. Let $n = 25$, $r = .8$, and $s_{b_1} = .71$. Write two or three sentences for each question as if you were explaining to a classmate.

  a. Is there a meaningful linear relation between units produced and years of experience. (*Note:* the given information is sufficient to make this test.)

  b. If there is a meaningful linear relation, estimate the increase in units produced for each additional year of experience. Give bounds and explain their meaning to your classmate.

  c. If a linear relation is not meaningful, describe what steps you would take to discern a meaningful relation between Y and X.

7. Assume you are the coordinator of English teachers in your school system with the responsibility of interpreting standardized (entrance) test scores to other teachers and to pupils. How would you interpret the following data as a generalization of the relation of test scores, X, to achievement, Y, in a senior English course? Suppose you have $\hat{Y} = 55 + .5X$ with $0 \leq X \leq 80$ points, and $.4 \leq B_1 \leq .6$ with 95% probability. Discuss the projected change in average achievement in the course for each additional point achieved on the test.

6/Essay Example

A teen-ager just beginning to smoke said, "Nobody young worries much about some disease they might get 40 or 50 years from now." A few million men and women aged 40–60 said, "I'd like to stop, but I can't." And tobacco companies, with hundreds of millions of dollars in advertising space, said, "Smoke!"

The Report of the Advisory Committee to the Surgeon General of the Public Health Service (The Surgeon General's Report) of 1964 included [6]:

> Cigarette smoking is causally related to lung cancer in men; the magnitude of the effect of cigarette smoking far outweighs all other factors. The data for women, though less extensive, points in the same direction.

The risk of developing lung cancer increases with duration of smoking and the number of cigarettes smoked per day, and is diminished by discontinuing smoking.

These four views, and others, exist today. How these attitudes and actions have affected our United States population is reflected in one way by smoking statistics [7].

Table 3/ Percentage of Current Smokers of Cigarettes (Regularly or Occasionally) by Sex and Age. U.S. Surveys: 1955 and 1966 by CPS (Current Population Surveys) and 1970 by NCSH (Survey Conducted for National Clearinghouse for Smoking and Health)[1]

	Male			Female		
Age	CPS 1955	CPS 1966	NCSH 1970	CPS 1955	CPS 1966	NCSH 1970
18–24	53.0	48.3	47.0[2]	33.3	34.7	31.1[2]
25–34	63.6	58.9	46.8	39.2	43.2	40.3
35–44	62.1	57.0	48.6	35.4	41.1	39.0
45–54	58.0	53.1	43.1	25.7	37.3	36.0
55–64	45.8	46.2	37.4	13.4	23.0	24.3
65+	25.8	24.6	23.7	4.7	8.1	11.8

[1] 1955 survey based on approximately 45,000 persons; 1966 survey based on approximately 35,000 persons, 1970 survey based on approximately 5,000 persons
[2] Estimated

Massive changes in smoking behavior have taken place among adults in recent years. (Observe that the patterns for male and female smoking are quite different.) These are, of course, reflections of industry advertising, personal concern for health, social pressures, and many other things. The analysis of United States mortality rates in future years should provide valuable information concerning smoking and death rates.

Our discussion is about the suggested "causal" relation between smoking and lung cancer. I want to emphasize the careful and extensive search of evidence made by the Committee before any conclusions were formulated. If the evidence was unclear, then a conclusion was not made.

The modern history of smoking and health began at the turn of this century. About 1900 an increase in cancer of the lung was noted, particularly

Regression, Paired Observations, and Linear Models

by vital statisticians. Since that time observations of thousands of patients and autopsy studies of smokers and nonsmokers indicate that many kinds of damage to body functions and to organs, cells, and tissues occur frequently and severely in smokers.

After 1930 a number of large scale medical studies were made on smoking. These were predominately of two types: (1) *retrospective studies* which concern data from the personal histories and medical and mortality records of individuals and, (2) *prospective studies,* in which persons chosen from some special group, such as a profession, were followed from their entry into the study for an indefinite time period, or until they died or were lost for other reasons. An account of several of these studies appears in [1]. Although these studies represent somewhat uncontrolled evidence, with many factors unmeasured and not controlled, the large numbers do show some patterns over time.

It is, of course, not reasonable to subject humans to experiments that might conceivably produce cancer or cause other physical harm. Subsequently only nonhuman animals have been required to smoke for extended periods and under strictly controlled conditions. In numerous studies, animals have been exposed to tobacco smoke and tars and to the various chemical compounds they contain.

Thus three lines of evidence were evaluated separately—clinical and autopsy studies, population studies, and animal experiments—and then considered together in drawing conclusions. When a relationship or an association between smoking and some physical condition was noted, the statistical significance of the association was assessed. But the Committee was aware that the establishment of a statistical association between the use of tobacco and a disease is insufficient for a judgment of a *causal* relation [6].

> Statistical methods cannot establish a proof of a causal relationship in an association. The causal significance of an association is a matter of judgment which goes beyond any statement of statistical probability. To judge or evaluate the causal significance of the association between the attribute or agent and the disease, or effect upon health, a number of criteria must be utilized, no one of which is an all-sufficient basis for judgment. These criteria include:
> a) The consistency of the association
> b) The strength of the association
> c) The specificity of the association
> d) The temporal relationship of the association
> e) The coherence of the association

It appears that various meanings and conceptions of the term *cause* were discussed "vigorously" by the Committee and that other terms including "factor" and "determinant" were considered as alternatives. Yet it was clearly put that [6],

> . . . no member of this committee used the word "cause" in an absolute sense in the area of this study. . . . All were thoroughly aware of the fact that there are series of events in occurrences and developments in these fields, and that the end results are the net effect of many actions and counteractions.
>
> Granted that these complexities were recognized, it is to be noted clearly that the Committee's considered decisions to use the words "a cause," or "a major cause," or "a significant cause," or "a causal association" in certain conclusions about smoking and health affirms their conviction.

I trust it is apparent that it was very difficult for these persons to judge a cause–effect relation. Furthermore the "strength of the association" (correlations) was only one of numerous instruments that they used; a high correlation, even $+1$ or -1, does *not* imply causation.

7/ Summary for Chapter 12

In regression we attempt to describe one variable through a functional relation to others. We begin with a data plot. From this visual inspection will usually suggest one or more reasonable patterns.

A quantitative evaluation of the data uses an assumed statistical model. We have discussed only the linear regression model. The statistical approach called *least squares* is used to estimate values for parameters in the (assumed) linear model. The objective of least squares is to identify that line which, in one sense, passes closest to all of the sample data points. If the error in describing these points is sufficiently small, i.e., if considering all sample points the line comes sufficiently close, then inference can be made about possible response values.

The Pearson coefficient of correlation measures the strength of the linear regression. This coefficient takes values between ± 1; that is, for sample values $-1 \leq r \leq +1$. The strength of a linear correlation is described by the coefficient of determination, $100r^2$. Correlation does not indicate causation, but rather indicates the amount of common description for two variables. Moreover a low linear correlation, near zero, means only that a simple linear regression is inappropriate and that other models should be tried. Again, data plots can be invaluable in deciding what to try next.

REFERENCES

[1] B. W. Brown, Jr., "Statistics, Scientific Method, and Smoking," in *Statistics: A Guide to the Unknown* edited by Judith M. Tanur, et al. (San Francisco: Holden-Day, 1972).

[2] Clelland, deCani and others, *Basic Statistics with Business Applications* (New York: John Wiley & Sons, 1966).

[3] Edward E. Cureton, in a letter to the editor, *The Statistician,* 25: 54–55, June, 1971.

[4] John Ingram, *Introductory Statistics* (Menlo Park, California: Cummings, 1974).

[5] Albert P. Shulte, "Points and Fouls in Basketball," in *Statistics by Example, Exploring Data,* edited by Fredrick Mosteller et al. (Reading, Massachusetts: Addison-Wesley, 1973).

[6] *Smoking and Health: Report of the Advisory Committee to the Surgeon General* (Washington, D.C.: U.S. Department of Health, Education, and Welfare, Public Health Service, Health Services and Mental Health Administration, 1964).

[7] *The Health Consequences of Smoking: A Report to the Surgeon General: 1971* (Washington, D.C.: U.S. Department of Health, Education, and Welfare, Public Health Service, Health Services and Mental Health Administration, 1971).

Review Problems

1. Examine the diagram below. Explain what is right and what is wrong with the following statements:
   a. $r \doteq -.98$
   b. The linear regression of Y on X is $\hat{Y} = 7 + 1X$.

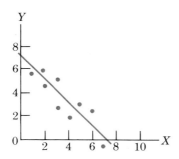

2. In a regression study Y = weight gain for beef cattle, X_1 = initial weight, and X_2 = pounds of feed per day. If $r_{YX_1} = .46$ and $r^2_{YX_2} = .3136$, which variable, X_1 or X_2, gives stronger linear description? By how much?

3. The following display is of observed data points and the estimated regression line. Something has gone wrong in my calculations, at what point should I recheck my math?

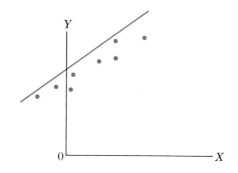

4. For the paired observations shown in the table:

X:	0	2	3	4	6
Y:	−1	0	1	2	3

   a. plot the data on XY coordinates.
   b. without making any calculations, place your (freehand) estimate of the regression line on the graph. Is the slope positive, negative, or nearly zero?
   c. the regression line, estimated by least squares, is $\hat{Y} = -1.1 + 0.7X$. Plot this line on your (already full) graph. How close does this line come to the one that you made for part b?

5. Using the following data, answer parts a, b, and c of question 4. In this case you will have to develop the equation for the least squares regression line.

X:	0	2	3	4	6
Y:	3	2	1	0	−1

6. Give a short answer of one or two sentences for the following questions:

   a. What is the essential difference between a situation where $r = -0.7$ and one where $r = +0.7$? Can you make better predictions in one case than in the other? Explain.
   b. An advertising agency selected a simple random sample of 250 families to study the age relation between husband and wife. For the sample $r = .88$, explain the practical meaning of this value.

7. The following table indicates the number of graduates from a single high school who went directly into college. Assume the class size was about the same for all years.

Year:	1	2	3	4	5	6
Number of persons:	121	138	115	162	160	174

   Plot the data. Does a linear trend appear reasonable? Estimate the trend line by a least squares analysis.

8. Treating the following data as a time series, find a linear trend equation and use it to predict the trend value for $X = 9.5$.

X:	4	5	6	7	8	9	10
Y:	4	6	8	12	18	24	26

9. For the sample data below, assume that the regression of Y on X is linear and establish the regression equation. Check reasonableness by plotting the data.

X:	1	3	4	6	8	9	11	14
Y:	1	2	4	4	5	7	8	9

Given:

$\sum X^2 = 524$ $\quad n\sum(X^2) - (\sum X)^2 = 1056$
$\sum (XY) = 364$ $\quad n\sum(XY) - \sum(X)\sum(Y) = 672$.

10. The least squares regression for the data in the table is described as $\hat{Y} = 4.70 + 4.10X$.

X:	1	2	3	4	5
Y:	9	13	18	18	27

a. Estimate Y for period $X = 6$.
b. Now treat this as a time series problem (i.e., equal spacing) coding origin at $X = 3$ and with whole-period increments. Again estimate Y for period $X = 6$. Compare your answer to that in Part a.
c. Suppose X denotes years and Y denotes per capita income. If, following a rather stable period (including years 1–5), in year $X = 6$ the economy experienced a recession, would you hesitate to make the predictions requested in Parts a and b? Explain why or why not.

11. Compute the coefficient of correlation for the given values. Now code the data using $X' = X/10$ and $Y' = Y/100$. Again compute r. Do you think any other coding could be done to simplify computing r?

X:	20	50	70	40	20
Y:	500	200	100	300	400

12. Ten students obtained these exam grades in a college chemistry course:

Student	Midterm	Final	Student	Midterm	Final
1	65	61	6	65	71
2	91	89	7	72	78
3	54	62	8	43	35
4	80	71	9	57	62
5	51	62	10	92	89

Compute the Pearson coefficient of correlation for this data given $\sum(XY) = 47{,}777$, $\sum(X^2) = 47{,}394$, and $\sum(Y^2) = 48{,}609$. Interpret the meaning of your value through $(100 \cdot r^2)$.

13. Suppose that the estimated regression of weights, Y, on heights, X, for adult males in one country is $\hat{Y} = 70 + 1.5X$ with $s_{b_1} = .25$.

   a. Use this information to make a point estimate for the average weight of 6 foot men in this country.

   b. What are the greatest and least values, u and l, for the *change* in weight for each extra inch in height assuming $n = 28$ and 95% probability?

14. For $n\sum X^2 - (\sum X)^2 = 28 \quad n\sum(XY) - \sum X \sum Y = 53$
    $\hat{Y} = -2.45 + 1.89X \quad n\sum Y^2 - (\sum Y)^2 = 102 \quad \overline{X} = 7, n = 7$
    $s_{b_1} \doteq .40$
    a. Compute r, the Pearson coefficient of correlation.
    b. Test $H_0: B_1 = 0$ against $H_A: B_1 \neq 0$ with $\alpha = .10$. State your conclusions.

Inference on Categorical Data and Frequencies

CHAPTER 13

This chapter fills several gaps remaining in our study of the basics of statistical inference. So far our work has been based on normally distributed random variables. Herein we discuss frequency data and values that can be categorized. This does *not,* however, assume an underlying normal probability distribution.

Inference is on discrete variables, but continuous random variables can be considered, too, by simply forming (discrete) classes on intervals of values. The discussion includes tests for binomial or extended multiple classification experiments. An important application is testing independence of (generally qualitative) classifications for categorical data.

Since rather minimal assumptions are required for the probability forms, these procedures are especially useful in the social sciences, in psychology, and generally in experiments where human behavior is studied. Moreover, nominally scaled data is sufficient for many of the tests; that is, procedures herein use frequencies in categories or name classes. Thus a nominal scale, that uses symbols to classify objects into groups, suffices for the calculations. Examples of nominal-scaled variables include zip codes, marital status, and sex classification.

The Chi-square statistic test can be used to find out if each face on these dice shows an equal number of times. This will allow us to determine if we have a balanced pair of dice.

1/ The Chi-Square Statistic

The basis for calculations is the multinomial experiment. This can be described as in tossing a die. For one toss, the object can show any one of six faces on top. These faces are the *multinomial,* or many named, classes in the experiment. Suppose, for example, $n = 300$ tosses. Of these the observed frequency of the one-face showing on top could be named f_1, or frequency of ones; f_2, the frequency of upturned two's, etc. The characteristic of interest is the balance of the die; does each face show on top about equally often? If so, then we should *expect* about one in six of the tosses to produce the one-face, say $E_1 = 300 \cdot 1/6 = 50$. An important assumption is independence of outcomes. This condition is assured as long as the die is reshaken so that an outcome does not depend on what has happened before. For a "fair die," and with independent tosses, the expected frequencies would be the same, 50, for all (six) multinomial possibilities.

As another example, if we rename the faces so we have just two possibilities, such as (1) the one-face and (2) other than the one-face, then the preceding reduces to a simple binomial-type experiment. Thus other than counting for the more numerous classes, the multinomial requires little more calculation than in binomial experiments.

The test statistic for multinomial experiments is an approximate chi-square form.

Rule

For inference on multinomial classification data we use the approximate chi-square form:

$$\chi^2_{df} = \sum^k \left[\frac{(f_i - E_i)^2}{E_i} \right]$$

where

$E_i = np_i$ = expected frequency in class i

k = the number of classes

f_i = observed frequency count for class i.

We do not assume any specific distribution, but rather that (1) the data can be divided into multinomial classes, and that (2) the sample is sufficiently large to assume χ^2_{df} has an approximate chi-square distribution. Of course this requires independence of outcomes to assure fixed class probabilities. In the test statistic, agreement of observed frequencies with hypothesized or expected values is substantiated by small squared differences, $(f_i - E_i)^2$. Subsequently, small calculated χ^2-values support a null hypothesis; disagreement of the observed and the expected results in a large positive χ^2 value.

Because of its form, this chi-square statistic cannot distinguish the direction of differences. That is, all differences are squared and so add to a positive value. Subsequently in hypothesis testing the alternative becomes does-not-equal, but at the same time the rejection region is forced to the upper tail. The test is structured in a very similar manner as F-test procedures in Chapter 11. For a less-than or a greater-than alternative hypothesis, a binomial, normal, or other procedure should be used.

The basic chi-square form, χ_1^2, is defined as the square of a Z-statistic. It should be no surprise then that the χ^2-distributions in Figure 1 appear much the shape of a distorted normal curve. The specific shape depends on the degrees of freedom, which in turn depend on the number of classes. The random variable takes only positive values, see Figure 1. Also the probability curve is positively skewed; that is, the distortion is one of asymmetry with less distortion from a bell-type curve occurring for the higher df curves.

Figure 1 / The chi-square distribution for various degrees of freedom

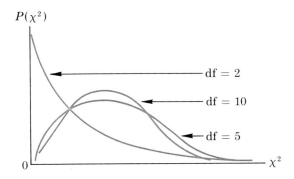

The percentage points in Table III of Appendix B are points from either tail of the chi-square distribution. See Figure 2. The column heading indicates the percent of measure falling below lower-tail percentage points (Table III-a, i.e., $P(\chi^2 < \chi_\alpha^2) = \alpha$) or above upper-tail percentage points (Table III-b, i.e., $P(\chi^2 > \chi_{1-\alpha}^2) = \alpha$).

The following examples illustrate the use of Table III. Check these with the table.

Inference on Categorical Data and Frequencies

Figure 2/ The chi-square probability distribution

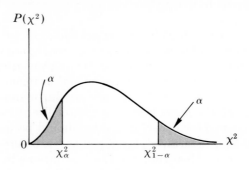

Example

For a test of hypothesis on equal proportions a multinomial classification experiment has 6 degrees of freedom. For this test with $\alpha = .05$, Table III-b gives $P(\chi^2 > 12.59) = .05$. Thus a calculated χ^2-value larger than 12.59 is required to reject "equal proportions" in favor of "unequal proportions."

The next example illustrates estimation, including both upper- and lower-tail chi-square percentage points.

Example

For a 90% probability interval on the scale for χ^2_{15} we observe from Table III that:

$\chi^2_{.05,15} = 7.26$ (lower-tail)
$\chi^2_{.95,15} = 25.00$ (upper-tail)
That is, $P(7.26 < \chi^2_{15} < 25.00) = .90$.

Observe that the nonsymmetry of the chi-square distribution makes interval estimates lopsided. (The mean for a chi-square distribution has a value equal to the degrees of freedom, here 15.)

The problems for this section consider the use of Table III. The next section will concern inference in binomial experiments, and there the chi-square, as well the binomial and other forms, compete as tools for inference.

Problems

1. Answer the following concerning the chi-square probability distribution. Use Table III.
   a. Determine $P(\chi^2_{21} > 8.897)$.
   b. If $P(\chi^2_{10} > 20.483) = \alpha$, find α.
   c. For $P(\chi^2_{.975,1} > C) = .025$, find C.

2. Show your understanding of Table III by answering the following questions:
   a. Find C and D for $P(C < \chi^2_5 < D) = .95$, so that 2.5% of the lack-of-confidence comes from either tail.
   b. $P(\chi^2_{11} > 3.053) = \alpha$; what is α?
   c. Find the 99th percentile point on the distribution of χ^2_{19}.

3. For a chi-square test using $\alpha = .01$ and with df = 13, the calculated $\chi^2 = 18.3$. Assume the chi-square approximation shown in this section so that the rejection region is entirely in the upper tail.
   a. Should the null hypothesis be rejected?
   b. Another chi-square test uses $\alpha = .025$ and 27 df. What must the calculated χ^2-value be in order to reject the (null) hypothesis of no difference?

4. Recall that $Z_{.025} = 1.96$. Now $Z^2_{.025} = (1.96)^2 = \chi^2_{\gamma,1}$. With the aid of Table III, if needed, identify the value for γ.

2/ Inference in Binomial Experiments

The binomial experiment was introduced in Chapter 5; then inference on p, the true proportion of successes, was discussed in Section 3 of Chapter 9. Therein we required $np \geq 5$ and $nq \geq 5$ as conditions suitable for the use of the normal distribution. Here we explore inference on p for *any* sample size. This discussion will utilize separately the binomial, normal, and chi-square statistics.

Example

A test on the advertising effectiveness of two displays was made as follows: each display was placed at the check-out counter of a single store and remained there for the same time period; also, a record was kept of the frequency of sales of the advertised product. One hypothesis is (the null) that the two displays gave equal sales of this product. Assuming that the stores handle comparable trade, were sales statistically equal? Use $\alpha = .05$ and sales counts for two weeks.

Display:	1	2	Total
Sales (in number of units):	82	118	200

Solution

Either a Z-statistic or the chi-square statistic can be applied. The chi-square procedure, being new, is discussed first. Since the test is for a difference in sales, a does-not-equal alternative is appropriate.

$H_0: p = 1/2$, p = true percentage of sales when display 1 is used
$H_A: p \neq 1/2$.

a. As a chi-square test:

Test: Reject H_0 if calculated $\chi^2 > 3.84$ ($\chi^2_{.95,1}$).

(For now df is one. There are two displays (classes) and our test requires one restriction, hence df $= 2 - 1 = 1$). Using the data given:

Observed, f_i	82	118
Expected, E_i	100	100

with the single restriction $\sum f_i = n = 200$, then

calculated $\chi^2 = \sum[(f_i - E_i)^2/E_i] = [(82 - 100)^2/100]$
$\qquad\qquad\qquad + [(118 - 100)^2/100] = 3.24 + 3.24 = 6.48$

So reject H_0, $\alpha = .05$.

b. As a Z-test:

Test: Reject H_0 if $|\text{calculated } Z| > 1.96$

The comparable Z-calculation for X = number of sales under display 1 is

$$\frac{82 - 100}{\sqrt{200 \cdot .5 \cdot .5}} = -18/\sqrt{50} \quad \text{(See Section 9.3)}$$

$$= -2.55$$

Again, reject H_0.

Notice that $Z^2 = \chi_1^2 = (-18/\sqrt{50})^2 = 6.48 \quad (> (1.96)^2 \doteq 3.84)$
$[(Z_{.025})^2 = \chi_{.95,1}^2]$

The data indicates a significant difference in sales for the two displays, $\alpha = .05$, at these stores. The conclusion is the same using the Z- or the chi-square procedure.

Conclusion

This illustrates that for two-tailed alternative tests on binomial p, with $np \geq 5, nq \geq 5$, we have a choice of using either the chi-square or the Z-test. For one-tailed tests with large sample size, our only choice is the Z-test.

For smaller samples the Z-approximation is ruled out; instead we consider exact binomial probabilities.

A coin is repeatedly tossed to determine whether it is balanced. Seven tails are observed in 25 independent tosses. Is this sufficient evidence to say, at $\alpha = .05$, that the coin is out of balance?

Example

Consider the exact binomial form. Here a difficulty arises. Since the binomial random variable is discrete, it admits at most a countable number of values. There is no guarantee that we can specify a test with exactly .05 α-risk. For example consider two possible tests for $H_0: p = .5$ against $H_A: p \neq .5$.

Solution

Test 1: Reject H_0 if $X \leq 7$ or $X \geq 18$ successes. Use Table V in Appendix B for $X = 7$ successes (tails) with $n = 25$ and $p = .5$.

$\alpha_1 = P(X = 0, 1, \ldots, 7, 18, 19, \ldots, 25)$
$= P(X \leq 7 \text{ or } X \geq 18/p = .5)$
$= 2(.0001 + .0004 + .0016 + .0053 + .0143)$
$= .0434 (< .05)$.

Test 2: Reject H_0 if $X \leq 8$ or $X \geq 17$.

$\alpha_2 = P(X \leq 8 \text{ or } X \geq 17/p = .05) = .0434 + 2(.0322)$
$= .1078 (> .05)$.

Since the observation $X = 7$ falls within the rejection region for both tests, we reject H_0 for $\alpha = .05$. The fact that one cannot arbitrarily set the α-risk makes this kind of test somewhat undesirable. Also in some cases, conflicting conclusions can result from tests with slightly higher or lower α-levels.

Conclusion

An alternative for the two-tailed test employs a form of the approximate chi-square statistic.

Inference on Categorical Data and Frequencies

Rule

For X_1 = the observed number of successes, let $X_2 = n - X_1$ be a count of nonsuccesses. Then $H_0: p = p_1$ = the true proportion of successes with $p_2 = 1 - p_1 = q$ = the true proportion of nonsuccesses. The chi-square form is:

$$\chi_1^2 = \frac{(X_1 - np_1)^2}{np_1} + \frac{(X_2 - np_2)^2}{np_2} = \sum_{i=1}^{2} \left[\frac{(X_i - np_i)^2}{np_i} \right].$$

This is the original approximate chi-square form (see p. 324) with $f_i = X_i$ and $E_i = np_i$. This is applied to the coin tossing experiment to test for a biased coin. Since either too many or too few tails would indicate an imbalance, the test alternative is does-not-equal.

Example

Again, is seven tails (successes) in 25 independent tosses of a single coin sufficient evidence to conclude that the coin is out of balance? As before $\alpha = .05$.

Solution

Let p_1 = the chance any toss produces a tail, and suppose $X_1 = 7$ tails in $n = 25$ tosses

$H_0: p_1 = .5, H_A: p_1 \neq .5$

Test: Reject H_0 if calculated $\chi^2 > 3.84$ ($= \chi^2_{.95,1}$)

Calculated $\chi^2 \doteq \left[\frac{(7 - 25 \cdot .5)^2}{25 \cdot .5} \right] + \left[\frac{(18 - 25 \cdot .5)^2}{25 \cdot .5} \right] = 4.84.$

Conclusion

As before the data indicates that this coin is out of balance, $\alpha = .05$. The direction of the imbalance is *not* determined by the test. Intuitively, however, there appear to be too few tails. New data could be gathered and an appropriate test performed, such as a one-tailed (less-than) binomial or Z-test.

Since the chi-square approximation is based upon a limiting form that is standard normal, the rule gives more exact results for larger samples. For an improved approximation, the (continuity) corrected chi-square form can be used for binomial or for other one degree of freedom tests [4]. The chi-square approximation is an alternative for tests on p in binomial experiments, but only with a does-not-equal alternative. The binomial test has the advantage in that it uses an exact probability distribution and it can be used for one-sided alternative tests.

There are sample size limitations in using the chi-square form. The following philosophy is used by many.

Inference in Binomial Experiments

Any expected frequency value of less than five denotes a short cell. *Rule*

The concern regarding a short cell is the matter of deciding how much information is required for meaningful statistical inference.

In many experiments projections from an expected frequency of four or fewer scores is questionable.*

Short cells should be treated using one of the following techniques, in order of preference:

1. Remove the short cell by taking sufficient sample to overcome the low expectation.
2. Correct by modification of observed counts using a correction rule.
3. Combine with other cells through logical combinations.
4. Remove by dropping the short cell and others that are no longer meaningful due to the cell exclusion.

Opinion polls may have short cells. The following example illustrates such a case.

Suppose a random selection of forty potential voters gave their preference for candidates in a congressional race. We hypothesize that the Independent candidate has 10% of the vote with the remainder evenly split between the others. We test $H_0: p_D = p_R = .45, p_I = .10$ $\Leftrightarrow p_D: p_R: p_I = .45: .45: .10$. Let $\alpha = .05$. *Example*

Voter Preference

Candidate	Number who favor
Democrat	21
Republican	13
Independent	6
Total	40

The source for problems, a low expected value, is in the Independent group, $E_I = np_I = (40) \cdot (.1) = 4.0$. This low expected value indicates a short cell. The correct approach for overcoming the short cell is a matter of opinion and should be determined by the most logical *Solution*

---

*Numerous statistics researchers contend that an expected value of 5 is an excessive restriction on the chi-square procedure, e.g., see [4]. The point is that at some point there is insufficient information to allow reasonable inference; I choose to go by the rule above.

approach under the circumstances. Certainly the preferred approach would be to obtain at least ten more opinions before an analysis is made. That is, $40 + 10 = 50$ opinions would yield $E_I = np_I = (50 \cdot .10) = 5$ —and this is minimal! In this case the cost of additional information, i.e., for interviewing ten or more voters, is reasonably low. The second table indicates counts, including 20 *more* opinions, for a total of 60 persons.

	Voter Preference-Extended
Candidate	Number who favor
Democrat	32
Republican	20
Independent	8
Total	60

Again $H_0 : p_D : p_R : p_I = .45 : .45 : .10$ with

$H_A : p_D : p_R : p_I \neq .45 : .45 : .10$.

The calculation of approximate χ^2 extends the preceding:

$$\text{calculated } \chi^2 = \frac{[32 - (60)(.45)]^2}{27} + \frac{[20 - (60)(.45)]^2}{27}$$

$$+ \frac{[8 - (60)(.10)]^2}{6} = 3.41$$

For df $= 3 - 1 = 2$, tabular $\chi^2_{.95,2} = 5.99$.

Conclusion

We cannot reject H_0, $\alpha = .05$. The sample shows that at this time there is no statistical difference in preference between the Democrat and Republican candidates, and the Independent has 10% of the vote.

As with other reasonably small sample tests, one should view the preceding "cannot-reject-the-hypothesis" conclusion with caution. There exists the possibility of a type 2 error, another reason that we should take larger samples.

In conclusion, caution is the best attitude toward treating short cells. Attempts to combine classes are dependent upon issues at that time and are very suspect if subjective placement is made. For example, are Independents closer in philosophy to Republicans—or to Democrats? Only after everything else fails should one consider leaving out meaningful information.

Problems

1. A binomial experiment including $n = 50$ trials produces $X = 24$ successes. Use the chi-square procedure to test the null hypothesis that $p = .40$. Let $\alpha = .05$ and use a does-not-equal alternative hypothesis. Use df = 1.

2. In a certain city 20 separate and independent attempts to capture narcotics dealers resulted in 17 individual arrests. For p = percent of attempts that result in an arrest, test $H_0: p = .7$ against $H_A: P > .7$ using $\alpha \doteq .05$. (*Hint:* in Table V, for $n = 20$ and $p = .7$, use $P(X \geq 17) = .1070$.)

3. A coin is tossed 20 times and shows 16 heads. The test is $H_0: p = .5$.
   a. Perform the test using the chi-square statistic with $\alpha = .05$.
   b. Compare the answers to Part a to a binomial test with the rejection region defined by $X = 0, 1, 2, 3, 4, 5, 15, 16, 17, 18, 19, 20$; i.e., $P(X \leq 5$ or $X \geq 15) = .0414$. Do you get the same conclusion? What is the α-risk level in the latter test?

4. An advertising firm is asked to test the hypothesis that more than 65% of all women in Des Moines prefer soap Brand A over Brand B. If 70 of 100 randomly chosen women expressed preference for A, establish and carry out a suitable test on the above contention. Use $\alpha = .10$. In deciding which test statistic to use, consider that the alternative hypothesis is one-tailed.

5. A random sample of 200 drivers who have taken a course in driver training contained 62 who had been involved in one or more accidents in the past five years. Can we reject the hypothesis that the proportion of accident-free drivers in this group is equal to 0.75? Use $\alpha = .10$ and your choice of either a Z or a chi-square test.

6. In a survey made at Trico State, six randomly-selected students were asked if they favored a change from the quarter system to a semester system. Five answered "no," the remainder answered "yes." Does the sample show that opinion is evenly divided on this issue?
   a. Why is the chi-square test inappropriate?
   b. If an increased sample gave a total of 31 out of 50 who were against the change, how would you answer the original question, $\alpha = .025$?

3/ Contingency Tables and Independence

In the last section we discussed tests on proportions for a single binomial distribution. That discussion is extended to k independent binomial distributions (Case 1), and then to multinomial cases (Case 2).

In Case 1, for k independent binomial random variables, we first consider random samples from k binomial distributions with parameters $n_1, p_1; n_2, p_2; \ldots; n_k, p_k$ and which yield respectively $X_1, X_2, \ldots, X_k$ successes. For sufficiently large samples, the distribution of

$$\frac{X_i - n_i p_i}{\sqrt{n_i p_i (1 - p_i)}}$$

can be approximated by the standard normal. When squared, the latter form has approximately the chi-square distribution. Using an addition rule for independent chi-square variables [3], their sum also has a chi-square distribution with df = the sum of the separate degrees of freedom. A common hypothesis is that the distributions have the same probability of success; that is, $p_1 = p_2 = \ldots = p_k = p$. Since p is unknown, it must be estimated. The resulting test has $k - 1$ degrees of freedom; that is, each sample for a binomial variable gets one degree of freedom (as in Section 2), but the one restriction, that the p_i are equal, results in a loss of one degree of freedom. Hence df $= k - 1$. The test statistic is given in the following rule.

Rule

To test $H_0: p_1 = p_2 = \ldots = p_k = p$ (unknown) for independent samples from k binomial distributions, use

$$\chi^2_{k-1} = \sum_{i=1}^{k} \left[\frac{(X_i - n_i p)^2}{n_i p} + \frac{(X'_i - n_i p')^2}{n_i p'} \right]$$

where

$$p = \frac{X_1 + X_2 + \cdots + X_k}{n_1 + n_2 + \cdots + n_k}$$

estimates the (common) probability of success

$p' = 1 - p$ estimates the (common) probability of nonsuccess

$X_i =$ the number of successes in a sample of n_i from distribution i

$X'_i = n_i - X_i =$ the observed number of failures in a sample of size n_i from distribution i.

Contingency Tables and Independence

This procedure is illustrated for a problem in which $k = 4$.

On the basis of the given data we want to test whether the proportion of nondefective parts produced is the same for all workers. Let $\alpha = .05$.

Example

		Nondefective Parts (success)	Defective Parts (nonsuccess)	Totals
w o r k e r	$i = 1$	$X_1 = 100$	$X_1' = 20$	120
	2	74	6	80
	3	106	14	120
	$4(k)$	64	16	80
	Total	344		400

$X_1 = 100$ indicates worker 1 produced 100 nondefective units, while $X_1' = 20$ means she produced 20 defective items. The true proportions are:

$p_1 =$ the proportion nondefective made by worker 1
$p_1' =$ the proportion defective for worker 1, etc.

The test is for an equal proportion of nondefective parts for all workers, $H_0: p_1 = p_2 = p_3 = p_4 \, (= p)$ against H_A: *at least* one P_i differs. The true proportion that is nondefective is estimated as

Solution

$$\hat{p} = \frac{100 + 74 + 106 + 64}{120 + 80 + 120 + 80} = \frac{344}{400} = .86, \quad \text{so}$$

$$\hat{p}' = 1 - \hat{p} = .14$$

Then $\quad n_1 \hat{p} = 120 \cdot .86 = 103.2 \quad (= E_1)$
$\quad\quad\quad n_1 \hat{p}' = 120 \cdot .14 = 16.8 \quad (= E_2)$
$\quad\quad\quad n_2 \hat{p} = 80 \cdot .86 = 68.8 \quad (= E_3)$, etc., give expected values.

The data is recorded in the following table. Expected values are recorded slightly above and to the right of the observed counts.

Worker	1	2	3	4
Nondefective	$100^{103.2}$	$74^{68.8}$	$106^{103.2}$	$64^{68.8}$
Defective	$20^{16.8}$	$6^{11.2}$	$14^{16.8}$	$16^{11.2}$

This yields:

$$\chi^2_{4-1} = \frac{(100 - 103.2)^2}{103.2} + \frac{(74 - 68.8)^2}{68.8} + \cdots + \frac{(16 - 11.2)^2}{11.2} = 6.45$$

For tabular $\chi^2_{.95,3} = 7.815$, we cannot reject the null hypothesis.

Inference on Categorical Data and Frequencies

Conclusion   This analysis indicates that the four workers are producing equal proportions of nondefective items.

This test is the only common one for testing the equality of proportions for $k > 2$ binomial distributions; a Z-form is available for tests on differences in proportions, $k = 2$, see [2].

Case 2 is an extension to the general two-way multinomial experiment. The chi-square procedure has the advantage over Z-testing in that it can be extended for $k > 2$. The extension is from binomial to multinomial classifications. This can be viewed as counts on classes where the classes are grouped in two directions, that is, for a two-way frequency distribution. Since computations are not made on the values of the levels of the classes, the classifications in a contingency table can be either qualitative or quantitative. The levels serve only as labels and so can be as inexact as you desire. However, they must be mutually exclusive. Contingency table procedures can be extended to three-way or n-way classifications, but the computations quickly become cumbersome. Consider the following for a two-way classification.

Example   A political scientist hypothesizes that there is some relation between work status and a woman's opinion about women's lib. The α-risk is set at .05. The population is the women at a large university, with a random sample of 700 opinions classed as follows:

		Opinion		
Work hours per week	In favor	No opinion	Against	Totals
none	80	124	16	220
1–9.9	75	95	10	180
10–19.9	61	73	6	140
20 or more	74	78	8	160
Totals	290	370	40	700

The question asked is, "Is there a dependence between opinion about women's lib and the number of hours of work?"

This is a question of determining whether the two classifications are independent. If so, comparing across the four work classes one should observe about equal numbers who "favor," about equal numbers with "no opinion," etc. Since the "total" numbers differ for every work class, proportions rather than counts are compared. If opinions are independent of

Contingency Tables and Independence

work group, the proportion would be the same across all of the groups. Thus the proportion of those who "favor" is estimated by the total number who "favor" divided by total n. Here $290/700 \doteq .414$. Then for example, the number *expected* to "favor" from work class "none" is the frequency times the estimated proportion, here $220 \cdot (290/700) \doteq 91.1$. This illustrates a generalization.

> Under the assumption of independence, the expected cell frequencies in a two-way contingency are estimated as the product of appropriate row total times column total divided by the sample size.

Rule

This gives a procedure for estimating expected values, E_i. It is in fact just what we did in estimating expected values for k-independent binomial variables. The extent of agreement between cell observed counts, X_i, and the expected values indicates whether the assumed independence actually exists. If so, we should find $E_i \doteq X_i$, hence overall, a small chi-square value.

The test critical value has not been specified, but again we extend the k-binomial distribution case. Here the row and column totals have specified values; this puts one restriction on each row and another on each column, hence $[(4-1)(3-1)] = 6$ degrees of freedom. The test is an approximate chi-square with six degrees of freedom. Now we have the equipment to approach the question on women's lib.

The hypotheses are

H_0: work hours and opinion concerning women's lib are independent
H_A: opinion concerning women's lib depends on hours worked

Solution

Cell expectations of "favor" for the remaining work groups are:

$220 \cdot (290/700) = 91.1$ (repeated)
$180 \cdot (.414) = 74.6$
$140 \cdot (.414) = 58.0$
and $290 - (91.1 + 74.6 + 58.0) = 66.3$.

Work hours per week	Opinion			Totals
	In favor	No opinion	Against	
none	$80^{91.1}$	$124^{116.3}$	$16^{12.6}$	220
1–9.9	$75^{74.6}$	$95^{95.1}$	$10^{10.3}$	180
10–19.9	$61^{58.0}$	$73^{74.0}$	$6^{8.0}$	140
20 or more	$74^{66.3}$	$78^{84.6}$	$8^{9.1}$	160
Totals	290	370	40	700

Inference on Categorical Data and Frequencies

Similarly, cell expectations are computed for the "no opinion" and the "against" cells. In all 12 values are needed. Then calculated

$$\chi^2 = \sum \left[\frac{(X_i - E_i)^2}{E_i} \right]$$

$$= \frac{(80 - 91.1)^2}{91.1} + \frac{(75 - 74.6)^2}{74.6} + \cdots + \frac{(8 - 9.1)^2}{9.1} \doteq 5.0.$$

Then for $\alpha = .05$, tabular $\chi^2_{.95,6} = 12.59$.

Conclusion

We cannot reject independence of opinion concerning women's lib and amount of work for these women. That is, statistically equal percentages "favor" the concept, equal (but likely different) percentages have "no opinion" of this issue, and the remaining percentage are "against" women's lib. The three percentages are *estimated* by $(290/700) \cdot 100 \doteq .414 \cdot (100) = 41.4\%$ who favor, $(370/700) \cdot (100) \doteq 52.9\%$ who have no opinion, and 5.7% who are against women's lib.

In the example, I indicated calculation of all 12 cell expected values. This is not necessary because the "observed" and "expected" totals should be equal for any row or for any column. Then a minimum of $[(r-1) \cdot (c-1)]$ cell expectations must be computed. Those remaining can be obtained by subtraction from a row or column total. This is a basis for the degrees of freedom in the test.

Rule

For a test of independence using an r (rows) by c (columns) contingency table, the number of degrees of freedom for the chi-square test is $[(r-1)(c-1)]$.

The rule is in agreement with the calculations for the preceding example. This rule can be used for the problems that follow, especially for Review Problems 7 and 11.

Problems

1. Use the two-way contingency table to answer these questions:
    a. What is the expected number of children that are adolescent and strong (assuming independence)?

b. What is the number of degrees of freedom for testing the null hypothesis, H_0: age and resistance are independent?

c. Suppose that the value for the usual statistic that is computed for this type of problem is 5.56. Do you reject the hypothesis given in Part b at the 5% level of significance? Why or why not?

Resistance to a disease	Infant 0–5	Early School 6–11	Adolescent 12–18
Weak	14	16	10
Strong	16	14	30

2. For the sample data given, test for independence of sex and car style preference. Use $\alpha = .05$ and state your conclusion.

Sex	Four-door	Car Style Hardtop	Station Wagon
Male	30	10	20
Female	20	10	10

3. The following data indicate the number of admissions to the pediatric section of a hospital for a randomly chosen week. Test the claim that admissions occurred uniformly throughout the week. That is, test $H_0: p_{Su} = p_M = \cdots = p_{Sa} = 1/7$. Use $\alpha = .025$.

Day	Su	M	T	W	Th	F	Sa	Total
Admissions	27	16	15	14	15	25	28	140

4. One-hundred boys and one-hundred girls were interviewed regarding their interest in a certain Saturday cartoon program. Sixty-five of the boys watch the program regularly. Eighty of the girls are regular viewers of this show. Is there a significant difference between the boys and girls in their viewing habits of this program, $\alpha = .05$?

5. A random sample of 350 women is selected and classified according to their drinking and smoking habits. Are drinking and smoking habits independent for this group? Use $\alpha = .01$, and calculated $\chi^2 = 57.25$.

Inference on Categorical Data and Frequencies

Drinking Habit	Smoking Habit		
	Light	Moderate	Heavy
None	51	38	37	14
Social	28	42	11	10
Heavy	21	40	12	46

6. A market research firm questions whether the inclusion of a small prize in a questionnaire would affect the number of responses to their survey. In a preliminary study 300 questionnaires, half with a prize and half without, were sent to a random selection of persons. This produced the values shown below. Is there sufficient evidence to conclude that response to the questionnaire depends upon the inclusion of a prize? Use $\alpha = .05$, and state the null hypothesis in terms of proportions.

Prize	Response	No Response	Total
Included	97	53	150
Not Included	80	70	150
Total	177	123	300

4/ Summary for Chapter 13

Tests on proportions for numerous-count data experiments, including the binomial, can be described through the multinomial distribution. The latter assumes a fixed number of classes and a fixed, but usually unequal, proportion of the population in each class. The random variable describes the number of (random) sample observations that fall into the separate classes. The appropriate statistic is a chi-square approximation.

This chi-square distribution is skewed with the amount of distortion depending on the number of cells. The chi-square form does not, however, distinguish the direction of differences. Subsequently it is used in testing with a does-not-equal alternative hypothesis. The exact binomial distribution and the Z-statistic can provide alternative test procedures.

The chi-square procedure can be used in tests for equality of proportions for two or more distributions. Also a two-way extension of the binomial classes is a basis for (multinomial) contingency tables. In a contingency table, the product [(number of rows − 1) · (number of columns − 1)] determines the degrees of freedom for tests of independence.

This work contains numerous first principles of statistics and by now, I'm sure we all recognize the need for the logical treatment of data. As often as possible, checks should be made concerning the "reasonableness" of procedures and results; intuition is often a valid indicator, as well as the more formal checks. Reasonable mastery of this work should provide you with an adequate background for a more comprehensive treatment of the statistical concepts frequently used in your discipline.

REFERENCES

[1] Allen L. Edwards, *Statistical Methods,* third edition (New York: Holt, Rinehart, and Winston, Inc., 1973).

[2] J. A. Ingram, *Introductory Statistics* (Menlo Park, California: Cummings, 1974).

[3] B. W. Lindgren, *Statistical Theory,* second edition (New York: Macmillan, 1968).

[4] Joseph A. Steger, editor, *Readings in Statistics for the Behavoral Scientist* (New York: Holt, Rinehart, and Winston, Inc., 1971).

Review Problems

1. Answer the following by using the chi-square distribution Table III:
   a. $P(\chi_5^2 < c) = .05$, find c.
   b. $P(\chi_{13}^2 > 15.984) = \alpha$, find α.
   c. If the chi-square test has 23 df and we want $\alpha = .01$, what calculated value is required to reject equality of proportions?

2. A binomial experiment of 400 independent trials showed 250 successes. Does this evidence other than $p = .60$? Use $\alpha = .05$.

3. A home center carries two major appliances—ranges and refrigerators. A one-month record of sales shows the following number of units of the major appliances sold by two salespeople. Does this record support the claim that each salesperson's ability to sell depends on the product he or she is selling? State recommendations at $\alpha = .025$.

Inference on Categorical Data and Frequencies

Salesperson	Ranges	Refrigerators
Ann	26	8
Steve	14	18

4. Suppose you are a purchasing agent and have received parts from two suppliers, A and B. Assuming that yours is an adequate random sample, which company would you choose to continue as sole supplier of the parts? Use $\alpha = .05$ and test $H_0: p_A = p_B$ where $p_A = $ percent defective received from Company A, etc. Assume that all other relevant factors such as cost, delivery, and so forth are equal. Begin with a check for short cells.

Company	Quantity Received	
	Defective	Nondefective
A	18	182
B	12	288

5. Let heavy drug users be classified as potheads (heavy marijuana smokers) or other. Test the claim of equal percentage of potheads among heavy drug users in these three cities. Use $\alpha = .05$ and check for short cells.

City	Potheads	Other
New York (NY)	12	4
Los Angeles (LA)	29	11
Miami (M)	15	9

6. The number of national conventions held in a large city was recorded over many months. Test the claim that the numbers have been evenly distributed over the week. Use $\alpha = .05$. Notice that df $= 7 - 1 = 6$.

Day	Sun	Mon	Tue	Wed	Thu	Fri	Sat	Total
Number of Meetings	27	16	15	15	14	25	28	140

7. A safety engineer for a manufacturer feels that there is a dependence between the hour and the day during which accidents occur. Do you agree? Use $\alpha = .05$ and the following record of the number and day of accidents. Note 15 accidents were observed and 9.6 accidents were expected for Monday from 9 to 10, etc. State your recommendations.

	Hour			
Day	9–10	11–12	2–3	4–5
Mon	$15^{9.6}$	$6^{9.0}$	$12^{10.3}$	$9^{12.1}$
Wed	$8^{7.8}$	$4^{7.3}$	$10^{8.3}$	$12^{10.6}$
Fri	$7^{12.6}$	$18^{11.7}$	$10^{13.4}$	$20^{17.3}$

8. Two youth organizations called Blue-Y and Red-Y compete with one another and so have exclusive memberships. If in one random sample, 82 youth belong to Blue-Y while 128 others belong to Red-Y, test $H_0: p = 1/2$ for $p =$ percentage of youth in Red- or Blue-Y that belong to Blue-Y. Use a less-than alternative hypothesis with $\alpha = .05$.

9. A manufacturer of cat foods has designed a new package for their product and has run the following test: Two stores were selected, one in a higher-income area and the other in a lower-income neighborhood. A display of cat food was arranged, consisting of packages of the old design and of the new. At the end of the test, 350 packages had been sold in Store A. Of these, 298 were the new packages. In Store B, 150 packages were sold of which 102 were the new packages. Is there a significant difference between the proportion of new packages sold in the two stores? Use $\alpha = .05$.

10. The following is a two-way classification by sex and marital status for the persons at a party. Test the hypothesis that 50% of the people in this group are now married. Use $\alpha = .05$. (*Suggestion:* use subclasses married and not married for marital status.)

Sex	Single	Married	Divorced
Male	11	22	7
Female	5	21	14

Inference on Categorical Data and Frequencies

11. A national retailing firm is studying the advertising media which most influence their customers. An analysis is made of findings from interviews with 180 randomly selected customers. Test for independence of media and income bracket. Use $\alpha = .01$. Would you make any further tests? As a preliminary procedure, check for short cells.

		Medium		
Income	TV	Local news	Magazines	Personal habit
Over $15,000	$10^{8.1}$	$17^{13.9}$	$16^{13.4}$	7
$8,000 to $15,000	$7^{12.9}$	$13^{22.2}$	$24^{21.3}$	36
Under $8,000	12	20	8	10

12. Mrs. Bothem has two sections of political science. Standardized practice tests were given to Section A, but not to Section B; otherwise the two groups were taught the same. At the end of the semester, 110 of 120 students in Section A passed the course and 8 of 80 students in Section B failed the course. State and make an appropriate test of independence at the 0.05 significance level.

13. The number of arrivals were recorded at a drive-in apothecary. Test the hypothesis that 40% of the arrivals are during 12–2 P.M. Use $\alpha = .01$.

Hours	10–12 A.M.	12–2 P.M.	2–4 P.M.
Arrivals	96	118	86

14. Test whether the proportion of sales was dependent upon a seasonal factor. Units are cases of liquor. Let $\alpha = .01$.

	Month of Purchase		
Number of cases	Nov	Dec	Jan
Retail	33	32	20
Wholesale	106	128	80

15. A certain drug is claimed to be effective in curing the common cold. In an experiment on 106 people with colds, half of them were given the drug and the others were given sugar pills. The patients were all told they were given a "drug" that hopefully would cure their colds. If 50 of those given the drug were helped while 44 of those given the sugar felt they were helped, make a test for independence of effect (helped–not helped) and treatment. Use $\alpha = .05$. State practical conclusions.

Appendix A/ Math Essentials

Many people who need statistics cannot take the time to get an extensive background in math, so this appendix offers some of the mathematics essentials necessary for our work. This is a compilation of the math tools that the author has found necessary in teaching statistics. The concepts are *not* restricted to any one math area, and other useful sources for increasing math background might include high school algebra texts or any of the several chapter references.

The concepts presented here, if understood and practiced, will allow you to apply the math necessary in this book in a comfortable fashion. A word of caution, math and statistics require a different kind of thinking than does a reading course such as history, basic education, or economics. A key point in learning a quantitative subject is to be certain that you understand each new concept *as it is presented*. This usually requires *working* enough problems to assure yourself that you understand the concepts and can apply them.

The material is divided into three sections: (1) math skills, (2) algebraic manipulations, and (3) statistical concepts. Hopefully you will encounter a number of concepts that you have learned before. Check on your understanding by working the problems in the problems section. Answers are provided for all of these problems.

1/ Math Skills

Math Symbols

A list of common math symbols along with their meanings appear in Table 1.

Table 1/ Common Math Symbols with Meanings

Symbol	Verbal Equivalent	Example
$\cdot$	is multiplied by	$3 \cdot 2$
$=$	is equal to	$3 = 1 + 2$
$\neq$	does not equal	$3 \neq 2$
$\doteq$	is approximately equal to	$2.98 \doteq 3.0$
$<$	is less than	$2 < 3$
$>$	is greater than	$3 > 2$
$\leq (\geq)$	is less (greater) than or equal to	$a \leq b, (b \geq a)$
$\|a\|$	the absolute (positive) value	$\|3\| = 3, \|-3\| = 3$
$\sqrt{b}$	the square root of $b (> 0$, i.e., requires that b is nonnegative	$\sqrt{4} = 2$
c^2	the square of c	$2^2 = 4$
$\ldots$	continuing this pattern	$1, 2, \ldots, 5$ implies $1, 2, 3, 4, 5$
$\approx$	is similar to	$a \approx b$
$\dfrac{a}{b}$ or a/b	a divided by b, requires $b \neq 0$	$6/2 = 3$

The symbols in Table 1 are used throughout the book and should become familiar with continued use.

Calculations with Zero and One

The following rules can serve as guides:

Rules	Examples
1. The product of a number and zero equals zero	$6 \cdot 0 = 0, -100 \cdot 0 = 0$
2. Zero raised to a positive power has the value zero	$0 \cdot 0 = 0^2 = 0^{100} = 0$
3. Division of zero by a nonzero number yields the quotient zero	$0/10 = 0/2 = 0$
4. Division by zero has no meaning; the result is undefined	$3/0$ is undefined
5. The value of a nonzero number raised to power zero is one	$-1^0 = 10^0 = 1$
6. The product of a number and one equals the number	$1 \cdot 3 = 3, 1 \cdot 0 = 0$
7. One raised to any power has the value one	$1^0 = 1^1 = 1^{-10} = 1$
8. A number divided by one has as its quotient the given number	$0/1 = 0, (1/2)/1 = 1/2$

Rule 4 has a unique consequence. In mathematics we can generally check a result by reversing the process. For example, the division $6/2 = 3$ is checked by multiplication, that is, by $3 \cdot 2 = 6$. But for $3/0 =$ "some number" there is no "number" that will, when multiplied by zero, give 3; i.e., by Rule 1 0 times some number does not equal 3 for any number. So we say division by 0 has no meaning and the quotient is undefined.

Setting Decimals

This includes considerations for multiplication, for powers, and for division.

> **Rule**
> In multiplying decimal numbers, multiply as though with whole numbers (integers), ignoring the decimal point. To set the decimal place in the answer count off as many places beginning from the right as there are decimal places in the numbers being multiplied.
>
> **Examples**
> $300 \cdot .03 = 9.00$ (2 decimals)
> $.003 \cdot .0003 = .0000009$ (7 decimals).

Powers for numbers can be considered a special case of multiplication, and the same rule applies.

Math Essentials

$(1.1)^2 = 1.1 \cdot 1.1 = 1.21$ (2 decimals)
$(.03)^3 = .03 \cdot .03 \cdot .03 = .000027$ (6 decimals)
[Or $(.03)^3 = (3/100)^3 = 3^3/100^3 = 27/1{,}000{,}000 = .000027$].

Rule

In division by decimals, first move the decimal point of the divisor to the right a sufficient number of places to give a whole number. Then move the decimal point in the numerator the same number of decimal positions in the same direction. Divide as with whole numbers.

Examples

$$\frac{3.4000}{.0034} = \frac{34000}{34.} = 1{,}000 \quad \text{(moved four decimal places)}$$

$$\frac{.540}{1.2} = \frac{.540 \times 10}{1.2 \times 10} = 5.40/12 = .45 \quad \text{(moved one decimal place).}$$

The addition or subtraction of decimal numbers is performed by placing the numbers in a column with the decimal points aligned, then proceed as usual with addition or subtraction.

Operations with Common (Ratio) Fractions

Common fractions are ratios of whole numbers (integers). These ratios, or so-called rational numbers, can contain three signs: (1) one for the numerator, (2) one for the denominator, and (3) one for the entire fraction.

	Signed Fractions		Realized Fractions	Example		
1.	$+\left(\dfrac{+a}{+b}\right) =$	$+\left(\dfrac{a}{b}\right) =$	$\dfrac{a}{b}$	$+\left(\dfrac{+2}{+3}\right) = \dfrac{2}{3}$		
2.	$-\left(\dfrac{+a}{+b}\right) =$	$-\left(\dfrac{a}{b}\right) =$	$-\dfrac{a}{b}$	$-\left(\dfrac{+2}{+3}\right) = -\dfrac{2}{3}$		
3.	$+\left(\dfrac{+a}{-b}\right) =$	$+\left(\dfrac{-a}{+b}\right) =$	$-\left(\dfrac{+a}{+b}\right) =$	$-\dfrac{a}{b}$	$+\left(\dfrac{+2}{-3}\right) =$	$+\left(\dfrac{-2}{+3}\right) = -\dfrac{2}{3}$
4.	$+\left(\dfrac{-a}{-b}\right) =$	$+\left(\dfrac{+a}{+b}\right) =$		$\dfrac{a}{b}$	$+\left(\dfrac{-2}{-3}\right) =$	$+\left(\dfrac{+2}{+3}\right) = \dfrac{2}{3}$
5.	$-\left(\dfrac{-a}{-b}\right) =$	$-\left(\dfrac{+a}{+b}\right) =$		$-\dfrac{a}{b}$	$-\left(\dfrac{-2}{-3}\right) =$	$-\left(\dfrac{+2}{+3}\right) = -\dfrac{2}{3}$

A fraction is *realized* when the first two signs are both made positive. The sign of the entire fraction is then determined, either positive or negative. The most common cases are illustrated for the positive integers a and b. Recall that denominators cannot be zero, $b \neq 0$.

> If a signed fraction contains an odd number of negative signs, its "realized" form is negative. Otherwise the fraction is positive.

Rule

Computations involving common fractions should use the "realized" form.

> Multiplying the numerator and denominator of a common fraction by the same number (excluding zero) does not change the value of the fraction since it is equivalent to multiplying the fraction by 1.

Rule

$$\frac{-2}{-3} = \frac{-2}{-3} \cdot \frac{-1}{-1} = \frac{2}{3}; \quad \frac{.034}{3.4} = \frac{.034}{3.4} \cdot \frac{1000}{1000} = \frac{34}{3400}.$$

Examples

This rule can also be used for adding or subtracting common fractions.

$$\frac{1}{2} - \frac{1}{3} = \left(\frac{1}{2} \cdot \frac{3}{3}\right) - \left(\frac{1}{3} \cdot \frac{2}{2}\right) = \frac{3}{6} - \frac{2}{6} = \frac{1}{6}.$$

Examples

The product $3 \cdot 2 = 6$ is called the *least common denominator*.

> The least common denominator is found by taking the product of all primes, each raised to its *highest* power, found in any of the denominators. (Primes are the nonfactorable whole numbers 1, 2, 3, 5, 7, 11, 13, ...)

Rule

$$\frac{1}{4} + \frac{1}{10} = \frac{1}{2 \cdot 2} + \frac{1}{5 \cdot 2} = \frac{1}{2 \cdot 2} \cdot \frac{5}{5} + \frac{1}{5 \cdot 2} \cdot \frac{2}{2} = \frac{5+2}{2 \cdot 2 \cdot 5} = \frac{7}{20}.$$

Example

Here the least common denominator is $2^2 \cdot 5 = 20$.

For the addition or subtraction of common fractions, all terms are first changed to have the same denominator. The least common denominator is the preferred form.

> The product of two or more common fractions is a fraction with a numerator that is the product of all the numerators and with a denominator the product of all the denominators.

Whole numbers are treated as fractions having a denominator 1.

Math Essentials

Examples

$$3 \cdot \frac{1}{4} \cdot \frac{5}{7} = \frac{3 \cdot 1 \cdot 5}{1 \cdot 4 \cdot 7} = \frac{15}{28}$$

$$\frac{2}{3} \cdot \frac{3}{4} = \frac{2 \cdot 3}{3 \cdot 4} = \frac{6}{12} = \frac{1}{2} \cdot \frac{\cancel{6}^1}{\cancel{6}_1} = \frac{1}{2}.$$

"Cancelling" is using the first multiplication rule in reverse. This can simplify the arithmetic in multiplying common fractions (such as above).

Examples

$$\frac{25}{81} \cdot \frac{27}{5} = \frac{5 \cdot 5}{3 \cdot 3 \cdot 3 \cdot 3} \cdot \frac{3 \cdot 3 \cdot 3}{5} = \frac{\cancel{5} \cdot 5}{\cancel{3^2} \cdot 3} \cdot \frac{\cancel{3^2}}{\cancel{5}} = \frac{5}{3} \cdot \frac{1}{1} = \frac{5}{3}$$

$$\frac{14}{12} \cdot \frac{27}{49} = \frac{2 \cdot 7}{2 \cdot 2 \cdot 3} \cdot \frac{3 \cdot 3 \cdot 3}{7 \cdot 7} = \frac{\cancel{2} \cdot \cancel{7} \cdot \cancel{3} \cdot 3 \cdot 3}{\cancel{2} \cdot 2 \cdot \cancel{3} \cdot \cancel{7} \cdot 7} = \frac{3^2}{2 \cdot 7} = \frac{9}{14}.$$

Rule

To divide a number by a ratio fraction, invert the divisor and multiply.

Examples

$$\frac{3}{\frac{1}{2}} = 3 \cdot \frac{2}{1} = \frac{3}{1} \cdot \frac{2}{1} = 6$$

$$\frac{\frac{1}{3}}{-\frac{3}{2}} = \frac{1}{3}\left(-\frac{2}{3}\right) = -\frac{2}{9}.$$

The statement that two common fractions are equal is also a proportion. Thus $a/b = c/d$ or $a:b = c:d$ is read "a is to b as c is to d."

Examples

$$\frac{1}{2} = \frac{2}{4}$$ is read "as 1 is to 2, so 2 is to 4"

$$-\frac{1}{3} = -\frac{3}{9} \text{ or } \frac{-1}{3} = \frac{-3}{9}$$ is read "as -1 is to 3, so -3 is to 9."

The positions held by a and d in $a/b = c/d$ are called the *extremes;* those of b and c the *means* (a name, not an average).

Rule

In a proportion, the product of the means equals the product of the extremes; that is, $a/b = c/d$ if and only if $b \cdot c = a \cdot d$.

Math Skills

$-\dfrac{1}{3} = -\dfrac{3}{9}$ or $\dfrac{-1}{3} = \dfrac{-3}{9}$ gives $3 \cdot (-3) = (-1) \cdot 9$ Examples

$\dfrac{X}{4} = \dfrac{1}{2}$ or $2(X) = 4 \cdot 1$ so that $X = 2$. Check: $\dfrac{2}{4} = \dfrac{1}{2}$.

Decimal and Common Fraction Conversion; Percentages

These are used for answers expressed as parts of one hundred. Conversion to percentages is best made from decimal form.

> To convert a decimal fraction to a percent, move the decimal point two places to the right and attach a percentage sign.   Rule

> .99 is 99 parts of 100 or 99%
> .002 becomes 0.2%; that is, .002 · 100 = 0.2%.   Examples

Expressing a decimal in percentage form is equivalent to multiplying the decimal number by 100 and attaching the percent (parts of 100) sign.

> A percent is changed to a decimal fraction by dividing by 100 (i.e., moving the decimal point two positions to the left) and dropping the percentage sign. This is a reversal of the decimal to percent procedure.   Rule

> 36% becomes 36/100 = .36
> 0.2% becomes 0.2/100 = .002.   Examples

A common fraction is changed to a percent by first changing to a decimal fraction, then to a percent.

> 3/4 (by dividing 3 by 4) = .75,   .75 · 100 = 75%
> $-1/8 = -.125$, or -12.5%
> $1/12 = .08\dot{3}$, or $8.\dot{3}\%$. The dot over the 3 in $.08\dot{3}$ indicates a repeating decimal, .0833333 . . . , then 8.33 . . . %.   Examples

353

Math Essentials

Exponents and Roots

In the form a^k, a is the base number and k is an exponent (or power). The basic rules regarding exponents are given below:

Rules	Examples
1. By definition $a^0 = 1$, $a^1 = a$, and $a^{-1} = \dfrac{1}{a}$ for any $a \neq 0$	$10^0 = 1;\ 10^1 = 10;\ 10^{-1} = \dfrac{1}{10}$
2. The addition or subtraction of numbers with exponents requires that the separate terms be evaluated first	$3^2 - 2^2 = (3 \cdot 3) - (2 \cdot 2) = 9 - 4 = 5$
3. The product of exponential numbers with the same base has that common base raised to the sum of the powers	$2^1 \cdot 2^2 \cdot 2^3 = 2^{1+2+3} = 2^6 = 64$
4. The quotient of exponential numbers with the same base is the common base raised to the difference (numerator − denominator) of the powers	$\dfrac{2^3}{2^2} = 2^{3-2} = 2$ $\dfrac{2^2}{2^3} = 2^{2-3} = 2^{-1} = \dfrac{1}{2}$
5. The product of powers having a different base requires that the separate terms first be evaluated and then the product taken	$2^3 \cdot 3^2 = 8 \cdot 9 = 72$ $2^{-1} \cdot 3 = \dfrac{1}{2} \cdot \dfrac{3}{1} = \dfrac{3}{2}$
6. Division of powers with different bases also requires that the separate terms be evaluated first	$\dfrac{3^2}{2^3} = \dfrac{9}{8}$ or 1.125

The expression $a^k = b$, for k a positive and for b a nonnegative number, is equivalent to $a = b^{\frac{1}{k}} = \sqrt[k]{b}$, or the kth root of b. The evaluation of roots, except for perfect roots, requires special methods. Our concern is with square roots, i.e., when $k = 2$.

Examples

Since $3^2 = 9$, then $3 = 9^{\frac{1}{2}}$ or $3 = \sqrt{9}$
$2^3 = 8$, then $2 = 8^{\frac{1}{3}}$ or $2 = \sqrt[3]{8}$.

(*Note:* For square roots the symbol $\sqrt{\ }$ is commonly used in place of $\sqrt[2]{\ }$.)

For ease of evaluation one can rationalize an expression to remove a radical from the denominator. This is done by multiplying both numerator and denominator by the same radical. The result generally is more easily evaluated than the original form.

$$\frac{3}{\sqrt{2}} = \frac{3}{\sqrt{2}} \cdot \frac{\sqrt{2}}{\sqrt{2}} = \frac{3}{2^{\frac{1}{2}}} \cdot \frac{\sqrt{2}}{2^{\frac{1}{2}}} = \frac{3\sqrt{2}}{2}$$

$$\frac{\sqrt{3}}{\sqrt{2}} = \frac{\sqrt{3}}{\sqrt{2}} \cdot \frac{\sqrt{2}}{\sqrt{2}} = \frac{\sqrt{3} \cdot \sqrt{2}}{2} = \frac{\sqrt{3 \cdot 2}}{2} = \frac{\sqrt{6}}{2}.$$

Examples

The next example displays an important arithmetic property for roots:

$$\sqrt{3} \cdot \sqrt{2} = \sqrt{3 \cdot 2} = \sqrt{6}.$$

The product of the square roots for two positive numbers equals the square root of their product.

Rule

Inequalities

These are expressed $a < b$; that is, "the value of a is less than that of b." Equivalently, $b > a$; "the value of b is greater than that of a." An inequality opens to the greater value.

The addition or subtraction of "equals" (equal amounts) to both sides of an inequality does not change the direction of the inequality.

Rule

$-2 < -1$ means the same as $-1 > -2$, and
$-2 + 2 < -1 + 2$, or $0 < 1$
$-2 - 2 < -1 - 2$, or $-4 < -3$.

Examples

When both sides of an inequality are multiplied or divided by the same *positive* number, the resulting inequality has the same direction as the original.

Rule

For $3 > 2$, $\quad 3 \cdot 5 > 2 \cdot 5$, $\quad$ or $\quad 15 > 10$, $\quad$ and
$3 \cdot \dfrac{1}{5} > 2 \cdot \dfrac{1}{5}$, $\quad$ or $\quad .6 > .4$.

Examples

If both sides of an inequality are multiplied or divided by the same *negative* number, the inequality is reversed.

Rule

For $2 < 3$, $\quad 2 \cdot (-2) > 3(-2)$ as $\quad -4 > -6$, and
$2 \div (-2) > 3 \div (-2)$ as $\quad -1 > -\dfrac{3}{2}.$

Examples

Math Essentials

Rule
If a and b are positive numbers with $a > b$, then $\dfrac{1}{a} < \dfrac{1}{b}$.

Example
$3 > 2$, so $1/3 < 1/2$ as $.33 < .50$.

The next rule is central in the study of statistics because we frequently multiply common fractions.

Rule
If two or more *positive* numbers each of a size less than 1 are multiplied, their product has a value less than that of any of the numbers.

Examples
$$\frac{1}{2} \cdot \frac{1}{3} = \frac{1}{6} \quad \left(\frac{1}{6} < \frac{1}{2}\right), \quad \text{and} \quad \left(\frac{1}{6} < \frac{1}{3}\right)$$

$$\frac{1}{2} \cdot \frac{1}{3} \cdot \frac{1}{4} = \frac{1}{2 \cdot 3 \cdot 4} = \frac{1}{24} \quad \left(\frac{1}{24} < \frac{1}{2}\right), \quad \left(\frac{1}{24} < \frac{1}{3}\right), \quad \text{and} \quad \left(\frac{1}{24} < \frac{1}{4}\right).$$

This rule can be used as one way to check for the reasonableness of probability answers (beginning with Chapter 4).

Rules for Rounding

Rules for rounding are required for consistency, especially in checking your work against the answer section. In general the number 3, 3.0, 3.00, 3.000, etc., will be considered accurate to an unspecified number of decimal positions. This is also the case for 200, 4, 9, 976, or -543, etc.; that is, for whole numbers. Otherwise decimal accuracy will be indicated.

Example
$1/3 = .333 \ldots$ will most commonly be rounded to two-decimal accuracy: $1/3 \doteq .33$, or $1/3$ is approximated by $.33$.

Generally, and unless otherwise indicated by the accuracy of given numbers, the author has adhered to two-decimal accuracy in calculating answers for problems and examples. Some problems allow slight variations in procedure and subsequently may produce slightly different answers.

Example
$$\frac{2}{1/3} = 2 \cdot \frac{3}{1} = 6 \text{ whereas}$$

$$\frac{2}{.33} = \frac{2.00}{.33} = \frac{200}{33} = 6.06$$

The more accurate answer is, of course, 6.

You should not be alarmed if frequently your answers are slighly different from those in the answer section. You might, however, check that you have used an accurate procedure. For example, the above illustrates that operations with ratio fractions can give greater accuracy than those with decimal approximations.

In the case of statistical tables (Appendix B) the decimal accuracy may vary from two to four decimal positions; for example, see page 385. Our tables are standard ones so this is a universal problem. Two rules are given for the most common uses of tabled values.

Rules

The accuracy used for tabled decimal numbers should be:

1. at most that given in the table if the number is to be used solely for comparison and will not be used in later calculations.
2. directed by the rules for decimal accuracy that follow if the numbers are to be used in later calculations.

The use of tabled values "solely for comparison" occurs in Chapter 9 and on in the form of "test values."

Several rules are given for determining decimal accuracy for answers using arithmetic operations.

Rule

For addition or subtraction the answer carries the same accuracy as the least accurate of the combined numbers.

Examples

Addition: $3.21 + 6.9 + 4 = 3.21 + 6.9 + 4.00 = 14.1\cancel{1}$
Subtraction: $6.98 - 7.2 = -0.2\cancel{2}$

Usually the numbers we treat will have the same decimal accuracy, then the common level dictates proper accuracy. In fact, all numbers in a sample should be recorded with the same accuracy. For example, grade point averages are often recorded with two decimal places, such as 2.74, 3.19, 3.50, etc.

Rule 1

For multiplication the answer has the same accuracy as the least accurate of the combined numbers so long as the product exceeds $+1$ or is more negative than -1.

Examples

$2361 \cdot 2.1 = 4958.1$
$.36 \cdot 3.6 = 1.\cancel{296} = 1.3$

Multiplication requires a special rule (Rule 2) to allow identification of products with a value near zero.

Math Essentials

Rule 2
For two or more numbers whose product is less than 1, but greater than -1, use one more decimal place in the product than appears in the most accurate of the original numbers.

Examples
$.2 \cdot .3 = .06 \qquad (.1)^2 = .1 \cdot .1 = .01$
$.23 \cdot .01 = .002$

Note: These last two products would be 0 (zero) by Rule 1. This is fine unless we need to divide by the product, which is a common occurrence.

The accuracy in a quotient (division) uses the same rule as for addition and subtraction. The accuracy is that of the least accurate of the numbers involved in the division.

Examples
$2 \div .2 = 2.00 \div .2 = 10.0$
1
$44.5 \div 2.11 = 21.09$

If you doubt a decimal answer, a safe approach is to check your work by using an opposite procedure; i.e., for the last example, $2.11(21.1) = 44.52$. Opposite procedures include addition with subtraction, multiplication with division, and squaring with square roots.

Common practice in rounding decimal numbers is if over halfway, round up whereas if less than halfway, round down. Then, for example, 3.64 rounds to 3.6, while 3.66 rounds to 3.7 for accuracy to the nearest tenth. A troublesome case arises when the number is exactly one-half of the way between two values.

Rule
ENGINEER'S RULE/ When the last digits in a decimal number are 5000 . . . , round to the nearest *even* digit at the desired level of accuracy.

For example, *30.25000* . . . , *600.500* . . . , and *55.99500* . . . are rounded to 30.2, 600 and 56.00, respectively. This procedure eliminates potential rounding biases as it sometimes leads to rounding up and at other times to rounding down. The rule applies *only* when the last digits are 5000 . . . . For example, *51.4999* . . . is rounded to 51, and 51.50500 . . . is rounded to 52 using conventional rounding methods.

Use of Squares and Square Root Tables

The use of Table VIII in Appendix B will be required early in our work. Squares are given for numbers, *n*, between 1 and 1,000 inclusive. For decimal numbers, the proper decimal position for squares can be found using Rule 1, page 357.

Examples

$$n = 349 \qquad n^2 = 121{,}801$$
$$n = 34.9 \qquad n^2 = 1218.01$$
$$n = .349 \qquad n^2 = .121801$$

Numerous methods exist for finding square roots. These include (1) longhand extraction, (2) using a calculator, (3) using a common logarithm method, or (4) obtaining the square root from a prepared table. The latter method is described here. It has, however, the definite disadvantage of limited accuracy.

Suppose we seek the square root for $n = 120$, i.e., $\sqrt{120}$. I begin with the approximations $10^2 = 100$ and $11^2 = 121$. Using the idea of an opposite process, my n is between 100 and 121 and so has come from squaring a number between 10 and 11. This unknown number, $\sqrt{n}$ (the square root of n) is over 10 and close to, but less than 11, i.e., $100 < n < 121$ so $10 < \sqrt{n} < 11$. A first approximation to the answer is 10; that is, I begin with a value that is less than, but near, the actual value. Having a first approximation and having set the decimal point, it remains to go to Table VIII on p. 398 to find the remaining digits. I find

n	n^2	$\sqrt{n}$	$\sqrt{10n}$
120	14400	10.95445	34.64102

The answer is 10.95445. The key to success in finding square roots is practice. Several examples are given for this purpose. Check your understanding of the process. Use Table VIII.

Example

n	First Approximation	$\sqrt{n}$
10	3.	3.162278
.011	.1	.1048809
1010.	30.	31.78050

Notice the last example. The first approximation need not be extremely close for locating the proper tabled value.

Square roots for decimal numbers between 0 and 1 can be tricky. For these I begin with an approximation. For example, .26 is just bigger than .25 and I know that $.25 = .5^2$ is a perfect square, hence $\sqrt{.25} = .5$, so $\sqrt{.26}$ is slightly larger than .5. Using Table VIII, $\sqrt{.26} = .5099$, or .51.

Math Essentials

Rule

To find the square root for a (positive) decimal number, (1) multiply the radical by $100 = 10^2$, or $10,000 = 10^4$, or etc., to obtain a number with a size between 1 and 1,000, (2) in Table VIII find the square root for the changed number, then (3) move the decimal one place to the left if your original multiplier was $100 = 10^2$, or two places to the left if $10,000 = 10^4$, etc. The result is the square root for the decimal number.

Several examples should clarify this rule:

Examples

n	(1) Whole Numbers	(2) Using Table VIII	(3) Decimal Correction	Answer $\sqrt{n}$
.0026	$\sqrt{.0026 \cdot 10,000} = \sqrt{26.}$	5.099	.05,099	.051
.026	$\sqrt{.026 \cdot 10,000} = \sqrt{260.}$	16.12	.16,12	.16
.26	$\sqrt{.26 \cdot 100} = \sqrt{26.}$	5.099	.5.099	.51
.25	$\sqrt{.25 \cdot 100} = \sqrt{25.}$	5.	.5.	.5

Note: For example, $\sqrt{.0026} = .051$ can be checked by squaring $(.051)^2 = .002601 \,(\doteq .0026)$.

The following table gives procedures for finding the square root for numbers (n) larger than 1. You should look at Table VIII to check these results.

n	Intermediate Steps Require Special Attention				Answer $\sqrt{n}$
2.57	$\sqrt{2.57} \doteq \sqrt{257 \times 10^{-2}}$	$= \sqrt{257}\sqrt{10^{-2}} \doteq 16.03 \cdot 10^{-2/2}$	$\doteq 16.0 \cdot \tfrac{1}{10}$		1.6
25.67	$\sqrt{25.67} \doteq \sqrt{25.7}$	$= \sqrt{2570 \cdot 10^{-2}}$	$\sqrt{(10 \cdot 257) \cdot 10^{-2}} \doteq 50.7 \cdot \tfrac{1}{10}$		5.1
256.7	$\sqrt{256.7} \doteq \sqrt{257}$	=			16.03
2567.3	$\sqrt{2567.3} \doteq \sqrt{2570}$	$= \sqrt{10 \cdot 2570}$	$= 50.69$		50.7

It can be helpful in the first practice to pick radicands that are near perfect squares. For example, I know that $\sqrt{.80} \doteq .9$ since $.9^2 = .81$, and that $\sqrt{145} \doteq 12$ as $12^2 = 144$. This way I have a check for the answers that I find using Table VIII, here .89 and 12.04, respectively.

Math operations involving square roots are often mistreated. Proper use is displayed below and some simplification is made in the operations for a, b positive.

Algebraic Manipulations

Math Operation	Simplification	Example
$\sqrt{a} + \sqrt{b}$	None	$\sqrt{4} + \sqrt{9} = 2 + 3 = 5$
$\sqrt{a} - \sqrt{b}$	None	$\sqrt{4} - \sqrt{9} = 2 - 3 = -1$
$\sqrt{a} \cdot \sqrt{b}$	$\sqrt{a \cdot b}$	$\sqrt{2} \cdot \sqrt{8} = \sqrt{2 \cdot 8} = \sqrt{16} = 4$
$\dfrac{\sqrt{a}}{\sqrt{b}}$	$\sqrt{\dfrac{a}{b}}$	$\dfrac{\sqrt{8}}{\sqrt{2}} = \sqrt{\dfrac{8}{2}} = \sqrt{4} = 2$

Square roots play an integral part in statistics. Understanding their use will pay high returns in correct answers.

2/ Algebraic Manipulations

Statistics is a mathematical science. As such, it uses concepts from numerous subdisciplines of math. The skills of algebra, especially solving word problems, are central in this work. Algebra first differs from arithmetic in that literal symbols are used to name "variable" or changeable quantities. Understanding and using symbolic forms is also essential in statistics.

Summation Notation

The summation symbol is the mathematicians' shorthand way of saying "sum a list of numbers." The symbol for summation, $\sum$, is the Greek capital letter *sigma*. It is used to shorten a statement of addition that incorporates numerous terms.

$$\underbrace{3 + 3 + 3 + \cdots + 3}_{100 \text{ terms}} = \sum_{}^{100} 3 = 100 \cdot 3 = 300$$

$$\sum_{i=1}^{5} X_i = X_1 + X_2 + X_3 + X_4 + X_5.$$

Examples

In the last example the symbol i is a counter also called the *index of summation*. We assume the counter goes $i = 1, 2, \ldots, n$ (in the above $n = 5$) even if the index is absent. We sum a first value, plus a second value, etc., through a last value, X_n. We will use $\sum X$ instead of $\sum X_i$.

Math Essentials

Example

For $X_1 = 0$, $X_2 = 2$, $X_3 = -1$, $X_4 = 3$, $X_5 = 2$,

$$\sum^5 X = \sum_{i=1}^5 X_i = X_1 + X_2 + \cdots + X_5 = 0 + 2 + (-1) + 3 + 2 = 6.$$

Parentheses, ( ), brackets, [ ], and braces, { }, are used to indicate that whatever is enclosed is to be treated as a single unit. These "grouping symbols" are often used in summations.

Example

For $X_1 = 1$, $X_2 = 0$, $X_3 = 2$,

$$\sum^3 (X + 2) = (X_1 + 2) + (X_2 + 2) + (X_3 + 2)$$
$$= (1 + 2) + (0 + 2) + (2 + 2) = 3 + 2 + 4 = 9.$$

Sometimes two or more of the grouping symbols will be used in a single summation. One of the most common uses is shown in the following:

Example

For $X_1 = 0$, $X_2 = 1$, $X_3 = 2$,

$$\sum^3 [(X - 1)^2] = [(X_1 - 1)^2] + [(X_2 - 1)^2] + [(X_3 - 1)^2]$$
$$= [(0 - 1)^2] + [(1 - 1)^2] + [(2 - 1)^2]$$
$$= [(-1)^2] + [0^2] + [1^2] = 1 + 0 + 1 = 2.$$

This shows the order of operations when terms are grouped by symbols.

Rule

In math expressions where symbols are used for grouping, work from inside the grouping symbols, moving outward.

In the last example we performed subtractions, then squared, and then summed. Observe that the work moved from inside to outside of the grouping symbols.

A number of rules can simplify the process of finding sums. The following are equivalent forms. In the rules, a, b, c . . . are generalizations for constants (fixed numbers); X, Y, Z are generalizations for variables (changeable quantities). The process in the right member is equivalent to, but generally is shorter than, that on the left.

Algebraic Manipulations

Rule	Example
1. $\sum_{}^{n}(a) = \overbrace{a + a + \cdots + a}^{n} = na$	$\sum_{}^{10}(2) = 10 \cdot 2 = 20$
2. $\sum_{}^{n}(bX) = bX_1 + bX_2 + \cdots + bX_n$	$\sum_{}^{3}(2X) = 2\left[\sum_{}^{3}(X)\right]$
$\quad = b(X_1 + X_2 + \cdots + X_n) = b\sum_{}^{n}(X)$	$= 2[0 + 1 + 2] = 6{*}$
3. $\sum_{}^{n}(cX + dY) = (cX_1 + dY_2) + (cX_2 + dY_2)$	$\sum_{}^{3}(2X + 3Y) =$
$\quad + \cdots + (cX_n + dy_n)$	$2\sum_{}^{3}(X) + 3\sum_{}^{3}(Y)$
$\quad = cX_1 + cX_2 + \cdots + cX_n + dY_1$	$= 2[0 + 1 + 2]$
$\quad + dY_2 + \cdots + dY_n$	$+ 3[-1 + 1 + 2]$
$\quad = c\sum_{}^{n}(X) + d\sum_{}^{n}(Y)$	$= 2 \cdot [3] + 3 \cdot [2] = 12$
4. $\sum_{}^{n}(X + e) = (X_1 + e) + (X_2 + e)$	$\sum_{}^{3}[X + (-1)] =$
$\quad + \cdots + (X_n + e)$	$\sum_{}^{3}(X) + 3(-1)$
$\quad = (X_1 + X_2 + \cdots + X_n)$	$= (0 + 1 + 2) - 3 = 0$
$\quad + \overbrace{e + e + \cdots + e}^{n}$	
$\quad = \sum_{}^{n}(X) + ne$	

*The values $X_1 = 0$, $X_2 = 1$, $X_3 = 2$, $Y_1 = -1$, $Y_2 = 1$, $Y_3 = 2$ are used in these examples.

Rectangular Coordinates

This is one way to display points in two dimensions. Figure 1 illustrates the rectangular coordinate system. The horizontal and vertical lines are called the X-axis and Y-axis, respectively. The axes intersect at right angles; they are perpendicular. The plane of the figure is divided into four parts, called *quadrants,* which are labeled I, II, III, and IV. Any position

Math Essentials

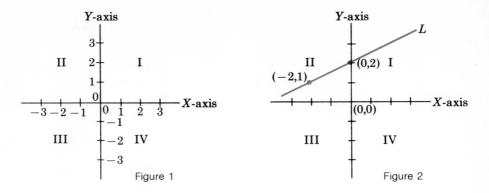

Figure 1

Figure 2

or point is identified by its *coordinates,* i.e., values projected to the coordinate axes. These are written as ordered pairs, (X, Y), with the intersection of the axes designating an *origin* or beginning position, $(0,0)$. An ordered pair lists first the *abscissa,* or X-value, and then the *ordinate,* or Y-value. Observe the position for $(0,2)$, then $(-2,1)$ in Figure 2. The point $(-2,1)$ is appropriately located in Quadrant II. This quadrant contains all points for which the X-coordinate (abscissa) is negative, but the Y-coordinate is positive. Similarly in Quadrant I, X and Y are both positive, etc.

The line segment L is a representation for many points; all of those in its path. Many other representations of numerous points, called *graphs,* can be made in two dimensions.

Linear Equations

Linear equations are relations that can be described on a rectangular coordinate system by lines. We define linear relations using rules (equations) like $Y = 2 + (1/2)X$. This specific form is displayed as the line L in Figure 2. In general, a special name—*domain*—is given to the collection of all values that can be taken by X in a relation. In Figure 2 X can take any real value so the domain is unrestricted. Similarly the values taken by Y are called the *range*. Here too the range is unlimited, but it is restricted by the relation. For $Y = 2 + (1/2)X$, the value $Y = 1$ can occur only when $X = -2$ (see Figure 2). We say the X-values are independent by choice, but that Y-values depend on the choice of a value for X. For example, the choice $X = -2$ requires $Y = 2 + 1/2(-2) = 1$.

A linear relation, generalized by $Y = a + bX$, plots as a straight line on the coordinate axes with $a = $ the point on the graph which has 0 as the value of its abscissa, that is, $(0,a)$ is the Y-intercept, and $b = $ the slope of the line. The slope is defined as change in Y divided by change in X.

Algebraic Manipulations

For any two points (X_1, Y_1) and (X_2, Y_2) with $X_1 \neq X_2$, then

$$b = \frac{Y_2 - Y_1}{X_2 - X_1}$$

is the slope for the line that includes these two points.

Rule

Lines like ╱ have a positive slope, while lines like ╲ have a negative slope.

In Figure 2 the line L contains the points $(-2, 1)$ and $(0, 2)$ and so has slope $b = (2 - 1)/[0 - (-2)] = 1/2$. Since the Y-intercept is at $Y = 2$, the relation describing the line is $Y = 2 + (1/2)X$. Observe that the slope (b) is positive, so the general direction of the line is ╱.

Example

The feature that distinguishes a line graph from other graphs is that the slope is constant between any two points in its path. The graph is "straight."

The relation $Y = -3X$ (or $Y = 0 - 3X$) describes a line, l, with a slope -3, therefore a line ╲. The Y-intercept value is zero; this line passes through the origin.

Example

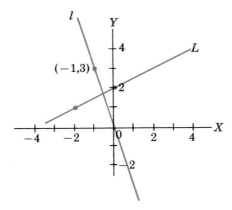

You will be asked to evaluate some linear relations, especially in Chapter 12. These will require solving for an "unknown." Several examples illustrate the correct manipulations.

1. The same number is added to (or subtracted from) each member (side) of an equation:

   $4 = 3 + X$ so $4 - 3 = (3 + X) - 3$ or $1 = X$

Examples

2. Each member (side) is multiplied by the same nonzero number:

$\frac{1}{2}X = 5$ so $2\left(\frac{1}{2}X = 5\right)$ and $\left(2 \cdot \frac{1}{2}\right)X = 2 \cdot 5$, or $X = 10$.

3. Each member is divided by the same nonzero number:

$2X = 8$ so $\frac{1}{2}(2X = 8)$ and $\frac{2}{2}X = \frac{8}{2}$, or $X = 4$

4. Each member is raised to the same power:

$X^{\frac{1}{2}} = 4$ so $(X^{\frac{1}{2}})^2 = 4^2$ and $X^{\frac{2}{2}} = X = 16$.

A final procedure allows any added or subtracted term to be moved from one side of an equality to the other side provided that its sign is changed. Zero takes the place of the lost term. This is another form of the first example shown above.

Examples

$X + 2 = 6$ so $X + 0 = 6 - 2, X = 4$
$Y + 2X - 2 = 3X$, so $Y + 0 = 3X - (2X - 2) = 3X - 2X + 2 = X + 2$

Practice by working the problems at the end of this section.

Word Problems

Word problems are the bigger part of the questions asked in this book. Word problem solving requires a different way of thinking in that you are expected to assimilate given information and from it formulate a quantitative problem. H. S. Bear [1] outlines a reasonable approach toward solving word problems.

> "Problems involving real quantities can frequently be solved by introducing symbols for the quantities and expressing the given relations in algebraic terms. A good systematic procedure to follow is given.
> 1. See what you are asked to find.
> 2. Introduce an unknown for this quantity.
> 3. Write the given facts as an equation or inequality in the unknown.
> 4. Solve the equation or inequality.
> 5. Check your solution in the stated problem."

Professor Bear has pointed out the need for an approach with some structure; all five steps are essential. I would, however, magnify step one. Try reading the problem three times. The first time read it for an overview;

Algebraic Manipulations

find out generally what the problem is about. Read the problem a second time to identify its type—costs, like price per pound and pounds; work, like times and jobs, etc. On the third reading, identify the unknown and what you are asked to find. The five steps can direct the solution.

Following are some typical algebra problems. I use these to illustrate the five steps applied to word problems. Related procedures are offered in Chapter 4, section 3 for the solution of statistics word problems.

Examples

If Kathy mailed three times as many letters at the first class rate of 10¢ as at the airmail rate of 13¢ (this is past history!), how many stamps of each type did she buy for a total of $4.30?

Solution

The five steps listed above are used. The problem concerns a total cost. Reread the problem. How is total cost determined?

1. We seek the total number of stamps she can buy for $4.30, requiring that there be three times as many 10¢ as 13¢ stamps.

2. I let $X =$ the number of 13¢ stamps. By the restrictions she wants 3 times as many 10¢ stamps as 13¢ stamps, so let $3X =$ the number of 10¢ stamps.

3. A proper relation to total cost is:

   10¢$(3X)$ + 13¢(X) = 430¢

   Note: This relation balances the cost for 10¢ stamps plus the cost for 13¢ stamps to equal the total cost.

4. The equation is $30X + 13X = 430$ then $43X = 430$ so $X = 10$, the number of 13¢ stamps, and $3X = 3(10) = 30$, the number of 10¢ stamps.

5. Checks include, for $X = 10$,
   a. $13(10) + 30(10) = 430$
      $130 + 300\ \ \ \ = 430$
   b. This gives three times as many 10¢ as 13¢ stamps.

A frequent mistake that is made is neglecting to use common units. Above we began with 30¢ and $4.30. It was necessary to use common units for a meaningful evaluation. My choice was cents.

The next section is related to the problems in Chapter 3, section 4.

Example

A family has a savings account of $5,000 at 6.25% interest. How much must they invest in a second account at 5.75% interest to accumulate a total of $500 interest for one year?

Solution

Again we seek a total:

1. We want the dollar amount which, if invested at 5.75%, will suffice to bring the *total* interest to $500.

2. Then, $X=$ the dollar amount invested at 5.75% is the unknown.

3. A total, for dollars of interest $=\sum[\text{Interest Rate} \cdot \text{Amount}]$, is $[.0625 \cdot \$5{,}000 + .0575X] = \$500$ (6.25% converted to a decimal number is .0625.)

4. So $.0575X = \$500 - (.0625 \cdot \$5{,}000) = \$500 - \$312.50 = \$187.50$
$X = \$187.50/.0575 = \$3{,}260.87$.

5. A check requires that $.0625 \cdot \$5{,}000 + .0575 \cdot \$3{,}260.87 = \$500.00$; $\$312.50 + 187.50 = \$500.00$.

Although the content of the two preceding examples is substantially different, the structure of the solutions is almost identical. You may want to review the algebraic-arithmetic operations for finding the unknowns. See p. 365.

In a third example there is a seeming lack of information. The solution takes some thought.

Two young men plan to buy a car to enter in a demolition derby and agree to split the cost evenly. After a third friend indicates his interest, they determine that their costs would be $33.3\dot{3}$ less for each of the original two. What is the price for this car?

Solution

We seek the price of the car. However, this does not enter into the calculations!

1. Let $X=$ the amount paid by each man *if two split the costs;* then *one* expression for the price of the car is $2X$.

2. The relation is a cost equality based on X. That is, total cost for two men = total cost for three men, or $2X = 3(X - \$33.3\dot{3})$.

3. Solving gives $2X = 3X - \$100$ or $X = \$100$ each (if two split the cost); the price of the car is $2X = 2(\$100) = \$200$.

4. As a check, $3(X - \$33.3\dot{3}) = 3(\$100 - \$33.3\dot{3}) = 3(\$66.67) = \$200$.

This example was appreciably more difficult than the first two because a proper relation is less obvious. All of the problems herein will provide information sufficient to determine an answer. The section material will generally help you in developing logical procedures to aid (Step 3) in formulating a relation. Problems 15–17 in the Problems section are offered to help you to develop problem solving skills.

3/ Statistical Concepts

This section introduces several topics that are founded on the mathematics, but that are central to our work. For example, the binomial theorem expansion of $(X + Y)^n$ is widely discussed in mathematics, and as a special case when $X + Y = 1$, is central in statistics.

The Binomial Theorem

This is a fundamental rule used in the mathematics. We begin with some notation, the *factorial* symbolism, used in one expression of this rule.

Rule

n-FACTORIAL, written $n!$, represents the product of the first n positive whole numbers.

Examples

$n! = n(n-1)(n-2) \ldots 3 \cdot 2 \cdot 1$ $1! = 1$
$6! = 6 \cdot 5 \cdot 4 \cdot 3 \cdot 2 \cdot 1 = 720$ $0! = 1$ (by definition)*
$4! = 4 \cdot 3 \cdot 2 \cdot 1 = 24$

This notation can be used to display the terms in the expansion of the binomial theorem. Several examples of this form, beginning with $n = 0$, follow:

$(X + Y)^0 = 1$ $\qquad 1$

$(X + Y)^1 = X + Y$ $\qquad X + Y$

$(X + Y)^2 = X^2 + \dfrac{2}{1}XY + \dfrac{2Y^2}{2 \cdot 1}$ $\qquad X^2 + 2XY + Y^2$

$(X + Y)^3 = X^3 + \dfrac{3}{1}X^2Y + \dfrac{3 \cdot 2 XY^2}{2 \cdot 1} + \dfrac{3 \cdot 2 \cdot 1 Y^3}{3 \cdot 2 \cdot 1} = X^3 + 3X^2Y + 3XY^2 + Y^3$

$\vdots$

---

*Defining $0! = 1$ leads to consistent results for calculations using factorials.

Math Essentials

$$(X + Y)^n = X^n + \frac{n}{1}X^{n-1}Y + \frac{n(n-1)}{2 \cdot 1}X^{n-2}Y^2 + \cdots$$

$$+ \frac{n(n-1)\cdots(n-r+1)}{r!}X^{n-r}Y^r + \cdots + \frac{n!}{n!}Y^n$$

The form is somewhat simplified by renaming the coefficients of the XY terms. The name *combination* identifies

$$\binom{n}{r} = \frac{n(n-1)\cdots(n-r+1)}{r!} = \frac{n!}{(n-r)!r!}.^*$$

This form indicates the count number of possibilities, assuming unspecified order, for groups of r items taken from a total of n, with $r \leq n$.

Examples

$$\binom{4}{3} = \frac{4 \cdot 3 \cdot (4 - 3 + 1)}{3 \cdot 2 \cdot 1} = \frac{4!}{(4-3)!\,3!} = \frac{4 \cdot 3 \cdot 2 \cdot 1}{1 \cdot 3 \cdot 2 \cdot 1} = 4$$

$$\binom{6}{4} = \frac{6!}{2!\,4!} = \frac{6 \cdot 5 \cdot 4 \cdot (6 - 4 + 1)}{4!} = \frac{6 \cdot 5 \cdot 4 \cdot 3}{4 \cdot 3 \cdot 2 \cdot 1} = 15$$

$$\binom{3}{3} = \frac{3 \cdot 2 \cdot 1}{3 \cdot 2 \cdot 1} = \frac{3!}{0!\,3!} = 1$$

The binomial form is central in Chapter 5. There we use $X = q$ and $Y = p$, where $q + p = 1$ so that $(q + p)^n = 1^n = 1$. This indicates that the total (probability) measure for the (statistical) binomial experiment is one. This can also be related to the expansion by the binomial theorem.

$$(q + p)^n = \binom{n}{0}p^0 q^n + \binom{n}{1}p^1 q^{n-1} + \binom{n}{2}p^2 q^{n-2} + \cdots$$

$$+ \binom{n}{r}p^r q^{n-r} + \cdots + \binom{n}{n}p^n q^0$$

The separate terms in this expansion give the point probabilities in a binomial experiment. The general term describes a binomial probability rule (see section 1 in Chapter 5). The coefficients are combinations. These appear, through $n = 5$, in Figure 3.

Pascal's triangle is extended to $n = 20$ in Table VI, Appendix B. These tables display: $1 = \binom{0}{0} = \binom{1}{0} = \binom{2}{0} = \binom{20}{0} = \cdots = \binom{1}{1} = \binom{2}{2} = \cdots = \binom{20}{20}$. Further, $\binom{1}{0} = 1, \binom{2}{1} = 2, \binom{3}{1} = 3, \ldots, \binom{n}{1} = n$. A reasonable exploration of the Pascal's triangle can help you to quickly determine numerous combination values.

---
*The form involving factorials appears in numerous math books.

Statistical Concepts

Figure 3/ Combination Counts or Pascal's Triangle

n	Binomial Expansion of $(q+p)^n$	Pascal's Triangle: Coefficients of terms
0	$(q+p)^0 = \binom{0}{0}$	1
1	$(q+p)^1 = \binom{1}{0}q + \binom{1}{1}p$	1 1
2	$(q+p)^2 = \binom{2}{0}q^2 + \binom{2}{1}qp + \binom{2}{2}p^2$	1 2 1
3	$(q+p)^3 = \binom{3}{0}q^3 + \binom{3}{1}q^2p + \binom{3}{2}qp^2 + \binom{3}{3}p^3$	1 3 3 1
4	$(q+p)^4 = \binom{4}{0}q^4 + \binom{4}{1}q^3p + \binom{4}{2}q^2p^2 + \binom{4}{3}qp^3 + \binom{4}{4}p^4$	1 4 6 4 1
5	$(q+p)^5 = \binom{5}{0}q^5 + \binom{5}{1}q^4p + \binom{5}{2}q^3p^2 + \binom{5}{3}q^2p^3 + \binom{5}{4}qp^4 + \binom{5}{5}p^5$	1 5 10 10 5 1

Many symmetries appear in the combination counts. These (symmetries) are generalized by $\binom{n}{r} = \binom{n}{n-r}$. This is required in Table VI, for example, to find $\binom{20}{16} = \binom{20}{4} = 4{,}845$. Similarly, to evaluate $\binom{15}{13} = \binom{15}{15-13} = \binom{15}{2} = 105$.

REFERENCES

[1] H. S. Bear, *Intermediate Algebra for College Students* (Menlo Park, California: Cummings, 1970).
[2] Charles F. Brumfield, R. E. Eicholz, M. E. Shanks, P. G. O'Daffer. *Principles of Arithmetic* (Reading, Massachusetts: Addison-Wesley, 1963).
[3] John A. Ingram, *Introductory Statistics* (Menlo Park, California: Cummings, 1974).
[4] Helen M. Walker, *Mathematics Essential for Elementary Statistics,* revised edition (New York: Holt, Rinehart, and Winston, 1966).

Problems

1. When possible evaluate each of the following; they all involve calculations with zero or one:

    a. $5 \cdot 0$
    b. 2^0
    c. $\dfrac{0}{1}$
    d. $\dfrac{2}{1}$
    e. 0^1
    f. 1^{-2}
    g. $\dfrac{1/10}{1}$
    h. $10 \cdot 1 \cdot 0$
    i. $\dfrac{2 \cdot 3}{4 \cdot 6 \cdot 0}$

2. Find the answers with proper decimal location for the following:

    a. $3.2 \cdot .001$
    b. $.01 \cdot .001$
    c. $(1.01)^2$
    d. $(.03)^2$
    e. $\dfrac{3400}{.0034}$
    f. $\dfrac{.54}{5.4}$

3. Perform the indicated arithmetic operations on common (ratio) fractions:

    a. $\dfrac{1}{3} - \dfrac{1}{4}$
    b. $\dfrac{1}{3} + \dfrac{1}{6}$
    c. $3 \cdot \dfrac{1}{6} \cdot \dfrac{2}{3}$
    d. $\dfrac{1}{3} \cdot \dfrac{1}{4} \cdot \dfrac{1}{2}$
    e. $-\dfrac{1}{2} \bigg/ -\dfrac{2}{3}$
    f. $\dfrac{6}{1/2}$

4. Using the concepts of ratios and proportions show that:

    a. $\dfrac{27}{171} = \dfrac{3}{19}$
    b. if $\dfrac{X}{6} = \dfrac{4}{15}$, then $X = 1.6$
    c. if $\dfrac{4}{7} = \dfrac{2}{X}$, then $X = 3.5$

5. Convert, as indicated, to either a ratio fraction, a decimal, or a percent:

    a. .875 to a percent
    b. 0.5% to a decimal
    c. .625 to a ratio fraction
    d. $\dfrac{2}{3}$ to a percent
    e. $\dfrac{5}{7}$ to a decimal
    f. 37.5% to a ratio fraction

Problems

6. Perform the indicated operations involving exponents or radicals:
   a. $2^3 - 3^2$
   b. $(-1)^2 \cdot (-1)^{-2}$
   c. $\dfrac{3^2}{2^{-1}}$
   d. $\dfrac{3^{-2}}{2^1}$
   e. Simplify $\dfrac{9}{\sqrt{3}}$
   f. Simplify $\dfrac{-\sqrt{9}}{3}$

7. Compute for these inequalities:
   a. Add $+2$ to both sides of $-3 < -1$.
   b. Subtract 2 from both sides of $3 > 1$.
   c. $1/2 > 1/3$ multiplied on both sides by -1.
   d. For $3 < 5$, divide each side by -2.
   e. Which is bigger, $(1/2)^2$ or $(1/2)^3$? Express as an inequality.

8. Indicate the number of decimals of accuracy for each answer.
   a. $3/1.5$
   b. $3.6 + 1.11 + 4.001$
   c. $6.921 - 5.8109$
   d. $6 \cdot .21$
   e. $(.5)^2$
   f. $12.5/10$

9. Use Table VIII to evaluate each of the following:
   a. $n = 251, n^2 = ?$
   b. $n = .251, n^2 = ?$
   c. $\sqrt{8.7}$
   d. $\sqrt{.87}$
   e. $\sqrt{.0087}$
   f. $\sqrt{.087}$

10. Evaluate these expressions. (*Hint:* All of the radicands can be made perfect squares, so Table VIII need not be used.)
    a. $\sqrt{9} + \sqrt{16}$
    b. $\sqrt{9} - \sqrt{16}$
    c. $\sqrt{5} \cdot \sqrt{5}$
    d. $\sqrt{.8} \cdot \sqrt{.2}$
    e. $\dfrac{\sqrt{32}}{\sqrt{8}}$
    f. $\dfrac{\sqrt{.27}}{\sqrt{3}}$

11. Perform each of the summations. Use the grouping symbols (parentheses) as is necessary. Let $X_1 = 0, X_2 = 1, X_3 = -1,$ and $X_4 = 2$.
    a. $\sum\limits^5 4$
    b. $\sum\limits^4 (X + 1)$
    c. $\sum\limits^4 [X - (1/2)]$
    d. $\sum\limits^4 [X - (1/2)]^2$
    e. $\sum\limits^4 (2X)$
    f. $\sum\limits^4 (3X - 1)$

12. Use the rectangular coordinates as described on p. 363 to answer the following:
    a. In which quadrant, I, II, III, or IV, is the point $(2, -1)$?
    b. Are the signs ($+$ or $-$) for both coordinates of points in Quadrant III always negative? (Exclude $(0,0)$ from those points in Quadrant III.)
    c. Are the points $(-2,1)$ and $(2,-1)$ in the same quadrant? In which quadrant(s) are these points?

13. Perform the designated operations for each linear relation:
    a. Graph on coordinate axes the line defined by $Y = 1 + 2X$.
    b. Graph on coordinate axes the line defined by $Y = 2X$.
    c. Find values for b_0 and b_1 for each of the following lines:

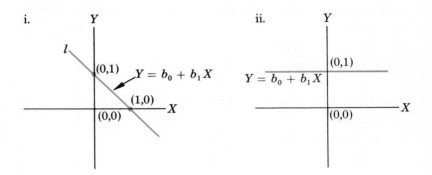

   d. In $Y = a + bX$, with a, b constants, explain how the direction of the line (slope) relates to the sign ($\pm$) for b.

14. Find the unknown, X, for each of the linear relations.
    a. $X + 1 = 2$
    b. $-(1/2)X = 3$
    c. $3X = 24$
    d. $X^{\frac{1}{3}} = 2$

Problems 15, 16, and 17 are word problems. Use the five steps outlined on p. 366 in solving them.

15. A board 9 ft long is to be cut into two pieces with one piece 3 ft shorter than the other. How long is each piece? (*Hint:* Let X = length of the longer piece.) Answer the same question, but for a board 7 ft 3 in long that is to be cut so that one piece is twice as long as the other.

16. An individual has a savings account of $1,000 invested at 6.0% interest. How much more must he invest in a second account at 5.0% interest to accumulate a total of $100 annually for the two accounts?

17. A household is insured for $31,250. If the cost per year per $1,000 of valuation for insurance is $2.85, what is the annual cost for this total policy?

18. Give a single numerical value for each expression:
    a. $3!$
    b. $0!$
    c. $5!$
    d. $6! - 5!$
    e. $3! \cdot 4!$
    f. $\dfrac{4!}{2!}$

19. Using the binomial formula, expand $(X + Y)^4$. Show for $X = 1/2$ and $Y = 1/2$ that your expansion of $(X + Y)^4$ sums to 1.

20. Use the Pascal's triangle (Table VI) to find:
    a. $\binom{10}{1}$
    b. $\binom{4}{3}$
    c. $\binom{8}{2}$
    d. $\binom{13}{11}$
    e. $\binom{15}{12} = \binom{15}{3}$
    f. $\binom{18}{16} - \binom{18}{1}$

Appendix B / Statistical Tables

Table I / The Standard Normal (Z) Distribution*

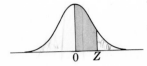

Z	.00	.01	.02	.03	.04	.05	.06	.07	.08	.09
0.0	.0000	.0040	.0080	.0120	.0160	.0199	.0239	.0279	.0319	.0359
0.1	.0398	.0438	.0478	.0517	.0557	.0596	.0636	.0675	.0714	.0753
0.2	.0793	.0832	.0871	.0910	.0948	.0987	.1026	.1064	.1103	.1141
0.3	.1179	.1217	.1255	.1293	.1331	.1368	.1406	.1443	.1480	.1517
0.4	.1554	.1591	.1628	.1664	.1700	.1736	.1772	.1808	.1844	.1879
0.5	.1915	.1950	.1985	.2019	.2054	.2088	.2123	.2157	.2190	.2224
0.6	.2257	.2291	.2324	.2357	.2389	.2422	.2454	.2486	.2517	.2549
0.7	.2580	.2611	.2642	.2673	.2704	.2734	.2764	.2794	.2823	.2852
0.8	.2881	.2910	.2939	.2967	.2995	.3023	.3051	.3078	.3106	.3133
0.9	.3159	.3186	.3212	.3238	.3264	.3289	.3315	.3340	.3365	.3389
1.0	.3413	.3438	.3461	.3485	.3508	.3531	.3554	.3577	.3599	.3621
1.1	.3643	.3665	.3686	.3708	.3729	.3749	.3770	.3790	.3810	.3830
1.2	.3849	.3869	.3888	.3907	.3925	.3944	.3962	.3980	.3997	.4015
1.3	.4032	.4049	.4066	.4082	.4099	.4115	.4131	.4147	.4162	.4177
1.4	.4192	.4207	.4222	.4236	.4251	.4265	.4279	.4292	.4306	.4319
1.5	.4332	.4345	.4357	.4370	.4382	.4394	.4406	.4418	.4429	.4441
1.6	.4452	.4463	.4474	.4484	.4495	.4505	.4515	.4525	.4535	.4545
1.7	.4554	.4564	.4573	.4582	.4591	.4599	.4608	.4616	.4625	.4633
1.8	.4641	.4649	.4656	.4664	.4671	.4678	.4686	.4693	.4699	.4706
1.9	.4713	.4719	.4726	.4732	.4738	.4744	.4750	.4756	.4761	.4767
2.0	.4772	.4778	.4783	.4788	.4793	.4798	.4803	.4808	.4812	.4817
2.1	.4821	.4826	.4830	.4834	.4838	.4842	.4846	.4850	.4854	.4857
2.2	.4861	.4864	.4868	.4871	.4875	.4878	.4881	.4884	.4887	.4890
2.3	.4893	.4896	.4898	.4901	.4904	.4906	.4909	.4911	.4913	.4916
2.4	.4918	.4920	.4922	.4925	.4927	.4929	.4931	.4932	.4934	.4936
2.5	.4938	.4940	.4941	.4943	.4945	.4946	.4948	.4949	.4951	.4952
2.6	.4953	.4955	.4956	.4957	.4959	.4960	.4961	.4962	.4963	.4964
2.7	.4965	.4966	.4967	.4968	.4969	.4970	.4971	.4972	.4973	.4974
2.8	.4974	.4975	.4976	.4977	.4977	.4978	.4979	.4979	.4980	.4981
2.9	.4981	.4982	.4982	.4983	.4984	.4984	.4985	.4985	.4986	.4986
3.0	.4987	.4987	.4987	.4988	.4988	.4989	.4989	.4989	.4990	.4990

*From Table A-1 of *Sturdy Statistics* by Mosteller & Rourke, Addison-Wesley Publishing Co., 1973 by permission of the publisher.

Table II / The *t*-Distribution*

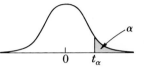

df	α					
	0.25	0.10	0.05	0.025	0.01	0.005
1	1.0000	3.0777	6.3138	12.7062	31.8207	63.6574
2	0.8165	1.8856	2.9200	4.3027	6.9646	9.9248
3	0.7649	1.6377	2.3534	3.1824	4.5407	5.8409
4	0.7407	1.5332	2.1318	2.7764	3.7469	4.6041
5	0.7267	1.4759	2.0150	2.5706	3.3649	4.0322
6	0.7176	1.4398	1.9432	2.4469	3.1427	3.7074
7	0.7111	1.4149	1.8946	2.3646	2.9980	3.4995
8	0.7064	1.3968	1.8595	2.3060	2.8965	3.3554
9	0.7027	1.3830	1.8331	2.2622	2.8214	3.2498
10	0.6998	1.3722	1.8125	2.2281	2.7638	3.1693
11	0.6974	1.3634	1.7959	2.2010	2.7181	3.1058
12	0.6955	1.3562	1.7823	2.1788	2.6810	3.0545
13	0.6938	1.3502	1.7709	2.1604	2.6503	3.0123
14	0.6924	1.3450	1.7613	2.1448	2.6245	2.9768
15	0.6912	1.3406	1.7531	2.1315	2.6025	2.9467
16	0.6901	1.3368	1.7459	2.1199	2.5835	2.9208
17	0.6892	1.3334	1.7396	2.1098	2.5669	2.8982
18	0.6884	1.3304	1.7341	2.1009	2.5524	2.8784
19	0.6876	1.3277	1.7291	2.0930	2.5395	2.8609
20	0.6870	1.3253	1.7247	2.0860	2.5280	2.8453
21	0.6864	1.3232	1.7207	2.0796	2.5177	2.8314
22	0.6858	1.3212	1.7171	2.0739	2.5083	2.8188
23	0.6853	1.3195	1.7139	2.0687	2.4999	2.8073
24	0.6848	1.3178	1.7109	2.0639	2.4922	2.7969
25	0.6844	1.3163	1.7081	2.0595	2.4851	2.7874
26	0.6840	1.3150	1.7056	2.0555	2.4786	2.7787
27	0.6837	1.3137	1.7033	2.0518	2.4727	2.7707
28	0.6834	1.3125	1.7011	2.0484	2.4671	2.7633
29	0.6830	1.3114	1.6991	2.0452	2.4620	2.7564
30	0.6828	1.3104	1.6973	2.0423	2.4573	2.7500
inf	0.6745	1.2816	1.6449	1.9600	2.3263	2.5758

*This table was abridged from the table of "Critical Values for Students' *t*-Distribution" in *Handbook of Statistical Tables* by D. B. Owen, Addison-Wesley Publishing Co., 1962 by permission of the publisher.

Table IIIA / The Chi-Square (χ^2) Distribution*

df	α					
	0.005	0.01	0.025	0.05	0.10	0.25
1	—	—	0.001	0.004	0.016	0.102
2	0.010	0.020	0.051	0.103	0.211	0.575
3	0.072	0.115	0.216	0.352	0.584	1.213
4	0.207	0.297	0.484	0.711	1.064	1.923
5	0.412	0.554	0.831	1.145	1.610	2.675
6	0.676	0.872	1.237	1.635	2.204	3.455
7	0.989	1.239	1.690	2.167	2.833	4.255
8	1.344	1.646	2.180	2.733	3.490	5.071
9	1.735	2.088	2.700	3.325	4.168	5.899
10	2.156	2.558	3.247	3.940	4.865	6.737
11	2.603	3.053	3.816	4.575	5.578	7.584
12	3.074	3.571	4.404	5.226	6.304	8.438
13	3.565	4.107	5.009	5.892	7.042	9.299
14	4.075	4.660	5.629	6.571	7.790	10.165
15	4.601	5.229	6.262	7.261	8.547	11.037
16	5.142	5.812	6.908	7.962	9.312	11.912
17	5.697	6.408	7.564	8.672	10.085	12.792
18	6.265	7.015	8.231	9.390	10.865	13.675
19	6.844	7.633	8.907	10.117	11.651	14.562
20	7.434	8.260	9.591	10.851	12.443	15.452
21	8.034	8.897	10.283	11.591	13.240	16.344
22	8.643	9.542	10.982	12.338	14.042	17.240
23	9.260	10.196	11.689	13.091	14.848	18.137
24	9.886	10.856	12.401	13.848	15.659	19.037
25	10.520	11.524	13.120	14.611	16.473	19.939
26	11.160	12.198	13.844	15.379	17.292	20.843
27	11.808	12.879	14.573	16.151	18.114	21.749
28	12.461	13.565	15.308	16.928	18.939	22.657
29	13.121	14.257	16.047	17.708	19.768	23.567
30	13.787	14.954	16.791	18.493	20.599	24.478
40	20.707	22.164	24.433	26.509	29.051	33.660
50	27.991	29.707	32.357	34.764	37.689	42.942
60	35.534	37.485	40.482	43.188	46.459	52.294
70	43.275	45.442	48.758	51.739	55.329	61.698
80	51.172	53.540	57.153	60.391	64.278	71.145
90	59.196	61.754	65.647	69.126	73.291	80.625
100	67.328	70.065	74.222	77.929	82.358	90.133

*This table was abridged from the table of "Critical Values for the Chi-Square Distribution" in D. B. Owen, *Handbook of Statistical Tables,* Addison-Wesley Publishing Co., 1962 by permission of the publisher.

Table IIIb / The Chi-Square (χ^2) Distribution (cont.)

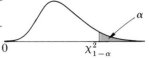

df	\multicolumn{6}{c}{$1 - \alpha$}
	0.75	0.90	0.95	0.975	0.99	0.995
1	1.323	2.706	3.841	5.024	6.635	7.879
2	2.773	4.605	5.991	7.378	9.210	10.597
3	4.108	6.251	7.815	9.348	11.345	12.838
4	5.385	7.779	9.488	11.143	13.277	14.860
5	6.626	9.236	11.071	12.833	15.086	16.750
6	7.841	10.645	12.592	14.449	16.812	18.548
7	9.037	12.017	14.067	16.013	18.475	20.278
8	10.219	13.362	15.507	17.535	20.090	21.955
9	11.389	14.684	16.919	19.023	21.666	23.589
10	12.549	15.987	18.307	20.483	23.209	25.188
11	13.701	17.275	19.675	21.920	24.725	26.757
12	14.845	18.549	21.026	23.337	26.217	28.299
13	15.984	19.812	22.362	24.736	27.688	29.819
14	17.117	21.064	23.685	26.119	29.141	31.319
15	18.245	22.307	24.996	27.488	30.578	32.801
16	19.369	23.542	26.296	28.845	32.000	34.267
17	20.489	24.769	27.587	30.191	33.409	35.718
18	21.605	25.989	28.869	31.526	34.805	37.156
19	22.718	27.204	30.144	32.852	36.191	38.582
20	23.828	28.412	31.410	34.170	37.566	39.997
21	24.935	29.615	32.671	35.479	38.932	41.401
22	26.039	30.813	33.924	36.781	40.289	42.796
23	27.141	32.007	35.172	38.076	41.638	44.181
24	28.241	33.196	36.415	39.364	42.980	45.559
25	29.339	34.382	37.652	40.646	44.314	46.928
26	30.435	35.563	38.885	41.923	45.642	48.290
27	31.528	36.741	40.113	43.194	46.963	49.645
28	32.620	37.916	41.337	44.461	48.278	50.993
29	33.711	39.087	42.557	45.722	49.588	52.336
30	34.800	40.256	43.773	46.979	50.892	53.672
40	45.616	51.805	55.758	59.342	63.691	66.766
50	56.334	63.167	67.505	71.420	76.154	79.490
60	66.981	74.397	79.082	83.298	88.379	91.952
70	77.577	85.527	90.531	95.023	100.425	104.215
80	88.130	96.578	101.879	106.629	112.329	116.321
90	98.650	107.565	113.145	118.136	124.116	128.299
100	109.141	118.498	124.342	129.561	135.807	140.169

Table IV / *F*-Distribution: Upper 10 percent points

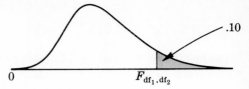

				Numerator degrees of freedom					
df_1 \\ df_2	1	2	3	4	5	6	7	8	9
1	39.863	49.500	53.593	55.833	57.240	58.204	58.906	59.439	59.858
2	8.5263	9.0000	9.1618	9.2434	9.2926	9.3255	9.3491	9.3668	9.3805
3	5.5383	5.4624	5.3908	5.3426	5.3092	5.2847	5.2662	5.2517	5.2400
4	4.5448	4.3246	4.1909	4.1072	4.0506	4.0097	3.9790	3.9549	3.9357
5	4.0604	3.7797	3.6195	3.5202	3.4530	3.4045	3.3679	3.3393	3.3163
6	3.7759	3.4633	3.2888	3.1808	3.1075	3.0546	3.0145	2.9830	2.9577
7	3.5894	3.2574	3.0741	2.9605	2.8833	2.8274	2.7849	2.7516	2.7247
8	3.4579	3.1131	2.9238	2.8064	2.7264	2.6683	2.6241	2.5893	2.5612
9	3.3603	3.0065	2.8129	2.6927	2.6106	2.5509	2.5053	2.4694	2.4403
10	3.2850	2.9245	2.7277	2.6053	2.5216	2.4606	2.4140	2.3772	2.3473
11	3.2252	2.8595	2.6602	2.5362	2.4512	2.3891	2.3416	2.3040	2.2735
12	3.1765	2.8068	2.6055	2.4801	2.3940	2.3310	2.2828	2.2446	2.2135
13	3.1362	2.7632	2.5603	2.4337	2.3467	2.2830	2.2341	2.1953	2.1638
14	3.1022	2.7265	2.5222	2.3947	2.3069	2.2426	2.1931	2.1539	2.1220
15	3.0732	2.6952	2.4898	2.3614	2.2730	2.2081	2.1582	2.1185	2.0862
16	3.0481	2.6682	2.4618	2.3327	2.2438	2.1783	2.1280	2.0880	2.0553
17	3.0262	2.6446	2.4374	2.3077	2.2183	2.1524	2.1017	2.0613	2.0284
18	3.0070	2.6239	2.4160	2.2858	2.1958	2.1296	2.0785	2.0379	2.0047
19	2.9899	2.6056	2.3970	2.2663	2.1760	2.1094	2.0580	2.0171	1.9836
20	2.9747	2.5893	2.3801	2.2489	2.1582	2.0913	2.0397	1.9985	1.9649
21	2.9610	2.5746	2.3649	2.2333	2.1423	2.0751	2.0233	1.9819	1.9480
22	2.9486	2.5613	2.3512	2.2193	2.1279	2.0605	2.0084	1.9668	1.9327
23	2.9374	2.5493	2.3387	2.2065	2.1149	2.0472	1.9949	1.9531	1.9189
24	2.9271	2.5383	2.3274	2.1949	2.1030	2.0351	1.9826	1.9407	1.9063
25	2.9177	2.5283	2.3170	2.1842	2.0922	2.0241	1.9714	1.9292	1.8947
26	2.9091	2.5191	2.3075	2.1745	2.0822	2.0139	1.9610	1.9188	1.8841
27	2.9012	2.5106	2.2987	2.1655	2.0730	2.0045	1.9515	1.9091	1.8743
28	2.8938	2.5028	2.2906	2.1571	2.0645	1.9959	1.9427	1.9001	1.8652
29	2.8870	2.4955	2.2831	2.1494	2.0566	1.9878	1.9345	1.8918	1.8568
30	2.8807	2.4887	2.2761	2.1422	2.0492	1.9803	1.9269	1.8841	1.8490
40	2.8354	2.4404	2.2261	2.0909	1.9968	1.9269	1.8725	1.8289	1.7929
60	2.7911	2.3933	2.1774	2.0410	1.9457	1.8747	1.8194	1.7748	1.7380
120	2.7478	2.3473	2.1300	1.9923	1.8959	1.8238	1.7675	1.7220	1.6842
∞	2.7055	2.3026	2.0838	1.9449	1.8473	1.7741	1.7167	1.6702	1.6315

df_2 \ df_1	10	12	15	20	24	30	40	60	120	∞
1	60.195	60.705	61.220	61.740	62.002	62.265	62.529	62.794	63.061	63.328
2	9.3916	9.4081	9.4247	9.4413	9.4496	9.4579	9.4662	9.4746	9.4829	9.4912
3	5.2304	5.2156	5.2003	5.1845	5.1764	5.1681	5.1597	5.1512	5.1425	5.1337
4	3.9199	3.8955	3.8704	3.8443	3.8310	3.8174	3.8036	3.7896	3.7753	3.7607
5	3.2974	3.2682	3.2380	3.2067	3.1905	3.1741	3.1573	3.1402	3.1228	3.1050
6	2.9369	2.9047	2.8712	2.8363	2.8183	2.8000	2.7812	2.7620	2.7423	2.7222
7	2.7025	2.6681	2.6322	2.5947	2.5753	2.5555	2.5351	2.5142	2.4928	2.4708
8	2.5380	2.5020	2.4642	2.4246	2.4041	2.3830	2.3614	2.3391	2.3162	2.2926
9	2.4163	2.3789	2.3396	2.2983	2.2768	2.2547	2.2320	2.2085	2.1843	2.1592
10	2.3226	2.2841	2.2435	2.2007	2.1784	2.1554	2.1317	2.1072	2.0818	2.0554
11	2.2482	2.2087	2.1671	2.1230	2.1000	2.0762	2.0516	2.0261	1.9997	1.9721
12	2.1878	2.1474	2.1049	2.0597	2.0360	2.0115	1.9861	1.9597	1.9323	1.9036
13	2.1376	2.0966	2.0532	2.0070	1.9827	1.9576	1.9315	1.9043	1.8759	1.8462
14	2.0954	2.0537	2.0095	1.9625	1.9377	1.9119	1.8852	1.8572	1.8280	1.7973
15	2.0593	2.0171	1.9722	1.9243	1.8990	1.8728	1.8454	1.8168	1.7867	1.7551
16	2.0281	1.9854	1.9399	1.8913	1.8656	1.8388	1.8108	1.7816	1.7507	1.7182
17	2.0009	1.9577	1.9117	1.8624	1.8362	1.8090	1.7805	1.7506	1.7191	1.6856
18	1.9770	1.9333	1.8868	1.8368	1.8103	1.7827	1.7537	1.7232	1.6910	1.6567
19	1.9557	1.9117	1.8647	1.8142	1.7873	1.7592	1.7298	1.6988	1.6659	1.6308
20	1.9367	1.8924	1.8449	1.7938	1.7667	1.7382	1.7083	1.6768	1.6433	1.6074
21	1.9197	1.8750	1.8271	1.7756	1.7481	1.7193	1.6890	1.6569	1.6228	1.5862
22	1.9043	1.8593	1.8111	1.7590	1.7312	1.7021	1.6714	1.6389	1.6041	1.5668
23	1.8903	1.8450	1.7964	1.7439	1.7159	1.6864	1.6554	1.6224	1.5871	1.5490
24	1.8775	1.8319	1.7831	1.7302	1.7019	1.6721	1.6407	1.6073	1.5715	1.5327
25	1.8658	1.8200	1.7708	1.7175	1.6890	1.6589	1.6272	1.5934	1.5570	1.5176
26	1.8550	1.8090	1.7596	1.7059	1.6771	1.6468	1.6147	1.5805	1.5437	1.5036
27	1.8451	1.7989	1.7492	1.6951	1.6662	1.6356	1.6032	1.5686	1.5313	1.4906
28	1.8359	1.7895	1.7395	1.6852	1.6560	1.6252	1.5925	1.5575	1.5198	1.4784
29	1.8274	1.7808	1.7306	1.6759	1.6465	1.6155	1.5825	1.5472	1.5090	1.4670
30	1.8195	1.7727	1.7223	1.6673	1.6377	1.6065	1.5732	1.5376	1.4989	1.4564
40	1.7627	1.7146	1.6624	1.6052	1.5741	1.5411	1.5056	1.4672	1.4248	1.3769
60	1.7070	1.6574	1.6034	1.5435	1.5107	1.4755	1.4373	1.3952	1.3476	1.2915
120	1.6524	1.6012	1.5450	1.4821	1.4472	1.4094	1.3676	1.3203	1.2646	1.1926
∞	1.5987	1.5458	1.4871	1.4206	1.3832	1.3419	1.2951	1.2400	1.1686	1.0000

Numerator degrees of freedom (column header); Denominator degrees of freedom (row header).

From *Biometrika Tables for Statisticians*, vol 2, E. S. Pearson and H. O. Hartley, Cambridge: The University Press, 1972 by permission of the Biometrika trustees.

Table IV / F-Distribution: Upper 5 percent points

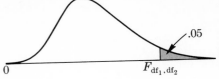

	Numerator degrees of freedom								
df_1 \ df_2	1	2	3	4	5	6	7	8	9
1	161.45	199.50	215.71	224.58	230.16	233.99	236.77	238.88	240.54
2	18.513	19.000	19.164	19.247	19.296	19.330	19.353	19.371	19.385
3	10.128	9.5521	9.2766	9.1172	9.0135	8.9406	8.8867	8.8452	8.8123
4	7.7086	9.9443	6.5914	6.3882	6.2561	6.1631	6.0942	6.0410	6.9988
5	6.6079	5.7861	5.4095	5.1922	5.0503	4.9503	4.8759	4.8183	4.7725
6	5.9874	5.1433	4.7571	4.5337	4.3874	4.2839	4.2067	4.1468	4.0990
7	5.5914	4.7374	4.3468	4.1203	3.9715	3.8660	3.7870	3.7257	3.6767
8	5.3177	4.4590	4.0662	3.8379	3.6875	3.5806	3.5005	3.4381	3.3881
9	5.1174	4.2565	3.8625	3.6331	3.4817	3.3738	3.2927	3.2296	3.1789
10	4.9646	4.1028	3.7083	3.4780	3.3258	3.2172	3.1355	3.0717	3.0204
11	4.8443	3.9823	3.5874	3.3567	3.2039	3.0946	3.0123	2.9480	2.8962
12	4.7472	3.8853	3.4903	3.2592	3.1059	2.9961	2.9134	2.8486	2.7964
13	4.6672	3.8056	3.4105	3.1791	3.0254	2.9153	2.8321	2.7669	2.7144
14	4.6001	3.7389	3.3439	3.1122	2.9582	2.8477	2.7642	2.6987	2.6458
15	4.5431	3.6823	3.2874	3.0556	2.9013	2.7905	2.7066	2.6408	2.5876
16	4.4940	3.6337	3.2389	3.0069	2.8524	2.7413	2.6572	2.5911	2.5377
17	4.4513	3.5915	3.1968	2.9647	2.8100	2.6987	2.6143	2.5480	2.4943
18	4.4139	3.5546	3.1599	2.9277	2.7729	2.6613	2.5767	2.5102	2.4563
19	4.3807	3.5219	3.1274	2.8951	2.7401	2.6283	2.5435	2.4768	2.4227
20	4.3512	3.4928	3.0984	2.8661	2.7109	2.5990	2.5140	2.4471	2.3928
21	4.3248	3.4668	3.0725	2.8401	2.6848	2.5727	2.4876	2.4205	2.3660
22	4.3009	3.4434	3.0491	2.8167	2.6613	2.5491	2.4638	2.3965	2.3419
23	4.2793	3.4221	3.0280	2.7955	2.6400	2.5277	2.4422	2.3748	2.3201
24	4.2597	3.4028	3.0088	2.7763	2.6207	2.5082	2.4226	2.3551	2.3002
25	4.2417	3.3852	2.9912	2.7587	2.6030	2.4904	2.4047	2.3371	2.2821
26	4.2252	3.3690	2.9752	2.7426	2.5868	2.4741	2.3883	2.3205	2.2655
27	4.2100	3.3541	2.9604	2.7278	2.5719	2.4591	2.3732	2.3053	2.2501
28	4.1960	3.3404	2.9467	2.7141	2.5581	2.4453	2.3593	2.2913	2.2360
29	4.1830	3.3277	2.9340	2.7014	2.5454	2.4324	2.3463	2.2783	2.2229
30	4.1709	3.3158	2.9223	2.6896	2.5336	2.4205	2.3343	2.2662	2.2107
40	4.0847	3.2317	2.8387	2.6060	2.4495	2.3359	2.2490	2.1802	2.1240
60	4.0012	3.1504	2.7581	2.5252	2.3683	2.2541	2.1665	2.0970	2.0401
120	3.9201	3.0718	2.6802	2.4472	2.2899	2.1750	2.0868	2.0164	1.9588
∞	3.8415	2.9957	2.6049	2.3719	2.2141	2.0986	2.0096	1.9384	1.8799

Numerator degrees of freedom

df_1 \ df_2	10	12	15	20	24	30	40	60	120	∞
1	241.88	243.91	245.95	248.01	249.05	250.10	251.14	252.20	253.25	254.31
2	19.396	19.413	19.429	19.446	19.454	19.462	19.471	19.479	19.487	19.496
3	8.7855	8.7446	8.7029	8.6602	8.6385	8.6166	8.5944	8.5720	8.5494	8.5264
4	5.9644	5.9117	5.8578	5.8025	5.7744	5.7459	5.7170	5.6877	5.6581	5.6281
5	4.7351	4.6777	4.6188	4.5581	4.5272	4.4957	4.4638	4.4314	4.3985	4.3650
6	4.0600	3.9999	3.9381	3.8742	3.8415	3.8082	3.7743	3.7398	3.7047	3.6689
7	3.6365	3.5747	3.5107	3.4445	3.4105	3.3758	3.3404	3.3043	3.2674	3.2298
8	3.3472	3.2839	3.2184	3.1503	3.1152	3.0794	3.0428	3.0053	2.9669	2.9276
9	3.1373	3.0729	3.0061	2.9365	2.9005	2.8637	2.8259	2.7872	2.7475	2.7067
10	2.9782	2.9130	2.8450	2.7740	2.7372	2.6996	2.6609	2.6211	2.5801	2.5379
11	2.8536	2.7876	2.7186	2.6464	2.6090	2.5705	2.5309	2.4901	2.4480	2.4045
12	2.7534	2.6866	2.6169	2.5436	2.5055	2.4663	2.4259	2.3842	2.3410	2.2962
13	2.6710	2.6037	2.5331	2.4589	2.4202	2.3803	2.3392	2.2966	2.2524	2.2064
14	2.6022	2.5342	2.4630	2.3879	2.3487	2.3082	2.2664	2.2229	2.1778	2.1307
15	2.5437	2.4753	2.4034	2.3275	2.2878	2.2468	2.2043	2.1601	2.1141	2.0658
16	2.4935	2.4247	2.3522	2.2756	2.2354	2.1938	2.1507	2.1058	2.0589	2.0096
17	2.4499	2.3807	2.3077	2.2304	2.1898	2.1477	2.1040	2.0584	2.0107	1.9604
18	2.4117	2.3421	2.2686	2.1906	2.1497	2.1071	2.0629	2.0166	1.9681	1.9168
19	2.3779	2.3080	2.2341	2.1555	2.1141	2.0712	2.0264	1.9795	1.9302	1.8780
20	2.3479	2.2776	2.2033	2.1242	2.0825	2.0391	1.9938	1.9464	1.8963	1.8432
21	2.3210	2.2504	2.1757	2.0960	2.0540	2.0102	1.9645	1.9165	1.8657	1.8117
22	2.2967	2.2258	2.1508	2.0707	2.0283	1.9842	1.9380	1.8894	1.8380	1.7831
23	2.2747	2.2036	2.1282	2.0476	2.0050	1.9605	1.9139	1.8648	1.8128	1.7570
24	2.2547	2.1834	2.1077	2.0267	1.9838	1.9390	1.8920	1.8424	1.7896	1.7330
25	2.2365	2.1649	2.0889	2.0075	1.9643	1.9192	1.8718	1.8217	1.7684	1.7110
26	2.2197	2.1479	2.0716	1.9898	1.9464	1.9010	1.8533	1.8027	1.7488	1.6906
27	2.2043	2.1323	2.0558	1.9736	1.9299	1.8842	1.8361	1.7851	1.7306	1.6717
28	2.1900	2.1179	2.0411	1.9586	1.9147	1.8687	1.8203	1.7689	1.7138	1.6541
29	2.1768	2.1045	2.0275	1.9446	1.9005	1.8543	1.8055	1.7537	1.6981	1.6376
30	2.1646	2.0921	2.0148	1.9317	1.8874	1.8409	1.7918	1.7396	1.6835	1.6223
40	2.0772	2.0035	1.9245	1.8389	1.7929	1.7444	1.6928	1.6373	1.5766	1.5089
60	1.9926	1.9174	1.8364	1.7480	1.7001	1.6491	1.5943	1.5343	1.4673	1.3893
120	1.9105	1.8337	1.7505	1.6587	1.6084	1.5543	1.4952	1.4290	1.3519	1.2539
∞	1.8307	1.7522	1.6664	1.5705	1.5173	1.4591	1.3940	1.3180	1.2214	1.0000

$F = \dfrac{s_1^2}{s_2^2} = \dfrac{S_1}{df_1} \bigg/ \dfrac{S_2}{df_2}$, where $s_1^2 = S_1/df_1$ and $s_2^2 = S_2/df_2$ are independent mean square estimators of a common variance σ^2, based on df_1 and df_2 degrees of freedom, respectively.

Table IV / F-Distribution: Upper 2.5 percent points

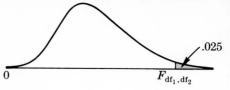

Numerator degrees of freedom

df_2 \ df_1	1	2	3	4	5	6	7	8	9
1	647.79	799.50	864.16	899.58	921.85	937.11	948.22	956.66	963.28
2	38.506	39.000	39.165	39.248	39.298	39.331	39.335	39.373	39.387
3	17.443	16.044	15.439	15.101	14.885	14.735	14.624	14.540	14.473
4	12.218	10.649	9.9792	9.6045	9.3645	9.1973	9.0741	8.9796	8.9047
5	10.007	8.4336	7.7636	7.3879	7.1464	6.9777	6.8531	6.7572	6.6811
6	8.8131	7.2599	6.5988	6.2272	5.9876	5.8198	5.6955	5.5996	5.5234
7	8.0727	6.5415	5.8898	5.5226	5.2852	5.1186	4.9949	4.8993	4.8232
8	7.5709	6.0595	5.4160	5.0526	4.8173	4.6517	4.5286	4.4333	4.3572
9	7.2093	5.7147	5.0781	4.7181	4.4844	4.3197	4.1970	4.1020	4.0260
10	6.9367	5.4564	4.8256	4.4683	4.2361	4.0721	3.9498	3.8549	3.7790
11	6.7241	5.2559	4.6300	4.2751	4.0440	3.8807	3.7586	3.6638	3.5879
12	6.5538	5.0959	4.4742	4.1212	3.8911	3.7283	3.6065	3.5118	3.4358
13	6.4143	4.9653	4.3472	3.9959	3.7667	3.6043	3.4827	3.3880	3.3120
14	6.2979	4.8567	4.2417	3.8919	3.6634	3.5014	3.3799	3.2853	3.2093
15	6.1995	4.7650	4.1528	3.8043	3.5764	3.4147	3.2934	3.1987	3.1227
16	6.1151	4.6867	4.0768	3.7294	3.5021	3.3406	3.2194	3.1248	3.0488
17	6.0420	4.6189	4.0112	3.6648	3.4379	3.2767	3.1556	3.0610	2.9849
18	5.9781	4.5597	3.9539	3.6083	3.3820	3.2209	3.0999	3.0053	2.9291
19	5.9216	4.5075	3.9034	3.5587	3.3327	3.1718	3.0509	2.9563	2.8801
20	5.8715	4.4613	3.8587	3.5147	3.2891	3.1283	3.0074	2.9128	2.8365
21	5.8266	4.4199	3.8188	3.4754	3.2501	3.0895	2.9686	2.8740	2.7977
22	5.7863	4.3828	3.7829	3.4401	3.2151	3.0546	2.9338	2.8392	2.7628
23	5.7498	4.3492	3.7505	3.4083	3.1835	3.0232	2.9023	2.8077	2.7313
24	5.7166	4.3187	3.7211	3.3794	3.1548	2.9946	2.8738	2.7791	2.7027
25	5.6864	4.2909	3.6943	3.3530	3.1287	2.9685	2.8478	2.7531	2.6766
26	5.6586	4.2655	3.6697	3.3289	3.1048	2.9447	2.8240	2.7293	2.6528
27	5.6331	4.2421	3.6472	3.3067	3.0828	2.9228	2.8021	2.7074	2.6309
28	5.6096	4.2205	3.6264	3.2863	3.0626	2.9027	2.7820	2.6872	2.6106
29	5.5878	4.2006	3.6072	3.2674	3.0438	2.8840	2.7633	2.6686	2.5919
30	5.5675	4.1821	3.5894	3.2499	3.0265	2.8667	2.7460	2.6513	2.5746
40	5.4239	4.0510	3.4633	3.1261	2.9037	2.7444	2.6238	2.5289	2.4519
60	5.2856	3.9253	3.3425	3.0077	2.7863	2.6274	2.5068	2.4117	2.3344
120	5.1523	3.8046	3.2269	2.8943	2.6740	2.5154	2.3948	2.2994	2.2217
∞	5.0239	3.6889	3.1161	2.7858	2.5665	2.4082	2.2875	2.1918	2.1136

Numerator degrees of freedom

df_2 \ df_1	10	12	15	20	24	30	40	60	120	∞
1	968.63	976.71	984.87	993.10	997.25	1001.4	1005.6	1009.8	1014.0	1018.3
2	39.398	39.415	39.431	39.448	39.456	39.465	39.473	39.481	39.490	39.498
3	14.419	14.337	14.253	14.167	14.124	14.081	14.037	13.992	13.947	13.902
4	8.8439	8.7512	8.6565	8.5599	8.5109	8.4613	8.4111	8.3604	8.3092	8.2573
5	6.6192	6.5245	6.4277	6.3286	6.2780	6.2269	6.1750	6.1225	6.0693	6.0153
6	5.4613	5.3662	5.2687	5.1684	5.1172	5.0652	5.0125	4.9589	4.9044	4.8491
7	4.7611	4.6658	4.5678	4.4667	4.4150	4.3624	4.3089	4.2544	4.1989	4.1423
8	4.2951	4.1997	4.1012	3.9995	3.9472	3.8940	3.8398	3.7844	3.7279	3.6702
9	3.9639	3.8682	3.7694	3.6669	3.6142	3.5604	3.5055	3.4493	3.3918	3.3329
10	3.7168	3.6209	3.5217	3.4185	3.3654	3.3110	3.2554	3.1984	3.1399	3.0798
11	3.5257	3.4296	3.3299	3.2261	3.1725	3.1176	3.0613	3.0035	2.9441	2.8828
12	3.3736	3.2773	3.1772	3.0728	3.0187	2.9633	2.9063	2.8478	2.7874	2.7249
13	3.2497	3.1532	3.0527	2.9477	2.8932	2.8372	2.7797	2.7204	2.6590	2.5955
14	3.1469	3.0502	2.9493	2.8437	2.7888	2.7324	2.6742	2.6142	2.5519	2.4872
15	3.0602	2.9633	2.8621	2.7559	2.7006	2.6437	2.5850	2.5242	2.4611	2.3953
16	2.9862	2.8890	2.7875	2.6808	2.6252	2.5678	2.5085	2.4471	2.3831	2.3163
17	2.9222	2.8249	2.7230	2.6158	2.5598	2.5020	2.4422	2.3801	2.3153	2.2474
18	2.8664	2.7689	2.6667	2.5590	2.5027	2.4445	2.3842	2.3214	2.2558	2.1869
19	2.8172	2.7196	2.6171	2.5089	2.4523	2.3937	2.3329	2.2696	2.2032	2.1333
20	2.7737	2.6758	2.5731	2.4645	2.4076	2.3486	2.2873	2.2234	2.1562	2.0853
21	2.7348	2.6368	2.5338	2.4247	2.3675	2.3082	2.2465	2.1819	2.1141	2.0422
22	2.6998	2.6017	2.4984	2.3890	2.3315	2.2718	2.2097	2.1446	2.0760	2.0032
23	2.6682	2.5699	2.4665	2.3567	2.2989	2.2389	2.1763	2.1107	2.0415	1.9677
24	2.6396	2.5411	2.4374	2.3273	2.2693	2.2090	2.1460	2.0799	2.0099	1.9353
25	2.6135	2.5149	2.4110	2.3005	2.2422	2.1816	2.1183	2.0516	1.9811	1.9055
26	2.5896	2.4908	2.3867	2.2759	2.2174	2.1565	2.0928	2.0257	1.9545	1.8781
27	2.5676	2.4688	2.3644	2.2533	2.1946	2.1334	2.0693	2.0018	1.9299	1.8527
28	2.5473	2.4484	2.3438	2.2324	2.1735	2.1121	2.0477	1.9797	1.9072	1.8291
29	2.5286	2.4295	2.3248	2.2131	2.1540	2.0923	2.0276	1.9591	1.8861	1.8072
30	2.5112	2.4120	2.3072	2.1952	2.1359	2.0739	2.0089	1.9400	1.8664	1.7867
40	2.3882	2.2882	2.1819	2.0677	2.0069	1.9429	1.8752	1.8028	1.7242	1.6371
60	2.2702	2.1692	2.0613	1.9445	1.8817	1.8152	1.7440	1.6668	1.5810	1.4821
120	2.1570	2.0548	1.9450	1.8249	1.7597	1.6899	1.6141	1.5299	1.4327	1.3104
∞	2.0483	1.9447	1.8326	1.7085	1.6402	1.5660	1.4835	1.3883	1.2684	1.0000

Denominator degrees of freedom

Table IV / F-Distribution: Upper 1 percent points

Numerator degrees of freedom

df_2 \ df_1	1	2	3	4	5	6	7	8	9
1	4052.2	4999.5	5403.4	5624.6	5763.6	5859.0	5928.4	5981.1	6022.5
2	98.503	99.000	99.166	99.249	99.299	99.333	99.356	99.374	99.388
3	34.116	30.817	29.457	28.710	28.237	27.911	27.672	27.489	27.345
4	21.198	18.000	16.694	15.977	15.522	15.207	14.976	14.799	14.659
5	16.258	13.274	12.060	11.392	10.967	10.672	10.456	10.289	10.158
6	13.745	10.925	9.7795	9.1483	8.7459	8.4661	8.2600	8.1017	7.9761
7	12.246	9.5466	8.4513	7.8466	7.4604	7.1914	6.9928	6.8400	6.7188
8	11.259	8.6491	7.5910	7.0061	6.6318	6.3707	6.1776	6.0289	5.9106
9	10.561	8.0215	6.9919	6.4221	6.0569	5.8018	5.6129	5.4671	5.3511
10	10.044	7.5594	6.5523	5.9943	5.6363	5.3858	5.2001	5.0567	4.9424
11	9.6460	7.2057	6.2167	5.6683	5.3160	5.0692	4.8861	4.7445	4.6315
12	9.3302	6.9266	5.9525	5.4120	5.0643	4.8206	4.6395	4.4994	4.3875
13	9.0738	6.7010	5.7394	5.2053	4.8616	4.6204	4.4410	4.3021	4.1911
14	8.8616	6.5149	5.5639	5.0354	4.6950	4.4558	4.2779	4.1399	4.0297
15	8.6831	6.3589	5.4170	4.8932	4.5556	4.3183	4.1415	4.0045	3.8948
16	8.5310	6.2262	5.2922	4.7726	4.4374	4.2016	4.0259	3.8896	3.7804
17	8.3997	6.1121	5.1850	4.6690	4.3359	4.1015	3.9267	3.7910	3.6822
18	8.2854	6.0129	5.0919	4.5790	4.2479	4.0146	3.8406	3.7054	3.5971
19	8.1849	5.9259	5.0103	4.5003	4.1708	3.9386	3.7653	3.6305	3.5225
20	8.0960	5.8489	4.9382	4.4307	4.1027	3.8714	3.6987	3.5644	3.4567
21	8.0166	5.7804	4.8740	4.3688	4.0421	3.8117	3.6396	3.5056	3.3981
22	7.9454	5.7190	4.8166	4.3134	3.9880	3.7583	3.5867	3.4530	3.3458
23	7.8811	5.6637	4.7649	4.2636	3.9392	3.7102	3.5390	3.4057	3.2986
24	7.8229	5.6136	4.7181	4.2184	3.8951	3.6667	3.4959	3.3629	3.2560
25	7.7698	5.5680	4.6755	4.1774	3.8550	3.6272	3.4568	3.3239	3.2172
26	7.7213	5.5263	4.6366	4.1400	3.8183	3.5911	3.4210	3.2884	3.1818
27	7.6767	5.4881	4.6009	4.1056	3.7848	3.5580	3.3882	3.2558	3.1494
28	7.6356	5.4529	4.5681	4.0740	3.7539	3.5276	3.3581	3.2259	3.1195
29	7.5977	5.4204	4.5378	4.0449	3.7254	3.4995	3.3303	3.1982	3.0920
30	7.5625	5.3903	4.5097	4.0179	3.6990	3.4735	3.3045	3.1726	3.0665
40	7.3141	5.1785	4.3126	3.8283	3.5138	3.2910	3.1238	2.9930	2.8876
60	7.0771	4.9774	4.1259	3.6490	3.3389	3.1187	2.9530	2.8233	2.7185
120	6.8509	4.7865	3.9491	3.4795	3.1735	2.9559	2.7918	2.6629	2.5586
∞	6.6349	4.6052	3.7816	3.3192	3.0173	2.8020	2.6393	2.5113	2.4073

Numerator degrees of freedom

df_2 \ df_1	10	12	15	20	24	30	40	60	120	∞
1	6055.8	6106.3	6157.3	6208.7	6234.6	6260.6	6286.8	6313.0	6339.4	6365.9
2	99.399	99.416	99.433	99.449	99.458	99.466	99.474	99.482	99.491	99.499
3	27.229	27.052	26.872	26.690	26.598	26.505	26.411	26.316	26.221	26.125
4	14.546	14.374	14.198	14.020	13.929	13.838	13.745	13.652	13.558	13.463
5	10.051	9.8883	9.7222	9.5526	9.4665	9.3793	9.2912	9.2020	9.1118	9.0204
6	7.8741	7.7183	7.5590	7.3958	7.3127	7.2285	7.1432	7.0567	6.9690	6.8800
7	6.6201	6.4691	6.3143	6.1554	6.0743	5.9920	5.9084	5.8236	5.7373	5.6495
8	5.8143	5.6667	5.5151	5.3591	5.2793	5.1981	5.1156	5.0316	4.9461	4.8588
9	5.2565	5.1114	4.9621	4.8080	4.7290	4.6486	4.5666	4.4831	4.3978	4.3105
10	4.8491	4.7059	4.5581	4.4054	4.3269	4.2469	4.1653	4.0819	3.9965	3.9090
11	4.5393	4.3974	4.2509	4.0990	4.0209	3.9411	3.8596	3.7761	3.6904	3.6024
12	4.2961	4.1553	4.0096	3.8584	3.7805	3.7008	3.6192	3.5355	3.4494	3.3608
13	4.1003	4.9603	3.8154	3.6646	3.5868	3.5070	3.4253	3.3413	3.2548	3.1654
14	3.9394	3.8001	3.6557	3.5052	3.4274	3.3476	3.2656	3.1813	3.0942	3.0040
15	3.8049	3.6662	3.5222	3.3719	3.2940	3.2141	3.1319	3.0471	2.9595	2.8684
16	3.6909	3.5527	3.4089	3.2587	3.1808	3.1007	3.0182	2.9330	2.8447	2.7528
17	3.5931	3.4552	3.3117	3.1615	3.0835	3.0032	2.9205	2.8348	2.7459	2.6530
18	3.5082	3.3706	3.2273	3.0771	2.9990	2.9185	2.8354	2.7493	2.6597	2.5660
19	3.4338	3.2965	3.1533	3.0031	2.9249	2.8442	2.7608	2.6742	2.5839	2.4893
20	3.3682	3.2311	3.0880	2.9377	2.8594	2.7785	2.6947	2.6077	2.5168	2.4212
21	3.3098	3.1730	3.0300	2.8796	2.8010	2.7200	2.6359	2.5484	2.4568	2.3603
22	3.2576	3.1209	2.9779	2.8274	2.7488	2.6675	2.5831	2.4951	2.4029	2.3055
23	3.2106	3.0740	2.9311	2.7805	2.7017	2.6202	2.5355	2.4471	2.3542	2.2558
24	3.1681	3.0316	2.8887	2.7380	2.6591	2.5773	2.4923	2.4035	2.3100	2.2107
25	3.1294	2.9931	2.8502	2.6993	2.6203	2.5383	2.4530	2.3637	2.2696	2.1694
26	3.0941	2.9578	2.8150	2.6640	2.5848	2.5026	2.4170	2.3273	2.2325	2.1315
27	3.0618	2.9256	2.7827	2.6316	2.5522	2.4699	2.3840	2.2938	2.1985	2.0965
28	3.0320	2.8959	2.7530	2.6017	2.5223	2.4397	2.3535	2.2629	2.1670	2.0642
29	3.0045	2.8685	2.7256	2.5742	2.4946	2.4118	2.3253	2.2344	2.1379	2.0342
30	2.9791	2.8431	2.7002	2.5487	2.4689	2.3860	2.2992	2.2079	2.1108	2.0062
40	2.8005	2.6648	2.5216	2.3689	2.2880	2.2034	2.1142	2.0194	1.9172	1.8047
60	2.6318	2.4961	2.3523	2.1978	2.1154	2.0285	1.9360	1.8363	1.7263	1.6006
120	2.4721	2.3363	2.1915	2.0346	1.9500	1.8600	1.7628	1.6557	1.5330	1.3805
∞	2.3209	2.1847	2.0385	1.8783	1.7908	1.6964	1.5923	1.4730	1.3246	1.0000

Denominator degrees of freedom

Table V / Binomial Probabilities

							p							
n	X	.05	.10	.20	.25	.30	.40	.50	.60	.70	.75	.80	.90	.95
1	0	.9500	.9000	.8000	.7500	.7000	.6000	.5000	.4000	.3000	.2500	.2000	.1000	.0500
	1	.0500	.1000	.2000	.2500	.3000	.4000	.5000	.6000	.7000	.7500	.8000	.9000	.9500
2	0	.9025	.8100	.6400	.5625	.4900	.3600	.2500	.1600	.0900	.0625	.0400	.0100	.0025
	1	.0950	.1800	.3200	.3750	.4200	.4800	.5000	.4800	.4200	.3750	.3200	.1800	.0950
	2	.0025	.0100	.0400	.0625	.0900	.1600	.2500	.3600	.4900	.5625	.6400	.8100	.9025
3	0	.8574	.7290	.5120	.4219	.3430	.2160	.1250	.0640	.0270	.0156	.0080	.0010	.0001
	1	.1354	.2430	.3840	.4219	.4410	.4320	.3750	.2880	.1890	.1406	.0960	.0270	.0071
	2	.0071	.0270	.0960	.1406	.1890	.2880	.3750	.4320	.4410	.4219	.3840	.2430	.1354
	3	.0001	.0010	.0080	.0156	.0270	.0640	.1250	.2160	.3430	.4219	.5120	.7290	.8574
4	0	.8145	.6561	.4096	.3164	.2401	.1296	.0625	.0256	.0081	.0039	.0016	.0001	
	1	.1715	.2916	.4096	.4219	.4116	.3456	.2500	.1536	.0756	.0469	.0256	.0036	.0005
	2	.0135	.0486	.1536	.2109	.2646	.3456	.3750	.3456	.2646	.2109	.1536	.0486	.0135
	3	.0005	.0036	.0256	.0469	.0756	.1536	.2500	.3456	.4116	.4219	.4096	.2916	.1715
	4		.0001	.0016	.0039	.0081	.0256	.0625	.1296	.2401	.3164	.4096	.6561	.8145
5	0	.7738	.5905	.3277	.2373	.1681	.0778	.0313	.0102	.0024	.0010	.0003		
	1	.2036	.3281	.4096	.3955	.3602	.2592	.1562	.0768	.0284	.0146	.0064	.0004	
	2	.0214	.0729	.2048	.2637	.3087	.3456	.3125	.2304	.1323	.0879	.0512	.0081	.0011
	3	.0011	.0081	.0512	.0879	.1323	.2304	.3125	.3456	.3087	.2637	.2048	.0729	.0214
	4		.0004	.0064	.0146	.0284	.0768	.1562	.2592	.3602	.3955	.4096	.3281	.2036
	5			.0003	.0010	.0024	.0102	.0313	.0778	.1681	.2373	.3277	.5905	.7738
6	0	.7351	.5314	.2621	.1780	.1176	.0467	.0156	.0041	.0007	.0002	.0001		
	1	.2321	.3543	.3932	.3560	.3025	.1866	.0938	.0369	.0102	.0044	.0015	.0001	
	2	.0305	.0984	.2458	.2966	.3241	.3110	.2344	.1382	.0595	.0330	.0154	.0012	.0001
	3	.0021	.0146	.0819	.1318	.1852	.2765	.3125	.2765	.1852	.1318	.0819	.0146	.0021
	4	.0001	.0012	.0154	.0330	.0595	.1382	.2344	.3110	.3241	.2966	.2458	.0984	.0305
	5		.0001	.0015	.0044	.0102	.0369	.0938	.1866	.3025	.3560	.3932	.3543	.2321
	6			.0001	.0002	.0007	.0041	.0156	.0467	.1176	.1780	.2621	.5314	.7351
7	0	.6983	.4783	.2097	.1335	.0824	.0280	.0078	.0016	.0002	.0001			
	1	.2573	.3720	.3670	.3115	.2471	.1306	.0547	.0172	.0036	.0013	.0004		
	2	.0406	.1240	.2753	.3115	.3177	.2613	.1641	.0774	.0250	.0115	.0043	.0002	
	3	.0036	.0230	.1147	.1730	.2269	.2903	.2734	.1935	.0972	.0577	.0287	.0026	.0002
	4	.0002	.0026	.0287	.0577	.0972	.1935	.2734	.2903	.2269	.1730	.1147	.0230	.0036
	5		.0002	.0043	.0115	.0250	.0774	.1641	.2613	.3177	.3115	.2753	.1240	.0406
	6			.0004	.0013	.0036	.0172	.0547	.1306	.2471	.3115	.3670	.3720	.2573
	7				.0001	.0002	.0016	.0078	.0280	.0824	.1335	.2097	.4783	.6983

(cont.)

Binomial Probabilities (cont.)

							p							
n	X	.05	.10	.20	.25	.30	.40	.50	.60	.70	.75	.80	.90	.95
8	0	.6634	.4305	.1678	.1001	.0576	.0168	.0039	.0007	.0001				
	1	.2793	.3826	.3355	.2670	.1976	.0896	.0312	.0079	.0012	.0004	.0001		
	2	.0515	.1488	.2936	.3115	.2965	.2090	.1094	.0413	.0100	.0038	.0011		
	3	.0054	.0331	.1468	.2076	.2541	.2787	.2188	.1239	.0467	.0231	.0092	.0004	
	4	.0004	.0046	.0459	.0865	.1361	.2322	.2734	.2322	.1361	.0865	.0459	.0046	.0004
	5		.0004	.0092	.0231	.0467	.1238	.2188	.2787	.2541	.2076	.1468	.0331	.0054
	6			.0011	.0038	.0100	.0413	.1094	.2090	.2965	.3115	.2936	.1488	.0515
	7			.0001	.0004	.0012	.0079	.0312	.0896	.1976	.2670	.3355	.3826	.2793
	8					.0001	.0007	.0039	.0168	.0576	.1001	.1678	.4305	.6634
9	0	.6302	.3874	.1342	.0751	.0404	.0101	.0020	.0003					
	1	.2985	.3874	.3020	.2253	.1556	.0605	.0176	.0035	.0004	.0001			
	2	.0628	.1722	.3020	.3003	.2668	.1612	.0703	.0212	.0039	.0012	.0003		
	3	.0077	.0446	.1762	.2336	.2668	.2508	.1641	.0743	.0210	.0087	.0028	.0001	
	4	.0006	.0074	.0661	.1168	.1715	.2508	.2461	.1672	.0735	.0389	.0165	.0008	
	5		.0008	.0165	.0389	.0735	.1672	.2461	.2508	.1715	.1168	.0661	.0074	.0006
	6		.0001	.0028	.0087	.0210	.0743	.1641	.2508	.2668	.2336	.1762	.0446	.0077
	7			.0003	.0012	.0039	.0212	.0703	.1612	.2668	.3003	.3020	.1722	.0628
	8				.0001	.0004	.0035	.0176	.0605	.1556	.2253	.3020	.3874	.2985
	9						.0003	.0020	.0101	.0404	.0751	.1342	.3874	.6302
10	0	.5987	.3487	.1074	.0563	.0282	.0060	.0010	.0001					
	1	.3151	.3874	.2684	.1877	.1211	.0403	.0098	.0016	.0001				
	2	.0746	.1937	.3020	.2816	.2335	.1209	.0439	.0106	.0014	.004	.0001		
	3	.0105	.0574	.2013	.2503	.2668	.2150	.1172	.0425	.0090	.0031	.0008		
	4	.0010	.0112	.0881	.1460	.2001	.2508	.2051	.1115	.0368	.0162	.0055	.0001	
	5	.0001	.0015	.0264	.0584	.1029	.2007	.2461	.2007	.1029	.0584	.0264	.0015	.0001
	6		.0001	.0055	.0162	.0368	.1115	.2051	.2508	.2001	.1460	.0881	.0112	.0010
	7			.0008	.0031	.0090	.0425	.1172	.2150	.2668	.2503	.2013	.0574	.0105
	8			.0001	.0004	.0015	.0106	.0439	.1209	.2335	.2816	.3020	.1937	.0746
	9					.0001	.0016	.0098	.0403	.1211	.1877	.2684	.3874	.3151
	10						.0001	.0010	.0060	.0282	.0563	.1074	.3487	.5987
11	0	.5688	.3138	.0859	.0422	.0198	.0036	.0005						
	1	.3293	.3835	.2362	.1549	.0932	.0266	.0054	.0007					
	2	.0867	.2131	.2953	.2581	.1998	.0887	.0269	.0052	.0005	.0001			
	3	.0137	.0710	.2215	.2581	.2568	.1774	.0806	.0234	.0037	.0011	.0002		
	4	.0014	.0158	.1107	.1721	.2201	.2365	.1611	.0701	.0173	.0064	.0017		
	5	.0001	.0025	.0388	.0803	.1321	.2207	.2256	.1471	.0566	.0268	.0097	.0003	

(cont.)

Binomial Probabilities (cont.)

								p						
n	X	.05	.10	.20	.25	.30	.40	.50	.60	.70	.75	.80	.90	.95
	6		.0003	.0097	.0268	.0566	.1471	.2256	.2207	.1321	.0803	.0388	.0025	.0001
	7			.0017	.0064	.0173	.0701	.1611	.2365	.2201	.1721	.1107	.0158	.0014
	8			.0002	.0011	.0037	.0234	.0806	.1774	.2568	.2581	.2215	.0710	.0137
	9				.0001	.0005	.0052	.0269	.0887	.1998	.2581	.2953	.2131	.0867
	10					.0007	.0054	.0266	.0932	.1549	.2362	.3835	.3293	
	11						.0005	.0036	.0198	.0422	.0859	.3138	.5688	
12	0	.5404	.2824	.0687	.0317	.0138	.0022	.0002						
	1	.3413	.3766	.2062	.1267	.0712	.0174	.0029	.0003					
	2	.0988	.2301	.2835	.2323	.1678	.0639	.0161	.0025	.0002				
	3	.0173	.0852	.2362	.2581	.2397	.1419	.0537	.0125	.0015	.0004	.0001		
	4	.0021	.0213	.1329	.1936	.2311	.2128	.1209	.0420	.0078	.0024	.0005		
	5	.0002	.0038	.0532	.1032	.1585	.2270	.1934	.1009	.0291	.0115	.0033		
	6		.0005	.0155	.0402	.0792	.1766	.2256	.1766	.0792	.0402	.0155	.0005	
	7			.0033	.0115	.0291	.1009	.1934	.2270	.1585	.1032	.0532	.0038	.0002
	8			.0005	.0024	.0078	.0420	.1208	.2128	.2311	.1936	.1329	.0213	.0021
	9			.0001	.0004	.0015	.0125	.0537	.1419	.2397	.2581	.2362	.0852	.0173
	10					.0002	.0025	.0161	.0639	.1678	.2323	.2835	.2301	.0988
	11						.0003	.0029	.0174	.0712	.1267	.2062	.3766	.3413
	12							.0002	.0022	.0138	.0317	.0687	.2824	.5404
13	0	.5133	.2542	.0550	.0238	.0097	.0013	.0001						
	1	.3512	.3672	.1787	.1029	.0540	.0113	.0016	.0001					
	2	.1109	.2448	.2680	.2059	.1388	.0453	.0095	.0012	.0001				
	3	.0214	.0997	.2457	.2517	.2181	.1107	.0349	.0065	.0006	.0001			
	4	.0028	.0277	.1535	.2097	.2337	.1845	.0873	.0243	.0034	.0009	.0002		
	5	.0003	.0055	.0691	.1258	.1803	.2214	.1571	.0656	.0142	.0047	.0011		
	6		.0008	.0230	.0559	.1030	.1968	.2095	.1312	.0442	.0186	.0058	.0001	
	7		.0001	.0058	.0186	.0442	.1312	.2095	.1968	.1030	.0559	.0230	.0008	
	8			.0011	.0047	.0142	.0656	.1571	.2214	.1803	.1258	.0691	.0055	.0003
	9			.0002	.0009	.0034	.0243	.0873	.1845	.2337	.2097	.1535	.0277	.0028
	10				.0001	.0006	.0065	.0349	.1107	.2181	.2517	.2457	.0997	.0214
	11					.0001	.0012	.0095	.0453	.1388	.2059	.2680	.2448	.1109
	12						.0001	.0016	.0113	.0540	.1029	.1787	.3672	.3512
	13							.0001	.0013	.0097	.0238	.0550	.2542	.5133
14	0	.4877	.2288	.0440	.0178	.0068	.0008	.0001						
	1	.3593	.3559	.1539	.0832	.0407	.0073	.0009	.0001					
	2	.1229	.2570	.2501	.1802	.1134	.0317	.0056	.0006					
	3	.0259	.1142	.2501	.2402	.1943	.0845	.0222	.0033	.0002				

(cont.)

Binomial Probabilities (cont.)

								p						
n	X	.05	.10	.20	.25	.30	.40	.50	.60	.70	.75	.80	.90	.95
	4	.0037	.0349	.1720	.2202	.2290	.1549	.0611	.0136	.0014	.0003			
	5	.0004	.0078	.0860	.1468	.1963	.2066	.1222	.0408	.0066	.0018	.0003		
	6		.0013	.0322	.0734	.1262	.2066	.1833	.0918	.0232	.0082	.0020		
	7		.0002	.0092	.0280	.0618	.1574	.2095	.1574	.0618	.0280	.0092	.0002	
	8			.0020	.0082	.0232	.0918	.1833	.2066	.1262	.0734	.0322	.0013	
	9			.0003	.0018	.0066	.0408	.1222	.2066	.1963	.1468	.0860	.0078	.0004
	10				.0003	.0014	.0136	.0611	.1549	.2290	.2202	.1720	.0349	.0037
	11					.0002	.0033	.0222	.0845	.1943	.2402	.2501	.1142	.0259
	12						.0006	.0056	.0317	.1134	.1802	.2501	.2570	.1229
	13						.0001	.0009	.0073	.0407	.0832	.1539	.3559	.3593
	14							.0001	.0008	.0068	.0178	.0440	.2288	.4877
15	0	.4633	.2059	.0352	.0134	.0047	.0005							
	1	.3658	.3432	.1319	.0668	.0305	.0047	.0005						
	2	.1348	.2669	.2309	.1559	.0916	.0219	.0032	.0003					
	3	.0307	.1285	.2501	.2252	.1700	.0634	.0139	.0016	.0001				
	4	.0049	.0428	.1876	.2252	.2186	.1268	.0417	.0074	.0006	.0001			
	5	.0006	.0105	.1032	.1651	.2061	.1859	.0916	.0245	.0030	.0007	.0001		
	6		.0019	.0430	.0917	.1472	.2066	.1527	.0612	.0116	.0034	.0007		
	7		.0003	.0138	.0393	.0811	.1771	.1964	.1181	.0348	.0131	.0035		
	8			.0035	.0131	.0348	.1181	.1964	.1771	.0811	.0393	.0138	.0003	
	9			.0007	.0034	.0116	.0612	.1527	.2066	.1472	.0917	.0430	.0019	
	10			.0001	.0007	.0030	.0245	.0916	.1859	.2061	.1651	.1032	.0105	.0006
	11				.0001	.0006	.0074	.0417	.1268	.2186	.2252	.1876	.0428	.0049
	12					.0001	.0016	.0139	.0634	.1700	.2252	.2501	.1285	.0307
	13						.0003	.0032	.0219	.0916	.1559	.2309	.2669	.1348
	14							.0005	.0047	.0305	.0668	.1319	.3432	.3658
	15								.0005	.0047	.0134	.0352	.2059	.4633
20	0	.3585	.1216	.0115	.0032	.0008								
	1	.3774	.2702	.0576	.0211	.0068	.0005							
	2	.1887	.2852	.1369	.0669	.0278	.0031	.0002						
	3	.0596	.1901	.2054	.1339	.0716	.0124	.0011						
	4	.0133	.0898	.2182	.1897	.1304	.0350	.0046	.0003					
	5	.0022	.0319	.1746	.2023	.1789	.0746	.0148	.0013					
	6	.0003	.0089	.1091	.1686	.1916	.1244	.0370	.0049	.0002				
	7		.0020	.0546	.1124	.1643	.1659	.0739	.0146	.0010	.0002			
	8		.0004	.0222	.0609	.1144	.1797	.1201	.0355	.0039	.0008	.0001		
	9		.0001	.0074	.0271	.0654	.1597	.1602	.0710	.0120	.0030	.0005		

(cont.)

Binomial Probabilities (cont.)

							p							
n	X	.05	.10	.20	.25	.30	.40	.50	.60	.70	.75	.80	.90	.95
	10			.0020	.0099	.0308	.1171	.1762	.1171	.0308	.0099	.0020		
	11			.0005	.0030	.0120	.0710	.1602	.1597	.0654	.0271	.0074	.0001	
	12			.0001	.0008	.0039	.0355	.1201	.1797	.1144	.0609	.0222	.0004	
	13				.0002	.0010	.0146	.0739	.1659	.1643	.1124	.0546	.0020	
	14					.0002	.0049	.0370	.1244	.1916	.1686	.1091	.0089	.0003
	15						.0013	.0148	.0746	.1789	.2023	.1746	.0319	.0022
	16						.0003	.0046	.0350	.1304	.1897	.2182	.0898	.0133
	17							.0011	.0124	.0716	.1339	.2054	.1901	.0596
	18							.0002	.0031	.0278	.0669	.1369	.2852	.1887
	19								.0005	.0068	.0211	.0576	.2702	.3774
	20									.0008	.0032	.0115	.1216	.3585
25	0	.2774	.0718	.0038	.0008	.0001								
	1	.3650	.1994	.0236	.0063	.0014								
	2	.2305	.2659	.0708	.0251	.0074	.0004							
	3	.0930	.2265	.1358	.0641	.0243	.0019	.0001						
	4	.0269	.1384	.1867	.1175	.0572	.0071	.0004						
	5	.0060	.0646	.1960	.1645	.1030	.0199	.0016						
	6	.0010	.0239	.1633	.1828	.1472	.0442	.0053	.0002					
	7	.0001	.0072	.1108	.1654	.1712	.0800	.0143	.0009					
	8		.0018	.0623	.1241	.1651	.1200	.0322	.0031	.0001				
	9		.0004	.0294	.0781	.1336	.1511	.0609	.0088	.0004				
	10		.0001	.0118	.0417	.0916	.1612	.0974	.0212	.0013	.0002			
	11			.0040	.0189	.0536	.1465	.1328	.0434	.0042	.0007	.0001		
	12			.0012	.0074	.0268	.1139	.1550	.0760	.0115	.0025	.0003		
	13			.0003	.0025	.0115	.0760	.1550	.1139	.0268	.0074	.0012		
	14			.0001	.0007	.0042	.0434	.1328	.1465	.0536	.0189	.0040		
	15				.0002	.0013	.0212	.0974	.1612	.0916	.0417	.0118	.0001	
	16					.0004	.0088	.0609	.1511	.1336	.0781	.0294	.0004	
	17					.0001	.0031	.0322	.1200	.1651	.1241	.0623	.0018	
	18						.0009	.0143	.0800	.1712	.1654	.1108	.0072	.0002
	19						.0002	.0053	.0442	.1472	.1828	.1633	.0239	.0010
	20							.0016	.0199	.1030	.1645	.1960	.0646	.0060
	21							.0004	.0071	.0572	.1175	.1867	.1384	.0269
	22							.0001	.0019	.0243	.0641	.1358	.2265	.0930
	23								.0004	.0074	.0251	.0708	.2659	.2305
	24									.0014	.0063	.0236	.1994	.3650
	25									.0001	.0008	.0038	.0718	.2774

Table VI / Binomial Coefficients (Pascal's Triangle)

n	$\binom{n}{0}$	$\binom{n}{1}$	$\binom{n}{2}$	$\binom{n}{3}$	$\binom{n}{4}$	$\binom{n}{5}$	$\binom{n}{6}$	$\binom{n}{7}$	$\binom{n}{8}$	$\binom{n}{9}$	$\binom{n}{10}$
0	1										
1	1	1									
2	1	2	1								
3	1	3	3	1							
4	1	4	6	4	1						
5	1	5	10	10	5	1					
6	1	6	15	20	15	6	1				
7	1	7	21	35	35	21	7	1			
8	1	8	28	56	70	56	28	8	1		
9	1	9	36	84	126	126	84	36	9	1	
10	1	10	45	120	210	252	210	120	45	10	1
11	1	11	55	165	330	462	462	330	165	55	11
12	1	12	66	220	495	792	924	792	495	220	66
13	1	13	78	286	715	1287	1716	1716	1287	715	286
14	1	14	91	364	1001	2002	3003	3432	3003	2002	1001
15	1	15	105	455	1365	3003	5005	6435	6435	5005	3003
16	1	16	120	560	1820	4368	8008	11440	12870	11440	8008
17	1	17	136	680	2380	6188	12376	19448	24310	24310	19448
18	1	18	153	816	3060	8568	18564	31824	43758	48620	43758
19	1	19	171	969	3876	11628	27132	50388	75582	92378	92378
20	1	20	190	1140	4845	15504	38760	77520	125970	167960	184756

Table VII / Random Numbers*

1368	9621	9151	2066	1208	2664	9822	6599	6911	5112
5953	5936	2541	4011	0408	3593	3679	1378	5936	2651
7226	9466	9553	7671	8599	2119	5337	5953	6355	6889
8883	3454	6773	8207	5576	6386	7487	0190	0867	1298
7022	5281	1168	4099	8069	8721	8353	9952	8006	9045
4576	1853	7884	2451	3488	1286	4842	7719	5795	3953
8715	1416	7028	4616	3470	9938	5703	0196	3465	0034
4011	0408	2224	7626	0643	1149	8834	6429	8691	0143
1400	3694	4482	3608	1238	8221	5129	6105	5314	8385
6370	1884	0820	4854	9161	6509	7123	4070	6759	6113
4522	5749	8084	3932	7678	3549	0051	6761	6952	7041
7195	6234	6426	7148	9945	0358	3242	0519	6550	1327
0054	0810	2937	2040	2299	4198	0846	3937	3986	1019
5166	5433	0381	9686	5670	5129	2103	1125	3404	8785
1247	3793	7415	7819	1783	0506	4878	7673	9840	6629
8529	7842	7203	1844	8619	7404	4215	9969	6948	5643
8973	3440	4366	9242	2151	0244	0922	5887	4883	1177
9307	2959	5904	9012	4951	3695	4529	7197	7179	3239
2923	4276	9467	9868	2257	1925	3382	7244	1781	8037
6372	2808	1238	8098	5509	4617	4099	6705	2386	2830
6922	1807	4900	5306	0411	1828	8634	2331	7247	3230
9862	8336	6453	0545	6127	2741	5967	8447	3017	5709
3371	1530	5104	3076	5506	3101	4143	5845	2095	6127
6712	9402	9588	7019	9248	9192	4223	6555	7947	2474
3071	8782	7157	5941	8830	8563	2252	8109	5880	9912
4022	9734	7852	9096	0051	7387	7056	9331	1317	7833
9682	8892	3577	0326	5306	0050	8517	4376	0788	5443
6705	2175	9904	3743	1902	5393	3032	8432	0612	7972
1872	8292	2366	8603	4288	6809	4357	1072	6822	5611
2559	7534	2281	7351	2064	0611	9613	2000	0327	6145
4399	3751	9783	5399	5175	8894	0296	9483	0400	2272
6074	8827	2195	2532	7680	4288	6807	3101	6850	6410
5155	7186	4722	6721	0838	3632	5355	9369	2006	7681
3193	2800	6184	7891	9838	6123	9397	4019	8389	9508
8610	1880	7423	3384	4625	6653	2900	6290	9286	2396

(cont.)

*From table of "Random Numbers" in *Handbook of Statistical Tables* by D. B. Owen, Addison-Wesley Publishing Co., 1962 by permission of the publisher.

Random Numbers (cont.)

4778	8818	2992	6300	4239	9595	4384	0611	7687	2088
3987	1619	4164	2542	4042	7799	9084	0278	8422	4330
2977	0248	2793	3351	4922	8878	5703	7421	2054	4391
1312	2919	8220	7285	5902	7882	1403	5354	9913	7109
3890	7193	7799	9190	3275	7840	1872	6232	5295	3148
0793	3468	8762	2492	5854	8430	8472	2264	9279	2128
2139	4552	3444	6462	2524	8601	3372	1848	1472	9667
8277	9153	2880	9053	6880	4284	5044	8931	0861	1517
2236	4778	6639	0862	9509	2141	0208	1450	1222	5281
8837	7686	1771	3374	2894	7314	6856	0440	3766	6047
6605	6380	4599	3333	0713	8401	7146	8940	2629	2006
8399	8175	3525	1646	4019	8390	4344	8975	4489	3423
8053	3046	9102	4515	2944	9763	3003	3408	1199	2791
9837	9378	3237	7016	7593	5958	0068	3114	0456	6840
2557	6395	9496	1884	0612	8102	4402	5498	0422	3335
2671	4690	1550	2262	2597	8034	0785	2978	4409	0237
9111	0250	3275	7519	9740	4577	2064	0286	3398	1348
0391	6035	9230	4999	3332	0608	6113	0391	5789	9926
2475	2144	1886	2079	3004	9686	5669	4367	9306	2595
5336	5845	2095	6446	5694	3641	1085	8705	5416	9066
6808	0423	0155	1652	7897	4335	3567	7109	9690	3739
8525	0577	8940	9451	6726	0876	3818	7607	8854	3566
0398	0751	8787	3043	5063	0617	1770	5048	7721	7032
3623	9636	3638	1406	5731	3978	8068	7238	9715	3363
0739	2644	4917	8866	3632	5399	5175	7422	2476	2607
6713	3041	8133	8749	8835	6745	3597	3476	3816	3455
7775	9315	0432	8327	0861	1515	2297	3375	3713	9174
8599	2122	6842	9202	0810	2936	1514	2090	3067	3574
7955	3759	5254	1126	5553	4713	9605	7909	1658	5490
4766	0070	7260	6033	7997	0109	5993	7592	5436	1727
5165	1670	2534	8811	8231	3721	7947	5719	2640	1394
9111	0513	2751	8256	2931	7783	1281	6531	7259	6993
1667	1084	7889	8963	7018	8617	6381	0723	4926	4551
2145	4587	8585	2412	5431	4667	1942	7238	9613	2212
2739	5528	1481	7528	9368	1823	6979	2547	7268	2467

(cont.)

Random Numbers (cont.)

8769	5480	9160	5354	9700	1362	2774	7980	9157	8788
6531	9435	3422	2474	1475	0159	3414	5224	8399	5820
2937	4134	7120	2206	5084	9473	3958	7320	9878	8609
1581	3285	3727	8924	6204	0797	0882	5945	9375	9153
6268	1045	7076	1436	4165	0143	0293	4190	7171	7932
4293	0523	8625	1961	1039	2856	4889	4358	1492	3804
6936	4213	3212	7229	1230	0019	5998	9206	6753	3762
5334	7641	3258	3769	1362	2771	6124	9813	7915	8960
9373	1158	4418	8826	5665	5896	0358	4717	8232	4859
6968	9428	8950	5346	1741	2348	8143	5377	7695	0685
4229	0587	8794	4009	9691	4579	3302	7673	9629	5246
3807	7785	7097	5701	6639	0723	4819	0900	2713	7650
4891	8829	1642	2155	0796	0466	2946	2970	9143	6590
1055	2968	7911	7479	8199	9735	8271	5339	7058	2964
2983	2345	0568	4125	0894	8302	0506	6761	7706	4310
4026	3129	2968	8053	2797	4022	9838	9611	0975	2437
4075	0260	4256	0337	2355	9371	2954	6021	5783	2827
8488	5450	1327	7358	2034	8060	1788	6913	6123	9405
1976	1749	5742	4098	5887	4567	6064	2777	7830	5668
2793	4701	9466	9554	8294	2160	7486	1557	4769	2781
0916	6272	6825	7188	9611	1181	2301	5516	5451	6832
5961	1149	7946	1950	2010	0600	5655	0796	0569	4365
3222	4189	1891	8172	8731	4769	2782	1325	4238	9279
1176	7834	4600	9992	9449	5824	5344	1008	6678	1921
2369	8971	2314	4806	5071	8908	8274	4936	3357	4441
0041	4329	9265	0352	4764	9070	7527	7791	1094	2008
0803	8302	6814	2422	6351	0637	0514	0246	1845	8594
9965	7804	3930	8803	0268	1426	3130	3613	3947	8086
0011	2387	3148	7559	4216	2946	2865	6333	1916	2259
1767	9871	3914	5790	5287	7915	8959	1346	5482	9251

Table VIII / Squares and Square Roots for Numbers 1 Through 1,000*

n	n^2	$\sqrt{n}$	$\sqrt{10n}$	n	n^2	$\sqrt{n}$	$\sqrt{10n}$
1	1	1.000 000	3.162 278	35	1 225	5.916 080	18.70829
2	4	1.414 214	4.472 136	36	1 296	6.000 000	18.97367
3	9	1.732 051	5.477 226	37	1 369	6.082 763	19.23538
4	16	2.000 000	6.324 555	38	1 444	6.164 414	19.49359
5	25	2.236 068	7.071 068	39	1 521	6.244 998	19.74842
6	36	2.449 490	7.745 967	40	1 600	6.324 555	20.00000
7	49	2.645 751	8.366 600	41	1 681	6.403 124	20.24846
8	64	2.828 427	8.944 272	42	1 764	6.480 741	20.49390
9	81	3.000 000	9.486 833	43	1 849	6.557 439	20.73644
10	100	3.162 278	10.00000	44	1 936	6.633 250	20.97618
11	121	3.316 625	10.48809	45	2 025	6.708 204	21.21320
12	144	3.464 102	10.95445	46	2 116	6.782 330	21.44761
13	169	3.605 551	11.40175	47	2 209	6.855 655	21.67948
14	196	3.741 657	11.83216	48	2 304	6.928 203	21.90890
15	225	3.872 983	12.24745	49	2 401	7.000 000	22.13594
16	256	4.000 000	12.64911	50	2 500	7.071 068	22.36068
17	289	4.123 106	13.03840	51	2 601	7.141 428	22.58318
18	324	4.242 641	13.41641	52	2 704	7.211 103	22.80351
19	361	4.358 899	13.78405	53	2 809	7.280 110	23.02173
20	400	4.472 136	14.14214	54	2 916	7.348 469	23.23790
21	441	4.582 576	14.49138	55	3 025	7.416 198	23.45208
22	484	4.690 416	14.83240	56	3 136	7.483 315	23.66432
23	529	4.795 832	15.16575	57	3 249	7.549 834	23.87467
24	576	4.898 979	15.49193	58	3 364	7.615 773	24.08319
25	625	5.000 000	15.81139	59	3 481	7.681 146	24.28992
26	676	5.099 020	16.12452	60	3 600	7.745 967	24.49490
27	729	5.196 152	16.43168	61	3 721	7.810 250	24.69818
28	784	5.291 503	16.73320	62	3 844	7.874 008	24.89980
29	841	5.385 165	17.02939	63	3 969	7.937 254	25.09980
30	900	5.477 226	17.32051	64	4 096	8.000 000	25.29822
31	961	5.567 764	17.60682				
32	1 024	5.656 854	17.88854				
33	1 089	5.744 563	18.16590				
34	1 156	5.830 952	18.43909				

(cont.)

*From *Handbook of Tables for Probability and Statistics*, Second Edition, William H. Beyer, editor, © The Chemical Rubber Co., 1968 by permission of the publisher.

Squares and Square Roots (cont.)

n	n^2	$\sqrt{n}$	$\sqrt{10n}$	n	n^2	$\sqrt{n}$	$\sqrt{10n}$
65	4 225	8.062 258	25.49510	100	10 000	10.00000	31.62278
66	4 356	8.124 038	25.69047	101	10 201	10.04988	31.78050
67	4 489	8.185 353	25.88436	102	10 404	10.09950	31.93744
68	4 624	8.246 211	26.07681	103	10 609	10.14889	32.09361
69	4 761	8.306 624	26.26785	104	10 816	10.19804	32.24903
70	4 900	8.366 600	26.45751	105	11 025	10.24695	32.40370
71	5 041	8.426 150	26.64583	106	11 236	10.29563	32.55764
72	5 184	8.485 281	26.83282	107	11 449	10.34408	32.71085
73	5 329	8.544 004	27.01851	108	11 664	10.39230	32.86335
74	5 476	8.602 325	27.20294	109	11 881	10.44031	33.01515
75	5 625	8.660 254	27.38613	110	12 100	10.48809	33.16625
76	5 776	8.717 798	27.56810	111	12 321	10.53565	33.31666
77	5 929	8.774 964	27.74887	112	12 544	10.58301	33.46640
78	6 084	8.831 761	27.92848	113	12 769	10.63015	33.61547
79	6 241	8.888 194	28.10694	114	12 996	10.67708	33.76389
80	6 400	8.944 272	28.28427	115	13 225	10.72381	33.91165
81	6 561	9.000 000	28.46050	116	13 456	10.77033	34.05877
82	6 724	9.055 385	28.63564	117	13 689	10.81665	34.20526
83	6 889	9.110 434	28.80972	118	13 924	10.86278	34.35113
84	7 056	9.165 151	28.98275	119	14 161	10.90871	34.49638
85	7 225	9.219 544	29.15476	120	14 400	10.95445	34.64102
86	7 396	9.273 618	29.32576	121	14 641	11.00000	34.78505
87	7 569	9.327 379	29.49576	122	14 884	11.04536	34.92850
88	7 744	9.380 832	29.66479	123	15 129	11.09054	35.07136
89	7 921	9.433 981	29.83287	124	15 376	11.13553	35.21363
90	8 100	9.486 833	30.00000	125	15 625	11.18034	35.35534
91	8 281	9.539 392	30.16621	126	15 876	11.22497	35.49648
92	8 464	9.591 663	30.33150	127	16 129	11.26943	35.63706
93	8 649	9.643 651	30.49590	128	16 384	11.31371	35.77709
94	8 836	9.695 360	30.65942	129	16 641	11.35782	35.91657
95	9 025	9.746 794	30.82207	130	16 900	11.40175	36.05551
96	9 216	9.797 959	30.98387	131	17 161	11.44552	36.19392
97	9 409	9.848 858	31.14482	132	17 424	11.48913	36.33180
98	9 604	9.899 495	31.30495	133	17 689	11.53256	36.46917
99	9 801	9.949 874	31.46427	134	17 956	11.57584	36.60601

(cont.)

Squares and Square Roots (cont.)

n	n^2	$\sqrt{n}$	$\sqrt{10n}$	n	n^2	$\sqrt{n}$	$\sqrt{10n}$
135	18 225	11.61895	36.74235	170	28 900	13.03840	41.23106
136	18 496	11.66190	36.87818	171	29 241	13.07670	41.35215
137	18 769	11.70470	37.01351	172	29 584	13.11488	41.47288
138	19 044	11.74734	37.14835	173	29 929	13.15295	41.59327
139	19 321	11.78983	37.28270	174	30 276	13.19091	41.71331
140	19 600	11.83216	37.41657	175	30 625	13.22876	41.83300
141	19 881	11.87434	37.54997	176	30 976	13.26650	41.95235
142	20 164	11.91638	37.68289	177	31 329	13.30413	42.07137
143	20 449	11.95826	37.81534	178	31 684	13.34166	42.19005
144	20 736	12.00000	37.94733	179	32 041	13.37909	42.30839
145	21 025	12.04159	38.07887	180	32 400	13.41641	42.42641
146	21 316	12.08305	38.20995	181	32 761	13.45362	42.54409
147	21 609	12.12436	38.34058	182	33 124	13.49074	42.66146
148	21 904	12.16553	38.47077	183	33 489	13.52775	42.77850
149	22 201	12.20656	38.60052	184	33 856	13.56466	42.89522
150	22 500	12.24745	38.72983	185	34 225	13.60147	43.01163
151	22 801	12.28821	38.85872	186	34 596	13.63818	43.12772
152	23 104	12.32883	38.98718	187	34 969	13.67479	43.24350
153	23 409	12.36932	39.11521	188	35 344	13.71131	43.35897
154	23 716	12.40967	39.24283	189	35 721	13.74773	43.47413
155	24 025	12.44990	39.37004	190	36 100	13.78405	43.58899
156	24 336	12.49000	39.49684	191	36 481	13.82027	43.70355
157	24 649	12.52996	39.62323	192	36 864	13.85641	43.81780
158	24 964	12.56981	39.74921	193	37 249	13.89244	43.93177
159	25 281	12.60952	39.87480	194	37 636	13.92839	44.04543
160	25 600	12.64911	40.00000	195	38 025	13.96424	44.15880
161	25 921	12.68858	40.12481	196	38 416	14.00000	44.27189
162	26 244	12 72792	40.24922	197	38 809	14.03567	44.38468
163	26 569	12.76715	40.37326	198	39 204	14.07125	44.49719
164	26 896	12.80625	40.49691	199	39 601	14.10674	44.60942
165	27 225	12.84523	40.62019	200	40 000	14.14214	44.72136
166	27 556	12.88410	40.74310	201	40 401	14.17745	44.83302
167	27 889	12.92285	40.86563	202	40 804	14.21267	44.94441
168	28 224	12.96148	40.98780	203	41 209	14.24781	45.05552
169	28 561	13.00000	41.10961	204	41 616	14.28286	45.16636

(cont.)

Squares and Square Roots (cont.)

n	n^2	$\sqrt{n}$	$\sqrt{10n}$	n	n^2	$\sqrt{n}$	$\sqrt{10n}$
205	42 025	14.31782	45.27693	240	57 600	15.49193	48.98979
206	42 436	14.35270	45.38722	241	58 081	15.52417	49.09175
207	42 849	14.38749	45.49725	242	58 564	15.55635	49.19350
208	43 264	14.42221	45.60702	243	59 049	15.58846	49.29503
209	43 681	14.45683	45.71652	244	59 536	15.62050	49.39636
210	44 100	14.49138	45.82576	245	60 025	15.65248	49.49747
211	44 521	14.52584	45.93474	246	60 516	15.68439	49.59839
212	44 944	14.56022	46.04346	247	61 009	15.71623	49.69909
213	45 369	14.59452	46.15192	248	61 504	15.74802	49.79960
214	45 796	14.62874	46.26013	249	62 001	15.77973	49.89990
215	46 225	14.66288	46.36809	250	62 500	15.81139	50.00000
216	46 656	14.69694	46.47580	251	63 001	15.84298	50.09990
217	47 089	14.73092	46.58326	252	63 504	15.87451	50.19960
218	47 524	14.76482	46.69047	253	64 009	15.90597	50.29911
219	47 961	14.79865	46.79744	254	64 516	15.93738	50.39841
220	48 400	14.83240	46.90416	255	65 025	15.96872	50.49752
221	48 841	14.86607	47.01064	256	65 536	16.00000	50.59644
222	49 284	14.89966	47.11688	257	66 049	16.03122	50.69517
223	49 729	14.93318	47.22288	258	66 564	16.06238	50.79370
224	50 176	14.96663	47.32864	259	67 081	16.09348	50.89204
225	50 625	15.00000	47.43416	260	67 600	16.12452	50.99020
226	51 076	15.03330	47.53946	261	68 121	16.15549	51.08816
227	51 529	15.06652	47.64452	262	68 644	16.18641	51.18594
228	51 984	15.09967	47.74935	263	69 169	16.21727	51.28353
229	52 441	15.13275	47.85394	264	69 696	16.24808	51.38093
230	52 900	15.16575	47.95832	265	70 225	16.27882	51.47815
231	53 361	15.19868	48.06246	266	70 756	16.30951	51.57519
232	53 824	15.23155	48.16638	267	71 289	16.34013	51.67204
233	54 289	15.26434	48.27007	268	71 824	16.37071	51.76872
234	54 756	15.29706	48.37355	269	72 361	16.40122	51.86521
235	55 225	15.32971	48.47680	270	72 900	16.43168	51.96152
236	55 696	15.36229	48.57983	271	73 441	16.46208	52.05766
237	56 169	15.39480	48.68265	272	73 984	16.49242	52.15362
238	56 644	15.42725	48.78524	273	74 529	16.52271	52.24940
239	57 121	15.45962	48.88763	274	75 076	16.55295	52.34501

(cont.)

Squares and Square Roots (cont.)

n	n^2	$\sqrt{n}$	$\sqrt{10n}$	n	n^2	$\sqrt{n}$	$\sqrt{10n}$
275	75 625	16.58312	52.44044	310	96 100	17.60682	55.67764
276	76 176	16.61325	52.53570	311	96 721	17.63519	55.76737
277	76 729	16.64332	52.63079	312	97 344	17.66352	55.85696
278	77 284	16.67333	52.72571	313	97 969	17.69181	55.94640
279	77 841	16.70329	52.82045	314	98 596	17.72005	56.03570
280	78 400	16.73320	52.91503	315	99 225	17.74824	56.12486
281	78 961	16.76305	53.00943	316	99 856	17.77639	56.21388
282	79 524	16.79286	53.10367	317	100 489	17.80449	56.30275
283	80 089	16.82260	53.19774	318	101 124	17.83255	56.39149
284	80 656	16.85230	53.29165	319	101 761	17.86057	56.48008
285	81 225	16.88194	53.38539	320	102 400	17.88854	56.56854
286	81 796	16.91153	53.47897	321	103 041	17.91647	56.65686
287	82 369	16.94107	53.57238	322	103 684	17.94436	56.74504
288	82 944	16.97056	53.66563	323	104 329	17.97220	56.83309
289	83 521	17.00000	53.75872	324	104 976	18.00000	56.92100
290	84 100	17.02939	53.85165	325	105 625	18.02776	57.00877
291	84 681	17.05872	53.94442	326	106 276	18.05547	57.09641
292	85 264	17.08801	54.03702	327	106 929	18.08314	57.18391
293	85 849	17.11724	54.12947	328	107 584	18.11077	57.27128
294	86 436	17.14643	54.22177	329	108 241	18.13830	57.35852
295	87 025	17.17556	54.31390	330	108 900	18.16590	57.44563
296	87 616	17.20465	54.40588	331	109 561	18.19341	57.53260
297	88 209	17.23369	54.49771	332	110 224	18.22087	57.61944
298	88 804	17.26268	54.58938	333	110 889	18.24829	57.70615
299	89 401	17.29162	54.68089	334	111 556	18.27567	57.79273
300	90 000	17.32051	54.77226	335	112 225	18.30301	57.87918
301	90 601	17.34935	54.86347	336	112 896	18.33030	57.96551
302	91 204	17.37815	54.95453	337	113 569	18.35756	58.05170
303	91 809	17.40690	55.04544	338	114 244	18.38478	58.13777
304	92 416	17.43560	55.13620	339	114 921	18.41195	58.22371
305	93 025	17.46425	55.22681	340	115 600	18.43909	58.30952
306	93 636	17.49286	55.31727	341	116 281	18.46619	58.39521
307	94 249	17.52142	55.40758	342	116 964	18.49324	58.48077
308	94 864	17.54993	55.49775	343	117 649	18.52026	58.56620
309	95 481	17.57840	55.58777	344	118 336	18.54724	58.65151

(cont.)

Squares and Square Roots (cont.)

n	n^2	$\sqrt{n}$	$\sqrt{10n}$	n	n^2	$\sqrt{n}$	$\sqrt{10n}$
345	119 025	18.57418	58.73670	380	144 400	19.49359	61.64414
346	119 716	18.60108	58.82176	381	145 161	19.51922	61.72520
347	120 409	18.62794	58.90671	382	145 924	19.54482	61.80615
348	121 104	18.65476	58.99152	383	146 689	19.57039	61.88699
349	121 801	18.68154	59.07622	384	147 456	19.59592	61.96773
350	122 500	18.70829	59.16080	385	148 225	19.62142	62.04837
351	123 201	18.73499	59.24525	386	148 996	19.64688	62.12890
352	123 904	18.76166	59.32959	387	149 769	19.67232	62.20932
353	124 609	18.78829	59.41380	388	150 544	19.69772	62.28965
354	125 316	18.81489	59.49790	389	151 321	19.72308	62.36986
355	126 025	18.84144	59.58188	390	152 100	19.74842	62.44998
356	126 736	18.86796	59.66574	391	152 881	19.77372	62.52999
357	127 449	18.89444	59.74948	392	153 664	19.79899	62.60990
358	128 164	18.92089	59.83310	393	154 449	19.82423	62.68971
359	128 881	18.94730	59.91661	394	155 236	19.84943	62.76942
360	129 600	18.97367	60.00000	395	156 025	19.87461	62.84903
361	130 321	19.00000	60.08328	396	156 816	19.89975	62.92853
362	131 044	19.02630	60.16644	397	157 609	19.92486	63.00794
363	131 769	19.05256	60.24948	398	158 404	19.94994	63.08724
364	132 496	19.07878	60.33241	399	159 201	19.97498	63.16645
365	133 225	19.10497	60.41523	400	160 000	20.00000	63.24555
366	133 956	19.13113	60.49793	401	160 801	20.02498	63.32456
367	134 689	19.15724	60.58052	402	161 604	20.04994	63.40347
368	135 424	19.18333	60.66300	403	162 409	20.07486	63.48228
369	136 161	19.20937	60.74537	404	163 216	20.09975	63.56099
370	136 900	19.23538	60.82763	405	164 025	20.12461	63.63961
371	137 641	19.26136	60.90977	406	164 836	20.14944	63.71813
372	138 384	19.28730	60.99180	407	165 649	20.17424	63.79655
373	139 129	19.31321	61.07373	408	166 464	20.19901	63.87488
374	139 876	19.33908	61.15554	409	167 281	20.22375	63.95311
375	140 625	19.36492	61.23724	410	168 100	20.24846	64.03124
376	141 376	19.39072	61.31884	411	168 921	20.27313	64.10928
377	142 129	19.41649	61.40033	412	169 744	20.29778	64.18723
378	142 884	19.44222	61.48170	413	170 569	20.32240	64.26508
379	143 641	19.46792	61.56298	414	171 396	20.34699	64.34283

(cont.)

Squares and Square Roots (cont.)

n	n^2	$\sqrt{n}$	$\sqrt{10n}$	n	n^2	$\sqrt{n}$	$\sqrt{10n}$
415	172 225	20.37155	64.42049	450	202 500	21.21320	67.08204
416	173 056	20.39608	64.49806	451	203 401	21.23676	67.15653
417	173 889	20.42058	64.57554	452	204 304	21.26029	67.23095
418	174 724	20.44505	64.65292	453	205 209	21.28380	67.30527
419	175 561	20.46949	64.73021	454	206 116	21.30728	67.37952
420	176 400	20.49390	64.80741	455	207 025	21.33073	67.45369
421	177 241	20.51828	64.88451	456	207 936	21.35416	67.52777
422	178 084	20.54264	64.96153	457	208 849	21.37756	67.60178
423	178 929	20.56696	65.03845	458	209 764	21.40093	67.67570
424	179 776	20.59126	65.11528	459	210 681	21.42429	67.74954
425	180 625	20.61553	65.19202	460	211 600	21.44761	67.82330
426	181 476	20.63977	65.26868	461	212 521	21.47091	67.89698
427	182 329	20.66398	65.34524	462	213 444	21.49419	67.97058
428	183 184	20.68816	65.42171	463	214 369	21.51743	68.04410
429	184 041	20.71232	65.49809	464	215 296	21.54066	68.11755
430	184 900	20.73644	65.57439	465	216 225	21.56386	68.19091
431	185 761	20.76054	65.65059	466	217 156	21.58703	68.26419
432	186 624	20.78461	65.72671	467	218 089	21.61018	68.33740
433	187 489	20.80865	65.80274	468	219 024	21.63331	68.41053
434	188 356	20.83267	65.87868	469	219 961	21.65641	68.48357
435	189 225	20.85665	65.95453	470	220 900	21.67948	68.55655
436	190 096	20.88061	66.03030	471	221 841	21.70253	68.62944
437	190 969	20.90454	66.10598	472	222 784	21.72556	68.70226
438	191 844	20.92845	66.18157	473	223 729	21.74856	68.77500
439	192 721	20.95233	66.25708	474	224 676	21.77154	68.84766
440	193 600	20.97618	66.33250	475	225 625	21.79449	68.92024
441	194 481	21.00000	66.40783	476	226 576	21.81742	68.99275
442	195 364	21.02380	66.48308	477	227 529	21.84033	69.06519
443	196 249	21.04757	66.55825	478	228 484	21.86321	69.13754
444	197 136	21.07131	66.63332	479	229 441	21.88607	69.20983
445	198 025	21.09502	66.70832	480	230 400	21.90890	69.28203
446	198 916	21.11871	66.78323	481	231 361	21.93171	69.35416
447	199 809	21.14237	66.85806	482	232 324	21.95450	69.42622
448	200 704	21.16601	66.93280	483	233 289	21.97726	69.49820
449	201 601	21.18962	67.00746	484	234 256	22.00000	69.57011

(cont.)

Squares and Square Roots (cont.)

n	n^2	$\sqrt{n}$	$\sqrt{10n}$	n	n^2	$\sqrt{n}$	$\sqrt{10n}$
485	235 225	22.02272	69.64194	520	270 400	22.80351	72.11103
486	236 196	22.04541	69.71370	521	271 441	22.82542	72.18033
487	237 169	22.06808	69.78539	522	272 484	22.84732	72.24957
488	238 144	22.09072	69.85700	523	273 529	22.86919	72.31874
489	239 121	22.11334	69.92853	524	274 576	22.89105	72.38784
490	240 100	22.13594	70.00000	525	275 625	22.91288	72.45688
491	241 081	22.15852	70.07139	526	276 676	22.93469	72.52586
492	242 064	22.18107	70.14271	527	277 729	22.95648	72.59477
493	243 049	22.20360	70.21396	528	278 784	22.97825	72.66361
494	244 036	22.22611	70.28513	529	279 841	23.00000	72.73239
495	245 025	22.24860	70.35624	530	280 900	23.02173	72.80110
496	246 016	22.27106	70.42727	531	281 961	23.04344	72.86975
497	247 009	22.29350	70.49823	532	283 024	23.06513	72.93833
498	248 004	22.31591	70.56912	533	284 089	23.08679	73.00685
499	249 001	22.33831	70.63993	534	285 156	23.10844	73.07530
500	250 000	22.36068	70.71068	535	286 225	23.13007	73.14369
501	251 001	22.38303	70.78135	536	287 296	23.15167	73.21202
502	252 004	22.40536	70.85196	537	288 369	23.17326	73.28028
503	253 009	22.42766	70.92249	538	289 444	23.19483	73.34848
504	254 016	22.44994	70.99296	539	290 521	23.21637	73.41662
505	255 025	22.47221	71.06335	540	291 600	23.23790	73.48469
506	256 036	22.49444	71.13368	541	292 681	23.25941	73.55270
507	257 049	22.51666	71.20393	542	293 764	23.28089	73.62065
508	258 064	22.53886	71.27412	543	294 849	23.30236	73.68853
509	259 081	22.56103	71.34424	544	295 936	23.32381	73.75636
510	260 100	22.58318	71.41428	545	297 025	23.34524	73.82412
511	261 121	22.60531	71.48426	546	298 116	23.36664	73.89181
512	262 144	22.62742	71.55418	547	299 209	23.38803	73.95945
513	263 169	22.64950	71.62402	548	300 304	23.40940	74.02702
514	264 196	22.67157	71.69379	549	301 401	23.43075	74.09453
515	265 225	22.69361	71.76350	550	302 500	23.45208	74.16198
516	266 256	22.71563	71.83314	551	303 601	23.47339	74.22937
517	267 289	22.73763	71.90271	552	304 704	23.49468	74.29670
518	268 324	22.75961	71.97222	553	305 809	23.51595	74.36397
519	269 361	22.78157	72.04165	554	306 916	23.53720	74.43118

(cont.)

Squares and Square Roots (cont.)

n	n^2	$\sqrt{n}$	$\sqrt{10n}$	n	n^2	$\sqrt{n}$	$\sqrt{10n}$
555	308 025	23.55844	74.49832	590	348 100	24.28992	76.81146
556	309 136	23.57965	74.56541	591	349 281	24.31049	76.87652
557	310 249	23.60085	74.63243	592	350 464	24.33105	76.94154
558	311 364	23.62202	74.69940	593	351 649	24.35159	77.00649
559	312 481	23.64318	74.76630	594	352 836	24.37212	77.07140
560	313 600	23.66432	74.83315	595	354 025	24.39262	77.13624
561	314 721	23.68544	74.89993	596	355 216	24.41311	77.20104
562	315 844	23.70654	74.96666	597	356 409	24.43358	77.26578
563	316 969	23.72762	75.03333	598	357 604	24.45404	77.33046
564	318 096	23.74868	75.09993	599	358 801	24.47448	77.39509
565	319 225	23.76973	75.16648	600	360 000	24.49490	77.45967
566	320 356	23.79075	75.23297	601	361 201	24.51530	77.52419
567	321 489	23.81176	75.29940	602	362 404	24.53569	77.58866
568	322 624	23.83275	75.36577	603	363 609	24.55606	77.65307
569	323 761	23.85372	75.43209	604	364 816	24.57641	77.71744
570	324 900	23.87467	75.49834	605	366 025	24.59675	77.78175
571	326 041	23.89561	75.65454	606	367 236	24.61707	77.84600
572	327 184	23.91652	75.63068	607	368 449	24.63737	77.91020
573	328 329	23.93742	75.69676	608	369 664	24.65766	77.97435
574	329 476	23.95830	75.76279	609	370 881	24.67793	78.03845
575	330 625	23.97916	75.82875	610	372 100	24.69818	78.10250
576	331 776	24.00000	75.89466	611	373 321	24.71841	78.16649
577	332 929	24.02082	75.96052	612	374 544	24.73863	78.23043
578	334 084	24.04163	76.02631	613	375 769	24.75884	78.29432
579	335 241	24.06242	76.09205	614	376 996	24.77902	78.35815
580	336 400	24.08319	76.15773	615	378 225	24.79919	78.42194
581	377 561	24.10394	76.22336	616	379 456	24.81935	78.48567
582	338 724	24.12468	76.28892	617	380 689	24.83948	78.54935
583	339 889	24.14539	66.35444	618	381 924	24.85961	78.61298
584	341 056	24.16609	76.41989	619	383 161	24.87971	78.67655
585	342 225	24.18677	76.48529	620	384 400	24.89980	78.74008
586	343 396	24.20744	76.55064	621	385 641	24.91987	78.80355
587	344 569	24.22808	76.61593	622	386 884	24.93993	78.86698
588	345 744	24.24871	76.68116	623	388 129	24.95997	78.93035
589	346 921	24.26932	76.74634	624	389 376	24.97999	78.99367

(cont.)

Squares and Square Roots (cont.)

n	n^2	$\sqrt{n}$	$\sqrt{10n}$	n	n^2	$\sqrt{n}$	$\sqrt{10n}$
625	390 625	25.00000	79.05694	660	435 600	25.69047	81.24038
626	391 876	25.01999	79.12016	661	436 921	25.70992	81.30191
627	393 129	25.03997	79.18333	662	438 244	25.72936	81.36338
628	394 384	25.05993	79.24645	663	439 569	25.74879	81.42481
629	395 641	25.07987	79.30952	664	440 896	25.76820	81.48620
630	396 900	25.09980	79.37254	665	442 225	25.78759	81.54753
631	398 161	25.11971	79.43551	666	443 556	25.80698	81.60882
632	399 424	25.13961	79.49843	667	444 889	25.82634	81.67007
633	400 689	25.15949	79.56130	668	446 224	25.84570	81.73127
634	401 956	25.17936	79.62412	669	447 561	25.86503	81.79242
635	403 225	25.19921	79.68689	670	448 900	25.88436	81.85353
636	404 496	25.21904	79.74961	671	450 241	25.90367	81.91459
637	405 769	25.23886	79.81228	672	451 584	25.92296	81.97561
638	407 044	25.25866	79.87490	673	452 929	25.94224	82.03658
639	408 321	25.27845	79.93748	674	454 276	25.96151	82.09750
640	409 600	25.29822	80.00000	675	455 625	25.98076	82.15838
641	410 881	25.31798	80.06248	676	456 976	26.00000	82.21922
642	412 164	25.33772	80.12490	677	458 329	26.01922	82.28001
643	413 449	25.35744	80.18728	678	459 684	26.03843	82.34076
644	414 736	25.37716	80.24961	679	461 041	26.05763	82.40146
645	416 025	25.39685	80.31189	680	462 400	26.07681	82.46211
646	417 316	25.41653	80.37413	681	463 761	26.09598	82.52272
647	418 609	25.43619	80.43631	682	465 124	26.11513	82.58329
648	419 904	25.45584	80.49845	683	466 489	26.13427	82.64381
649	421 201	25.47548	80.56054	684	467 856	26.15339	82.70429
650	422 500	25.49510	80.62258	685	469 225	26.17250	82.76473
651	423 801	25.51470	80.68457	686	470 596	26.19160	82.82512
652	425 104	25.53429	80.74652	687	471 969	26.21068	82.88546
653	426 409	25.55386	80.80842	688	473 344	26.22975	82.94577
654	427 716	25.57342	80.87027	689	474 721	26.24881	83.00602
655	429 025	25.59297	80.93207	690	476 100	26.26785	83.06624
656	430 336	25.61250	80.99383	691	477 481	26.28688	83.12641
657	431 649	25.63201	81.05554	692	478 864	26.30589	83.18654
658	432 964	25.65151	81.11720	693	480 249	26.32489	83.24662
659	434 281	25.67100	81.17881	694	481 636	26.34388	83.30666

(cont.)

Squares and Square Roots (cont.)

n	n^2	$\sqrt{n}$	$\sqrt{10n}$	n	n^2	$\sqrt{n}$	$\sqrt{10n}$
695	483 025	26.36285	83.36666	730	532 900	27.01851	85.44004
696	484 416	26.38181	83.42661	731	534 361	27.03701	85.49854
697	485 809	26.40076	83.48653	732	535 824	27.05550	85.55700
698	487 204	26.41969	83.54639	733	537 289	27.07397	85.61542
699	488 601	26.43861	83.60622	734	538 756	27.09243	85.67380
700	490 000	26.45751	83.66600	735	540 225	27.11088	85.73214
701	491 401	26.47640	83.72574	736	541 696	27.12932	85.79044
702	492 804	26.49528	83.78544	737	543 169	27.14774	85.84870
703	494 209	26.51415	83.84510	738	544 644	27.16616	85.90693
704	495 616	26.53300	83.90471	739	546 121	27.18455	85.96511
705	497 025	26.55184	83.96428	740	547 600	27.20294	86.02325
706	498 436	26.57066	84.02381	741	549 081	27.22132	86.08136
707	499 849	26.58947	84.08329	742	550 564	27.23968	86.13942
708	501 264	26.60827	84.14274	743	552 049	27.25803	86.19745
709	502 681	26.62705	84.20214	744	553 536	27.27636	86.25543
710	504 100	26.64583	84.26150	745	555 025	27.29469	86.31338
711	505 521	26.66458	84.32082	746	556 516	27.31300	86.37129
712	506 944	26.68333	84.38009	747	558 009	27.33130	86.42916
713	508 369	26.70206	84.43933	748	559 504	27.34959	86.48699
714	509 796	26.72078	84.49852	749	561 001	27.36786	86.54479
715	511 225	26.73948	84.55767	750	562 500	27.38613	86.60254
716	512 656	26.75818	84.61678	751	564 001	27.40438	86.66026
717	514 089	26.77686	84.67585	752	565 504	27.42262	86.71793
718	515 524	26.79552	84.73488	753	567 009	27.44085	86.77557
719	516 961	26.81418	84.79387	754	568 516	27.45906	86.83317
720	518 400	26.83282	84.85281	755	570 025	27.47726	86.89074
721	519 841	26.85144	84.91172	756	571 536	27.49545	86.94826
722	521 284	26.87006	84.97058	757	573 049	27.51363	87.00575
723	522 729	26.88866	85.02941	758	574 564	27.53180	87.06320
724	524 176	26.90725	85.08819	759	576 081	27.54995	87.12061
725	525 625	26.92582	85.14693	760	577 600	27.56810	87.17798
726	527 076	26.94439	85.20563	761	579 121	27.58623	87.23531
727	528 529	26.96294	85.26429	762	580 644	27.60435	87.29261
728	529 984	26.98148	85.32292	763	582 169	27.62245	87.34987
729	531 441	27.00000	85.38150	764	583 696	27.64055	87.40709

(cont.)

Squares and Square Roots (cont.)

n	n^2	$\sqrt{n}$	$\sqrt{10n}$	n	n^2	$\sqrt{n}$	$\sqrt{10n}$
765	585 225	27.65863	87.46428	800	640 000	28.28427	89.44272
766	586 756	27.67671	87.52143	801	641 601	28.30194	89.49860
767	588 289	27.69476	87.57854	802	643 204	28.31960	89.55445
768	589 824	27.71281	87.63561	803	644 809	28.33725	89.61027
769	591 361	27.73084	87.69265	804	646 416	28.35489	89.66605
770	592 900	27.74887	87.74964	805	648 025	28.37252	89.72179
771	594 441	27.76689	87.80661	806	649 636	28.39014	89.77750
772	595 984	27.78489	87.86353	807	651 249	28.40775	89.83318
773	597 529	27.80288	87.92042	808	652 864	28.42534	89.88882
774	599 076	27.82086	87.97727	809	654 481	28.44293	89.94443
775	600 625	27.83882	88.03408	810	656 100	28.46050	90.00000
776	602 176	27.85678	88.09086	811	657 721	28.47806	90.05554
777	603 729	27.87472	88.14760	812	659 344	28.49561	90.11104
778	605 284	27.89265	88.20431	813	660 969	28.51315	90.16651
779	606 841	27.91057	88.26098	814	662 596	28.53069	90.22195
780	608 400	27.92848	88.31761	815	664 225	28.54820	90.27735
781	609 961	27.94638	88.37420	816	665 856	28.56571	90.33272
782	611 524	27.96426	88.43076	817	667 489	28.58321	90.38805
783	613 089	27.98214	88.48729	818	669 124	28.60070	90.44335
784	614 656	28.00000	88.54377	819	670 761	28.61818	90.49862
785	616 225	28.01785	88.60023	820	672 400	28.63564	90.55385
786	617 796	28.03569	88.65664	821	674 041	28.65310	90.60905
787	619 369	28.05352	88.71302	822	675 684	28.67054	90.66422
788	620 944	28.07134	88.76936	823	677 329	28.68798	90.71935
789	622 521	28.08914	88.82567	824	678 976	28.70540	90.77445
790	624 100	28.10694	88.88194	825	680 625	28.72281	90.82951
791	625 681	28.12472	88.93818	826	682 276	28.74022	90.88454
792	627 264	28.14249	88.99438	827	683 929	28.75761	90.93954
793	628 849	28.16026	89.05055	828	685 584	28.77499	90.99451
794	630 436	28.17801	89.10668	829	687 241	28.79236	91.04944
795	632 025	28.19574	89.16277	830	688 900	28.80972	91.10434
796	633 616	28.21347	89.21883	831	690 561	28.82707	91.15920
797	635 209	28.23119	89.27486	832	692 224	28.84441	91.21403
798	636 804	28.24889	89.33085	833	693 889	28.86174	91.26883
799	638 401	28.26659	89.38680	834	695 556	28.87906	91.32360

(cont.)

Squares and Square Roots (cont.)

n	n^2	$\sqrt{n}$	$\sqrt{10n}$	n	n^2	$\sqrt{n}$	$\sqrt{10n}$
835	697 225	28.89637	91.37833	870	756 900	29.49576	93.27379
836	698 896	28.91366	91.43304	871	758 641	29.51271	93.32738
837	700 569	28.93095	91.48770	872	760 384	29.52965	93.38094
838	702 244	28.94823	91.54234	873	762 129	29.54657	93.43447
839	703 921	28.96550	91.59694	874	763 876	29.56349	93.48797
840	705 600	28.98275	91.65151	875	765 625	29.58040	93.54143
841	707 281	29.00000	91.70605	876	767 376	29.59730	93.59487
842	708 964	29.01724	91.76056	877	769 129	29.61419	93.64828
843	710 649	29.03446	91.81503	878	770 884	29.63106	93.70165
844	712 336	29.05168	91.86947	879	772 641	29.64793	93.75500
845	714 025	29.06888	91.92388	880	774 400	29.66479	93.80832
846	715 716	29.08608	91.97826	881	776 161	29.68164	93.86160
847	717 409	29.10326	92.03260	882	777 924	29.69848	93.91486
848	719 104	29.12044	92.08692	883	779 689	29.71532	93.96808
849	720 801	29.13760	92.14120	884	781 456	29.73214	94.02127
850	722 500	29.15476	92.19544	885	783 225	29.74895	94.07444
851	724 201	29.17190	92.24966	886	784 996	29.76575	94.12757
852	725 904	29.18904	92.30385	887	786 769	29.78255	94.18068
853	727 609	29.20616	92.35800	888	788 544	29.79933	94.23375
854	729 316	29.22328	92.41212	889	790 321	29.81610	94.28680
855	731 025	29.24038	92.46621	890	792 100	29.83287	94.33981
856	732 736	29.25748	92.52027	891	793 881	29.84962	94.39280
857	734 449	29.27456	92.57429	892	795 664	29.86637	94.44575
858	736 164	29.29164	92.62829	893	797 449	29.88311	94.49868
859	737 881	29.30870	92.68225	894	799 236	29.89983	94.55157
860	739 600	29.32576	92.73618	895	801 025	29.91655	94.60444
861	741 321	29.34280	92.79009	896	802 816	29.93326	94.65728
862	743 044	29.35984	92.84396	897	804 609	29.94996	94.71008
863	744 769	29.37686	92.89779	898	806 404	29.96665	94.76286
864	746 496	29.39388	92.95160	899	808 201	29.98333	94.81561
865	748 225	29.41088	93.00538	900	810 000	30.00000	94.86833
866	749 956	29.42788	93.05912	901	811 801	30.01666	94.92102
867	751 689	29.44486	93.11283	902	813 604	30.03331	94.97368
868	753 424	29.46184	93.16652	903	815 409	30.04996	95.02631
869	755 161	29.47881	93.22017	904	817 216	30.06659	95.07891

(cont.)

Squares and Square Roots (cont.)

n	n^2	$\sqrt{n}$	$\sqrt{10n}$	n	n^2	$\sqrt{n}$	$\sqrt{10n}$
905	819 025	30.08322	95.13149	940	883 600	30.65942	96.95360
906	820 836	30.09983	95.18403	941	885 481	30.67572	97.00515
907	822 649	30.11644	95.23655	942	887 364	30.69202	97.05668
908	824 464	30.13304	95.28903	943	889 249	30.70831	97.10819
909	826 281	30.14963	95.34149	944	891 136	30.72458	97.15966
910	828 100	30.16621	95.39392	945	893 025	30.74085	97.21111
911	829 921	30.18278	95.44632	946	894 916	30.75711	97.26253
912	831 744	30.19934	95.49869	947	896 809	30.77337	97.31393
913	833 569	30.21589	95.55103	948	898 704	30.78961	97.36529
914	835 396	30.23243	95.60335	949	900 601	30.80584	97.41663
915	837 225	30.24897	95.65563	950	902 500	30.82207	97.46794
916	839 056	30.26549	95.70789	951	904 401	30.83829	97.51923
917	840 889	30.28201	95.76012	952	906 304	30.85450	97.57049
918	842 724	30.29851	95.81232	953	908 209	30.87070	97.62172
919	844 561	30.31501	95.80449	954	910 116	30.88689	97.67292
920	846 400	30.33150	95.91663	955	912 025	30.90307	97.72410
921	848 241	30.34798	95.96874	956	913 936	30.91925	97.77525
922	850 084	30.36445	96.02083	957	915 849	30.93542	97.82638
923	851 929	30.38092	96.07289	958	917 764	30.95158	97.87747
924	853 776	30.39737	96.12492	959	919 681	30.96773	97.92855
925	855 625	30.41381	96.17692	960	921 600	30.98387	97.97959
926	857 476	30.43025	96.22889	961	923 521	31.00000	98.03061
927	859 329	30.44667	96.28084	962	925 444	31.01612	98.08160
928	861 184	30.46309	96.33276	963	927 369	31.03224	98.13256
929	863 041	30.47950	96.38465	964	929 296	31.04835	98.18350
930	864 900	30.49590	96.43651	965	931 225	31.06445	98.23441
931	866 761	30.51229	96.48834	966	933 156	31.08054	98.28530
932	868 624	30.52868	96.54015	967	935 089	31.09662	98.33616
933	870 489	30.54505	96.59193	968	937 024	31.11270	98.38699
934	872 356	30.56141	96.64368	969	938 961	31.12876	98.43780
935	874 225	30.57777	96.69540	970	940 900	31.14482	98.48858
936	876 096	30.59412	96.74709	971	942 841	31.16087	98.53933
937	877 969	30.61046	96.79876	972	944 784	31.17691	98.59006
938	879 844	30.62679	96.85040	973	946 729	31.19295	98.64076
939	881 721	30.64311	96.90201	974	948 676	31.20897	98.69144

(cont.)

Squares and Square Roots (cont.)

n	n^2	$\sqrt{n}$	$\sqrt{10n}$	n	n^2	$\sqrt{n}$	$\sqrt{10n}$
975	950 625	31.22499	98.74209	990	980 100	31.46427	99.49874
976	952 576	31.24100	98.79271	991	982 081	31.48015	99.54898
977	954 529	31.25700	98.84331	992	984 064	31.49603	99.59920
978	956 484	31.27299	98.89388	993	986 049	31.51190	99.64939
979	958 441	31.28898	98.94443	994	988 036	31.52777	99.69955
980	960 400	31.30495	98.99495	995	990 025	31.54362	99.74969
981	962 361	31.32092	99.04544	996	992 016	31.55947	99.79980
982	964 324	31.33688	99.09591	997	994 009	31.57531	99.84989
983	966 289	31.35283	99.14636	998	996 004	31.59114	99.89995
984	968 256	31.36877	99.19677	999	998 001	31.60696	99.94999
985	970 225	31.38471	99.24717	1000	1 000 000	31.62278	100.00000
986	972 196	31.40064	99.29753				
987	974 169	31.41656	99.34787				
988	976 144	31.43247	99.39819				
989	978 121	31.44837	99.48848				

Selected Answers to Problems

1. 2, 3, 3   3. a. $\overline{X}$ or median   b. mode or median   c. mode or median   d. median   e. $\overline{X}$ or median   f. $\overline{X}$ 5. a. 396   b. 33; $\Sigma (X - \overline{X}) = (26 - 33) + (34 - 33) + \cdots + (32 - 33) = 0$ 7. 1.03, 1.02, 1.02 (meters)

Chapter 2 / Section 1, page 13

1. a. 9.5   b. 22.5   c. 15.2, 3.9   3. 1.5, 3.6; checks 1, 4   5. 304, 2.4; 4 + 300, 2.4. Subtract the least value from each score, then add this back to get $\overline{X}$. s will not require correction.   7. 6.2′, .4′   9. .03 centimeters

Chapter 2 / Section 2, page 21

1. a. A histogram for take-home pay   b. 189.95, 199.95   c. 200.0, 209.9   d. 214.95   e. Frequency curve for take-home pay

Chapter 2 / Section 3, page 28

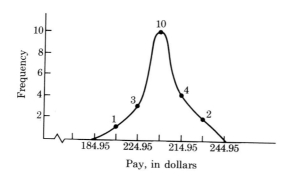

3.

Number of Children	Frequency
0	104
1	123
2	287
3	98
4 or more	88

413

Selected Answers to Problems

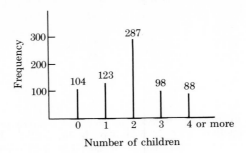

A line diagram

5. More leisure time for all days except for Sunday. Saturday leisure time is now the same as for Sunday.

Chapter 3 / Section 1, page 42

1. a. 3.00, 5.99, 2.995, 5.995   b. 3.00   c. 11.57

3.

Class	Frequency
3–6	1
7–10	2
11–14	1
15–18	4
19–22	4
23–26	2
27–30	1

$n = 15$

5. a.

Class	Frequency
$200.00–209.99	1
210.00–219.99	3
220.00–229.99	10
230.00–239.99	4
240.00–249.99	2

$n = 20$

b. 16   c. 14   d. 10%   7. a. 59.5, 69.5   b. 6   c. $\frac{11}{60}$, or $18\frac{1}{3}\%$ d. $\frac{52}{60}$, or $86\frac{2}{3}\%$

Selected Answers to Problems

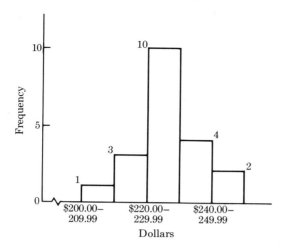

Histogram on weekly pay, in dollars

1. $\overline{X} \doteq 11.2$, $s^2 \doteq 23.09$, $s \doteq 4.8$ $\left(\text{check: } s \approx \dfrac{20-0}{5} = 4\right)$, mode $\doteq 13$

Chapter 3 / Section 2, page 47

3. $s^2 \doteq .5$, $s \doteq .7$ $\left(\text{check } s \approx \dfrac{5.4-1.5}{4} \doteq 1.0\right)$, mode $\doteq 3.7$

5. $\overline{X} \doteq 77.7¢$, $s^2 \doteq 13.7$, $s \doteq 3.7¢$
Answers are based on classes 68–69, 70–71, etc.

1.

Percentile	By Inspection	By Calculation
P_{31}	70	69.7
P_{55}	78	76.0
$Q_2 = P_{50}$	74	74.4 (See Figure 2, p. 50)
$D_9 = P_{90}$	87	88.4
$Q_3 = P_{75}$	83	82.3

Chapter 3 / Section 3, page 53

3. a. See page 416.
b. P_{50}, est. = 135, P_{50} = 132.3   c. P_{25}, est. = 110, P_{25} = 114.4; P_{75}, est. = 150, P_{75} = 151.2

Selected Answers to Problems

3. a.

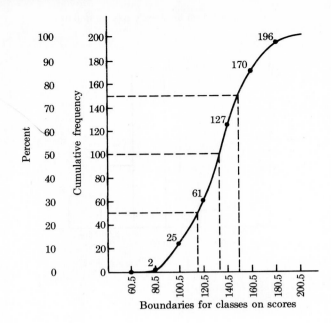

5. a.

Class	Frequency	Cumulative "less than" Frequencies		
110–119	10	Less than 110	0
120–129	15	"	120	10
130–139	25	"	130	25
140–149	90	"	140	50
150–159	70	"	150	140
160–169	110	"	160	210
170–179	40	"	170	320
180–189	30	"	180	360
190–199	10	"	190	390
Total	400	"	200	400

b. median = $P_{50} \approx 160$, Percentile rank $\approx 85\%$ c. median = $P_{50} = 158.1$, Percentile rank = 85.5

Chapter 3 / Section 4, page 60

1. 2.7   3. a. $17\frac{1}{4}\%$ b. $34.50   5. 108; his "buying power" has just stayed even

7.

Year	CPI Value
1970	100
1971	104.3
1972	107.7
1973	111.6
1974	127.0
1975	137.0

Based on the CPI, the cost of consumer goods rose 27% from 1970 to 1974.
9. a. $I_0 = 100(\%)$, $I_1 = 113.7$, $I_2 = 98.0$ b. $I_0 = 100(\%)$, $I_1 = 112.5$, $I_2 = 95.7$ c. By simple aggregates, down 2%; by simple relatives, down 4.3%

Review Problems page 62

1. $\bar{X} = 12$ (check: midrange $= 12$), median $= 11.5$, mode $=$ none, $s^2 = 16$, $s = 4$ (check: range/2 $= 5$)   3. 18.0, 17.3, 4.5, 3.1, 1.76   5. -1.8, -1.8, 3.4, 1.5, 1.24   7. median $= P_{50} = 2.80$, 2.75, .28, .53, (check: range/4 $= .5$), 2.8, 3.15   9. 24.5, 36, 6 (check: range/4 $= 6.75$), 24.5   11. a. Change scales to nearly a 1 to 1 ratio, including values 0, 0.5, 1.0, ..., 5.0 on the vertical scale.   b. Label the vertical axis, e.g. "sales, $100,000."   c. Label the figure, "A Line Graph for Sales, 1970–1975."   13. a. about 4 years of age b. 50 in c. 40 in d. 4 years of age

15. a.

Month	Value
January	100(%)
April	$103\frac{1}{3}$
August	$106\frac{2}{3}$

b. up to $6\frac{2}{3}\%$ 17. a. $P_{90} = 173.5$ b. $Q_3 = P_{75} = 160.4$
c. Percentile rank for 135 is 16.2 d. Q_4 19. 11.3 seconds. Newscasters often quote the statistic that is the single best time.

Chapter 4 / Section 1, page 77

1. a. $\{1,2\}$ b. $\{1\}$ c. S d. $\{3,4\}$ e. $\{3,4\}$ 3. a. $U = \{(A,M), (E,M), (F,M)\}$, $V = \{(A,B), (E,B), (F,B), (I,B)\}$ b. $(T \text{ or } U) = \{(A,B), (A,M), (A,P), (E,M), (F,M)\}$, $V' = \{(A,M), (E,M), (F,M), (A,P), (E,P)\}$, $(U \text{ or } V)' = \{(A,P), (E,P)\}$, $(U \text{ or } V') = \{(A,M), (E,M), (F,M), (A,P), (E,P)\}$ c. T and U, T or V, U' d. no, yes   5. $N(S) = 15$, $N(A \text{ or } B) = 9$, $N(A' \text{ or } B') = 13$, $N(A) = 6$, $N(A \text{ and } B') = 4$, $N(A \text{ or } B)' = 6$, $N(A \text{ and } B) = 2$ See page 418.

7.

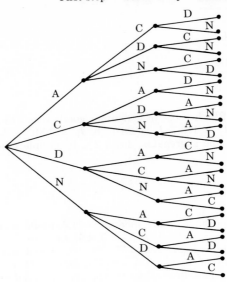

First stop Second stop Third stop

Chapter 4 / Section 2, page 86

1. .004, or 0.4%   3. a. 1 b. .5 c. .6 d. .9 e. .4 Answers are generally different because the assignment of probabilities to simple events is different.
5. a. 12/16 b. 1/2 c. 1/16 d. 1 e. 5/16   7. a. 1 b. 6/15 c. 2/15 d. 9/15 e. 4/15 f. 13/15 g. 6/15   9. S = separation, D = divorce, $P(S) = .10$, $P(S \text{ and } D) = .04$, $P(D) = .08$, $P(S \text{ or } D) = .10 + .08 - .04 = .14$

Chapter 4 / Section 3, page 94

1. a. .3 b. .5 c. .7 d. .1 e. .2 f. 1/3
3.

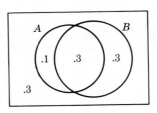

a. F(.7) b. F c. T d. F(.3)   5. U = under 21, S = salaried, $P(U) = .30$, $P(S \text{ and } U) = .42$, $P(S) = .74$, $P(U \text{ or } S) = .30 + .74 - .42 = .62$
7. A = first is defective, B = second is defective, $P(A) = \frac{2}{6}$, $P(B/A) = \frac{1}{5}$, $P(A \text{ and } B) = \frac{2}{6} \cdot \frac{1}{5} = \frac{1}{15}$, or $6\frac{2}{3}\%$ 9. $P(A \text{ or } B) = .70$, $P(A \text{ or } B) = 1 - P[(A \text{ or } B)'] = 1 - P(C) = .70$

Selected Answers to Problems

Chapter 5 / Section 1, page 106

1. a. $\sum P(X) = 1$ b. Both, as $P(3) = \frac{3}{2}$ defies $0 \le P(X) \le 1$
3. a. 1, 1, 1; 1, 1, 1   b. 3, 4, 20; 3, 4, 20   c. 1, 1; 6, 6; 15, 15   d. 1, 1
   e. n, n f. 12, 1, 153, 15504
5. a.

X	$P(X)$
0	.6561
1	.2916
2	.0486
3	.0036
4	.0001
	1.0000

b. $P(X \ge 2) = .0523$

7. a. $P(X \ge 4) = .0067$ b. .0512   c. $.2^3 \cdot .8^2 = .00512$ 9. a. 1/36; Yes
   b. $2(1/52)(1/52) = 1/1352$; Yes   c. $2(1/52)(1/51) = 1/1326$; No   d. .001 or 0.1%

Chapter 5 / Section 2, page 113

1. X = number of red delicious trees   a. $P(2) \doteq .36$
b.

X	$P(X)$
0	10/84
1	40/84
2	30/84
3	4/84

3.

X	$P(X)$	
0	.2	= 0.2
1	.2(.8)	= 0.16
2	$.2(.8)^2$	= 0.128
3	$.2(.8)^3$	= 0.1024
4	$.2(.8)^4$	= 0.0819
5	$.2(.8)^5$	= 0.0655
6	$.2(.8)^6$	= 0.0524
.	.	.
.	.	.
.	.	.

$P(X > 6) \doteq 0.2098$

5. a. .4   b. .8   c. .4   d. 0   7. a. 1/3   b. .3

Chapter 5 / Section 3, page 118

1. a. 2, 2, 1.2, 1.1
b.

X	$P(X)$	$X \cdot P(X)$	$(X - \mu)$	$[(X - \mu)^2 \cdot P(X)]$
0	.0778		-2	
1	.2592		-1	
2	.3456		0	
3	.2304		1	
4	.0768		2	
5	.0102		3	
		$\mu = 2$		$\sigma^2 = 1.2$
				$\sigma \doteq 1.1$

419

Selected Answers to Problems

3.

X	P(X)	X · P(X)	(X − .80)	(X − .80)²	[(X − .80)² · P(X)]
0	.36	0	−.8	.64	.2304
1	.48	.48	.2	.04	.0192
2	.16	.32	1.2	1.44	.2304

Total
$n = 2, p = .4$ $.80 = \mu$ $\sigma^2 = .4800$
   $\sigma \doteq .69$

5.

X	P(X)	X · P(X)	(X − μ)	(X − μ)²	[(X − μ)² · P(X)]
0	10	0	−4/3		
1	40	40	−1/3		
2	30	60	2/3		
3	4	12	5/3		

(with 84 shown as totals for P(X) and X·P(X) columns)

$$\mu = \frac{112}{84} = \frac{4}{3}$$
$$= 1.21$$

$\sigma^2 = 420/756 = .55$
$\sigma \doteq .75$

7. a.

X	P(X)	X · P(X)	(X − 3.1)	[(X − 3.1)² · P(X)]
1	.10	.1	−2.1	.441
2	.20	.4	−1.1	.242
3	.30	.9	−.1	.003
4	.30	1.2	.9	.243
5	.10	.5	1.9	.361

$\mu = 33.1$ $\sigma^2 = 1.290$
   $\sigma = \sqrt{1.29} \doteq 1.1$

b. .6, 1.0, 0.8

Chapter 6 / Section 1, page 129

1. a. 0, 100/12, 2.9 b. 1/2, 0, 1/2
3. a.

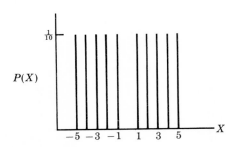

b. 1/2, 1/10, 4/10 c. The distributions cover the same space [−5,5] (except for the discreteness). Probabilities differ due to discrete or continuous distribution.   5. a. no b. 1/3 c. 2/7

Selected Answers to Problems

Chapter 6 / Section 2, page 135

1. a. 34.13%

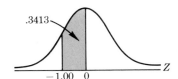

d. 5.05%

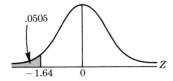

b. 95%

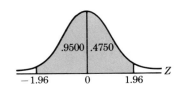

e. 2.5%

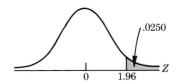

c. 14.88%

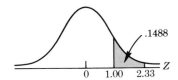

f. 86.41%

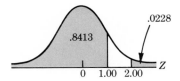

3. a. $Z = -1.30$

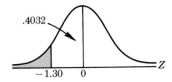

c. $Z = -.67$

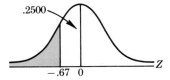

b. $Z = \pm 1.40$

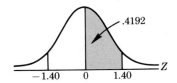

d. $Z = -1.42$

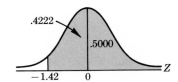

Selected Answers to Problems

e. $Z = 3.08$

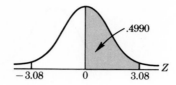

5. a. .01

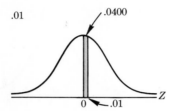

d. 2.15

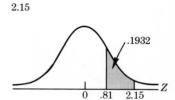

b.

b. $\doteq 1.00$

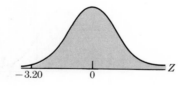

e. .2283

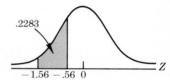

c. .9803

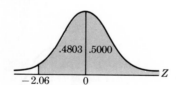

f. $P_{60} \doteq 0.25$ on Z-scale; $Z = 0.30$ is further to the right (on Z-scale), so is bigger and encompasses more area.

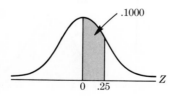
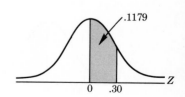

Selected Answers to Problems

Chapter 6 / Section 3, page 143

1. a. $Z = \dfrac{2-4}{1} = -2$

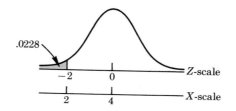

b. $Z = \dfrac{-54 - (-60)}{6} = 1$, $P(X \leq -54) = P(Z \leq 1) = \Phi(1) = .8413$

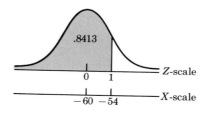

c. $Z = \dfrac{5-10}{.8} = -6.25$

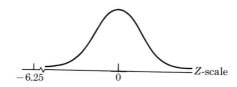

d. $P(-2 < X < 3) = P(-2 < Z < .5) = .6687$

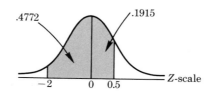

e. $P(X < -1.5 \text{ or } X > 0) = P(Z < -.25 \text{ or } Z > .5) = .7048$

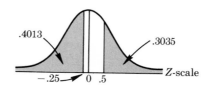

Selected Answers to Problems

3. $P(X < 4.0) = P(Z < -2.0) = .0228$
5. a. $P(X < 48.0 \text{ or } X > 50.00) = .095$, or 9.5%   b. 76¢ per 100 boxes
7. a. $P(Z < .7) = {}_F(.7) = .7580$, answer: 76 (rounded)   b. $P(Z < .25) = \Phi(.25) = .5987 \doteq .6$, so $.25 = \dfrac{X - 386}{20}$, answer: $X = P_{60} \doteq 391$ c. $\Phi(.67) = P(Z < .67) = .7486 \doteq .75$, so $.67 = \dfrac{X - 386}{20}$, answer: $P_{75} \doteq 399$. Using the symmetry of the normal curve $P(Z < -.67) = .2514 \doteq .25$ so $-.67 = \dfrac{X - 386}{20}$, answer: $P_{25} \doteq 373$

d. Third quarter (see part c); $P_{75} \doteq 399$, and median $= \mu = 386$.
e. $P(Z < -.8) = \Phi(-.8) = 1 - \Phi(.8) = 1 - .7881 = .2119$

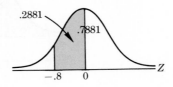

1. a. .2025   b. Approximation .2144; the approximation is off by 1.19% in absolute difference   3. Let X = number of Florida mosquitos killed $= 0, 1, 2, \ldots, 100$ so $P(X > 85) \doteq P(Z > 1.38) = .0838$

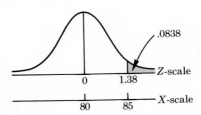

5. $Z = +.12$, $P(X \leq 20) \doteq .5478$ 7. For X = number of special lunches prepared $= 0, 1, 2 \ldots, 450$ $.1469 = P(Z > 1.05) \doteq P(X > 165)$

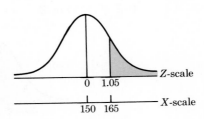

Selected Answers to Problems

Review Problems
page 152

1. a. .1446   b. $P(.4 < Z < 1) = .1859$ c. 2.10   d. ± 0.3 e. $P(Z > 1.45) = .0735$
f. $\Phi(.52) \doteq .70$, so $.52 \doteq \dfrac{X - 0}{2}$, $X \doteq 1.04$ 3. $P(X > 25) \doteq P(Z > 1.38) = .0838$

5. a. $\mu = -1$, $\sigma^2 = 10$, $\sigma = 3.16$ b. 2/3   7. a. $\binom{9}{3} = 84$ b. $\binom{2}{1}\binom{7}{2} = 42$
c. 1/2   9. a. $6/45 = .13$ b. $\binom{2}{2}(.4)^2(.6)^0 = .16$ c. $12/45 = .27$ 11. a. 69
(rounded)   b. $P_{45} \doteq \Phi(-.12) = 169$ where $-.12 = \dfrac{X - 170}{10}$ gives $P_{45} = X$
13. $P(Z < -1.50) = .0668$ 15. a. Binomial; $P(X \le 2) = .8370$ b. Normal approximation; $P(X \le 100) \doteq P(Z < 1.32) = .9066$ 17. $P(X = 1, 2, \text{ or } 3) = \dfrac{91}{216} = .42$, X = number of spots, has geometric distribution   19. $P(X < 12) \doteq P(Z < -2.12) = .017$ 21. $P(X < 60) \doteq P(Z < 1.90) = .9713$

23.

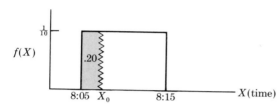

$X_0 = 8{:}07$ P.M.

25.

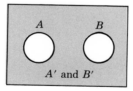

No

Chapter 7 / Section 1,
page 160

1. a. 780   b. 79,800   3. a. Parameters: $N = 642$ = number of registered voters, p = percent who voted in this election, μ = mean age for all registered voters, $p = .623$, $\hat{\mu} = 28.4$.   b. Likely not, there are one or more reasons why some did not vote   5. a. 9/15   b. $P(X \ge 1) = 1 - P(X = 0) = .96$

Chapter 7 / Section 2,
page 167

1. $\mu_{\bar{X}} = 50$, $\sigma_{\bar{X}} = 20/\sqrt{100} = 2$ 3. $\sigma = \sigma_X$ refers to the distribution of "individual" values; $\sigma_{\bar{X}}$ refers to the distribution of sample mean values. For $\dfrac{n}{N} \le .05$, $\sigma_{\bar{X}} = \sigma/\sqrt{n}$.

Selected Answers to Problems

5.

Sample	$\overline{X}$		Sampling Distribution			
		$\overline{X}$	$P(\overline{X})$	$[\overline{X}P(\overline{X})]$	$(\overline{X} - \mu)$	$[(\overline{X} - \mu)^2 \cdot P(\overline{X})]$
A, B, C	2	2	.1		−1	.9
A, B, D	2	$\frac{7}{3}$	.1		$-\frac{2}{3}$	.4
A, B, E	$\frac{7}{3}$	$\frac{8}{3}$	.2		$-\frac{1}{3}$	.2
A, B, F	$\frac{8}{3}$	3	.2		0	0.9
A, C, D	$\frac{7}{3}$	$\frac{10}{3}$	.2		$\frac{1}{3}$	.2
A, C, E	$\frac{8}{3}$	$\frac{11}{3}$	.1		$\frac{2}{3}$	.4
A, C, F	3	4	.1		1	.9
A, D, E	$\frac{8}{3}$				$\mu_{\overline{X}} = 3.0$	$\sigma_{\overline{X}}^2 = \frac{3.0}{9} = .33$
A, D, F	3					$\sigma_{\overline{X}} = .58$
A, E, F	$\frac{10}{3}$					
B, C, D	$\frac{8}{3}$		Comparison to Problem 4 a. $\mu_{\overline{X}}$ is unchanged			
B, C, E	3		b. For bigger n, $\sigma_{\overline{X}}$ is decreased			
B, C, F	$\frac{10}{3}$					
B, D, E	3					
B, D, F	$\frac{10}{3}$					
B, E, F	$\frac{11}{3}$					
C, D, E	$\frac{10}{3}$					
C, D, F	$\frac{11}{3}$					
C, E, F	4					
D, E, F	4					

Chapter 7 / Section 3, page 173

1. $P(\overline{X} > 7.0) = .0475$

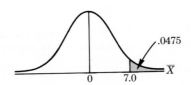

Note: $\sigma_{\overline{x}} = .3$

3. $P(\overline{X} < 32) = P(Z < 1.37) = .9147$ 5. a. $P(X > \$1.30) = P(Z > .4) = .3446$,
X_1 = amounts spent; has normal distribution   b. $P(1.20 < \overline{X} < 1.30)$
$= P(-2.67 < Z < 4.00) = .9962$

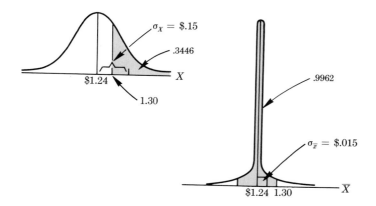

Note: $\sigma_{\bar{x}} = \dfrac{\sigma}{\sqrt{n}} = \dfrac{15¢}{\sqrt{100}} = 1.5¢$ i.e., the deviation unit in part a is $\sqrt{100} =$ 10 times the size of that in part b.

1. $42.13 \pm \left(\dfrac{1.96 \cdot \$6.50}{\sqrt{100}}\right) = \$42.13 \pm \$1.27, \$40.86 < \mu < \$43.40$ 3. 60 ± 3.7 mph. Interpretation: 95% of all intervals computed in this way will include the true mean speed. This single interval gives estimates 56.3 mph $< \mu <$ 63.7 mph. b. 60 ± 4.8 mph, 95% confidence: $u - l = 63.7 - 56.3 = 7.4$ units, 99% confidence: $u - l = 64.8 - 55.2 = 9.6$ units. For a fixed sample size, a higher confidence requires a longer interval (span) of possibilities.   5. $\overline{X} = 0.8$
a. $s^2 = \dfrac{46}{49} = .9388$, $s = .97$ b. using $\overline{X}$ and s from Part a, $0.8 \pm .3$ siblings, $0.5 < \mu < 1.1$ siblings

Chapter 8 / Section 1, page 184

1. $\$1,320.00 \pm \left(\dfrac{1.96 \cdot 280.00}{\sqrt{101}}\right)$, or $\$1,320.00 \pm \$54.61$ 3. $n = (20.64)^2 \doteq 427$
5. a. $n = 139$ loads (rounded up)   b. 25.2 tons ± 0.1 tons $\to 25.1 < \mu < 25.3$ tons/load   c. $3000 \cdot 25.2 = 75,600$ tons

Chapter 8 / Section 2, page 190

1. a. $\hat{p} = 576/900 = .64$ b. $\sigma_{\hat{p}} = .02$ c. $E = 1.96\,\sigma_{\hat{p}} = .04$ d. $\hat{p} \pm E$ $= .64 \pm .04$, so $.60 < p < .68$ 3. $n = 2171.56 \to$ take $n = 2172$
5. $.75 \pm .09$ 7. a. Yes   b. 30% to 34%   c. \$5,544

Chapter 8 / Section 3, page 195

Selected Answers to Problems

Chapter 9 / Section 1, page 207

1. $H_0: \mu = \$4.75$, $H_A: \mu < \$4.75$ 3. Calculated $Z = -1.75$, decision: cannot reject H_0. Conclusion: the evidence does not support the claim that these workers are underpaid, $\alpha = .025$.   5. a. H_0: the vaccine is effective   b. H_0: the vaccine is effective

Chapter 9 / Section 2, page 212

1. a. $H_0: \mu = 10$, $H_A: \mu > 10$ b. Test: reject H_0 if calculated $Z > 1.64$, do not reject H_0 if calculated $Z < 1.64$ c. Calculated $Z = 6$ d. Decision: Reject H_0. Conclusion: the sessions increased the tolerance score above 10 points for these persons, $\alpha = .05$.   3. $H_0: \mu = \$38,950$, $H_A: \mu \neq \$38,950$, rejection region: calculated $Z < -1.96$ or calculated $Z > +1.96$, calculated $Z = 4.05$, decision: reject H_0. Conclusion: valuation in the University area is not statistically the same as the mean valuation, $\alpha = .05$.   5. $H_A: \mu < \$320$, calculated $Z = -2.8$. Food costs were "substantially" decreased, $\alpha = .05$.

Chapter 9 / Section 3, page 217

1. $H_0: p = .65$, p = percent of junior-high youth with attitude against littering, $H_A: p \neq .65$, test: reject H_0 if |calculated Z| > 2.05, decision: cannot reject H_0. Conclusion: essentially 65% of the junior-high youth had an attitude against littering.   3. $H_A: p < .80$, calculated $Z = -1.25$, decision: cannot reject H_0. Conclusion: the claim is upheld, $\alpha = .075$. The cancerous growth was stopped on 80% or more of these mice.   5. $H_A: p > 1/2$, p = true percent who favor the proposed change, test: reject H_0 if calculated $Z > 1.64$, Calculated $Z = 1.40$, decision: cannot reject H_0.

Review Problems page 220

1. a. $P(16.0 < X < 20.0) = P(-2.30 < Z < -0.30) = .3714$ b. $P(\overline{X} \leq 21.0) = P(Z \leq 1.60) = \Phi(1.60) = .9452$ 3. a. 20.0 ± 1.31 b. $4 \cdot 36 = 144$ (Again, see the Nielsen Company pictures in Chapter 1)   5. $H_A: \mu > 66.8$, test: reject H_0 if calculated $Z > 1.64$, calculated $Z = 4.0$, decision: reject H_0, $\alpha = .05$.   7. a. .10   b. .30   9. a. $n = 62$ b. $E = (1.96 \cdot \$40)/\sqrt{100} = \$7.84$ 11. .7888 (note that $\sigma_{\bar{x}} = 4.0$)   13. Calculated $Z = 2.50$. Implement the advanced science program.   15. $P(\overline{X} > 72) = P(Z > 1.50) = .0668$ 17. $H_A: p < 1/2$, p = percentage of peanuts, calculated $Z = -2.04$, decision: reject H_0, $\alpha = .025$. Conclusion: less than half of the nuts are peanuts.   19. $n \leq 269$ students. Note: since we used $\hat{p} = 1/2$, we should expect $n = 269$ to be quite sufficient; possibly even a smaller sample would do.   21. a. $\mu_{\overline{X}} = \sum [\overline{X} P(\overline{X})] = 9.0$, $\sigma_{\overline{X}} = \sqrt{\sum (\overline{X} - \mu_{\overline{X}})^2 \cdot P(\overline{X})} \doteq 1.7$ b. .3, 0, .4, .8   23. $H_A: \mu < 85$, calculated $Z = -2.33$, decision: reject H_0, $\alpha = .05$ 25. $\overline{X} = \dfrac{1920}{60} = 32$, $s = \sqrt{\dfrac{4860}{59}} = 9.08$, $29.7 < \mu < 34.3$

Chapter 10 / Section 1, page 230

1. a. 1.71   b. 1.76   c. 2.13   d. 2.49   e. 1.71   f. 1.32   3. a. $t_{.05, 21} = -1.72$ b. 95%   c. 1.81   d. $\alpha = .01$ 5. a. Go to the row labeled 17, column headed 0.25. Since 25% of the area is above 0.6892 (and so 75% is below), this is P_{75}.   b. For a two-tailed test use the column headed $.05/2 = .025$. For the row df = 12 this column entry is 2.18.   c. This is a matter of space. Values for $\alpha = .01$, .02, ..., .99 would take many pages.   7. a. 4   b. 14   c. 0   d. 20

Selected Answers to Problems

Chapter 10 / Section 2, page 235

1. $18.9 < \mu < 21.1$ years 3. $n = 44$ 5. a. 29.35 ± 9.13 b. $n = 76$ (rounded-up) 7. Use $Z = 1.96$ instead of $t_{.025, 15} = 2.13$. It is narrower because the error, E, is determined through a bigger divisor.

Chapter 10 / Section 3, page 239

1. $\overline{X} = 4.5$, $s = 1.52$, $H_A: \mu < 5.0$, calculated $t = -.81$, μ is *not* less than 5.0 3. $H_A: \mu < 4{,}000$ pounds. Conclusion: his claim is discredited by this test. 5. $H_A: \mu > 12.62$, test: reject H_0 if calculated $t > 1.83$, Calculated $t = 2.38$ for $\overline{X} = 14$, $s^2 = 3\,1/3$, $s = 1.83$. Possible type 2 Error: to decide that the additive doesn't improve mileage, when in fact it does.

Chapter 10 / Section 4, page 244

1. $H_A: \mu_d < 0$ (for d = after-before), test $t = -2.82$, calculated $t = -2.31$, decision: cannot reject H_0. Conclusion: according to this test the diet did not produce substantial weight loss, $\alpha = .01$. 3. $H_A: \mu_d > 0$ (d = new-old), calculated $t = 3.80$. Conclusion: the new layout affords faster processing, $\alpha = .05$. 5. a. $H_A: \mu_d > 30$ (d = increase in reading score), test: reject H_0 if calculated $t > 2.13$ ($t_{.025, 15}$). Conclusion: the average increase for these second graders does not exceed 32 points. b. $32 \pm (2.13 \cdot 1.38)$. The true average increase in reading score for second graders at this school is estimated to be somewhere between 29.1 and 34.9 points.

Chapter 11 / Section 1, page 256

1. $H_A: \mu_1 \neq \mu_2 (\mu_1 - \mu_2 \neq 0)$, calculated $t = 3.61$ for $s_p^2 = 26.3$. Conclusion: average achievement on this test is not the same for students in these two American History classes. 3. a. $s_A^2 = 3.07$, $s_B^2 = .92$ b. $s_p^2 = 1.85$ 5. Test $t = 1.72$, calculated $t = 1.17$ for $s_p^2 = 35.8$, cannot reject equal means 7. $\overline{X}_1 = 5.22$, $s_1^2 = 2.44$, $\overline{X}_2 = 7.00$, $s_2^2 = 2.00$, calculated $t = 2.97$ for $s_p^2 = 2.25$. Conclusion: there is a higher average productivity for those chosen by a job test over those chosen by the quick interview selection, $\alpha = .025$.

Chapter 11 / Section 2, page 262

1. a. $\doteq 4.20$ b. $\doteq 5.67$ c. 4.30 d. $\doteq 2.16$ e. ≈ 1.40 f. 4052.2 3. For higher degrees of freedom (both numerator and denominator) *upper-tail* percentage points are closer to zero.

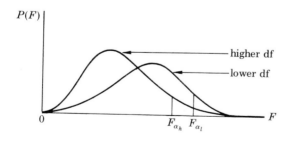

5. S_1^2, s_2^2 are each computed like $\dfrac{\Sigma(X - \overline{X})^2}{n - 1}$ and so are zero or larger. The quotient of nonnegative numbers is nonnegative.

Selected Answers to Problems

Chapter 11 / Section 3, page 267

1.

Source	df	Sum of squares	Mean squares	F
Among groups	3	120	40	4.21
Within groups	8	76	9.5	
Total	11	196	$F_{.01,3,8} = 7.59$	

Cannot reject equal means, $\alpha = .01$.

3. Calculated $F = 2.59$, $F_{.05,4,55} = 2.52$, tabular $F_{.025,4,55} = 3.01$. By choosing an α-level after calculating F a significant difference can usually be found, but this is *not* statistics.   5. $H_0: \mu_C = \mu_E$, $H_A: \mu_C \neq \mu_E$ a. Test: reject H_0 if $|\text{calculated } t| > 2.10(t_{.025,18})$, calculated $t = -2.27$ for $s_p^2 = 3.9$, decision: reject H_0, $\alpha = .05$. Conclusion: mean time to recall is not the same.   b. Use the same H_0 and H_A as in part a, Within MS $= \dfrac{s_1^2 + s_2^2}{2} = \dfrac{7.8}{2} = 3.9$

Source	df	Sum of squares	Mean squares	F
Among	1	20	20	5.13
Within	18		3.9	
Total	19		$F_{.05,1,18} = 4.41$	

Decision: reject H_0, $\alpha = .05$.

Conclusion: mean time to recall is not the same.

Chapter 11 / Section 4, page 273

1. $H_0: \mu_A = \mu_B = \mu_C = \mu_D$

Source	df	Sum of squares	Mean squares	F
Among groups	3	126.19	42.06	3.33
Within groups	11	138.81	12.62	
Total	14	265	$F_{.01,3,11} = 6.22$	

Decision: cannot reject (4) equal means.

3. $H_0: \mu_1 = \mu_2 = \mu_3$

Source	df	Sum of squares	Mean squares	F
Among groups	2	2,533.5	1,266.75	31.25
Within groups	9	364.75	40.53	
Total	11	2,898.25	$F_{.025,2,9} = 5.71$	

Reject H_0, $\alpha = .025$. Mean IQ's are not the same.

Selected Answers to Problems

5. $H_0: \mu_C = \mu_1 = \mu_2$, H_A: at least one $\mu \neq$ the others

Source	df	Sum of squares	Mean squares	F
Among groups	2	10.71	5.36	2.82
Within groups	11	21.00	1.91	
Total	13	31.71	$F_{.05,2,11} = 3.98$	

Cannot reject H_0. The run times have statistically equal means, $\alpha = .05$.

Review Problems page 276

1. a. .0207 b. 1.77 c. .95 d. 10.97 e. .05 f. 3.12
3.

Source	df	Sum of squares	Mean squares	F
Treatments	2	198	99	12.25
Within	24	194	8.08	
Total	26	392	$F_{.05,2,24} = 3.40$	

Reject H_0: the means are not all equal.   5. a. $H_A: \sigma_1^2 \neq \sigma_2^2$, test: Reject H_0 if calculated $F > 5.60$ ($F_{.025,8,6}$), calculated $F = 3.2$, cannot reject H_0, $\alpha = .05$.   b. $H_0: \mu_1 = \mu_2$, $H_A: \mu_1 \neq \mu_2$, calculated $t = 2.22$. Conclusion: the means are not equal.

c.

Source	df	Sum of squares	Mean squares	F
Universities	1	7.72	7.72	4.89
Within	14	22.10	1.58	
Total	15	29.82	$(t^2_{.025,14}) = F_{.05,1,14} = 4.60$	

Same conclusion as in Part b [Note: (calculated t)² = $(2.22)^2 = 4.93 \doteq$ calculated $F = 4.89$, with a slight error due to rounding off in calculations.]
7. Calculated $F = 4.42$, Test $F_{.025,15,11} = 3.33$ 9. Using total SS − Within SS = Groups SS gives us $1397 - 1244.3 = 152.7$, while the direct calculation gives Groups SS = $5(26.4 - 31.7)^2 + \cdots + 5(36.4 - 31.7)^2 = ?$ Some math is incorrect; in this case, correct the total SS to 1497.0.   11. $H_A: \mu_d > 0$, $d =$ Rating 1 − Rating 2, calculated $t = 2.67$, decision: reject H_0, $\alpha = .05$. The evidence supports a higher first rating.   13. $H_0: \mu = 10.0$ beats/minute
a. $H_A: \mu \neq 10.0$ beats/minute, calculated $t = 2.38$, decision: reject H_0, $\alpha = .05$. Conclusion: This group reacts differently to the drug.   b. $10.1 < \mu < 15.9$. The mean increase is estimated at between 10.1 and 15.9 beats/minute.

Selected Answers to Problems

15. a. $H_0: \mu_A = \mu_B = \mu_C$

Source	df	Sum of squares	Mean squares	F
Brands	2	170	85	27.4
Error	9	28	3.1	
Total	11	198	$F_{.05,2,9} = 4.26$	

Reject H_0, average mileage is not the same for all brands, A, B, and C.
b. For Brand C: $26.4 < \mu_C < 33.4$ miles/gallon.
17. Calculated $Z = 2.47$. Conclusion: Average wage for newly trained elementary education majors is statistically greater than the average paid to new secondary education graduates, $\alpha = .05$. 19. $H_0: \mu_1 - \mu_2 = \$.50$ ($\mu_1 = \mu_2 + \$.50$), $H_A: \mu_1 - \mu_2 > \$.50 (\mu_1 > \mu_2 + \$.50)$, test: Reject H_0 if calculated $t > 2.13$ ($t_{.025,15}$), $s_p^2 = .10$. Conclusion: the mean wage paid city employees in City 1 is not over $.50 per hour above the mean wage for City 2 employees. A possible type 1 error: To decide the difference exceeds $.50 when actually it does not.

Chapter 12 / Section 1, page 286

1.

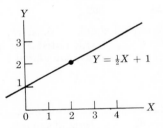

3. A plot of entrance scores against English grades

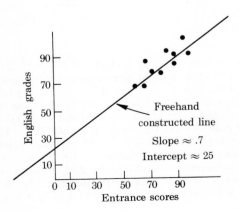

Selected Answers to Problems

5. a. A   b. A, B   c. $Y_i = c + dX_i + fX_i^2 + e_i$. The points display curvature, but they do not fall exactly in a quadratic pattern, so there is also statistical error (e_i).

1. a. XY data plots with freehand regression line

Chapter 12 / Section 2, page 291

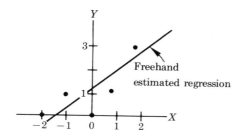

b. $\hat{Y} = 1 + .6x$ 3. $\hat{Y} = 28.4 + .67X$. My first estimates, by graphing, were 25 and 0.7.

5. a. Data plot of percent absent (Y) by weeks (X)

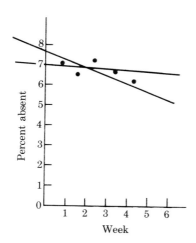

It is quite difficult to tell with so few points (e.g., observe the two lines above); which gives a better fit?   b. $\hat{Y} = 7.2 - .2X$ c. $\hat{Y} = 7.2 - .2(5) = 6.2$ $Y - \hat{Y} = 6 - 6.2 = -.2$ looks fairly close   d. Due to the small sample, statistical testing is questionable. The slope is *negative* but small. The relation is not yet clearly observed.

Selected Answers to Problems

Chapter 12 / Section 3, 1.
page 298

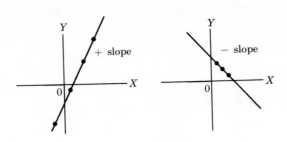

These are a perfect positive and a perfect negative relation, respectively.

3. a. Data plot of yield versus fertilizer

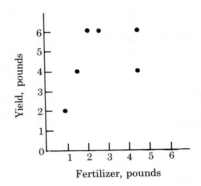

There does *not* appear to be a strong correlation.   b. $r = .05$, $100r^2 = 0.25\%$ explains less than 1% of the variation in growth.   5. $r_{1Y} = .69$, $r_{2Y} = -.66$, so Variable 1 has a stronger relation to Y.   7. $r = .98$

Chapter 12 / Section 4, 1. $\hat{Y} = 4.90 + 0.14X_c$, $5.32 million (origin = period 3, X_c-units = 1 period)
page 303
3. $\hat{Y} = 6.6 - .2X_c$ (origin = week 3, X_c = whole units), $\hat{Y}_6 = 6.6 - .2(3) = 6.0(\%)$. (Note: Problem 5 in Section 2 has its origin at time 0, so there the Y-intercept is 7.2.) We're assuming that this linear trend continues through week 6. As the hint suggests it is unlikely that such a strong reduction would persist.   5. $\hat{Y} = 8.0 + 1.1X_c$ (origin = 11 - 12, $X_c - \frac{1}{2}$ units). Moving toward either extreme, say 0 or 50, a linear relation would not be reasonable.

Chapter 12 / Section 5, 1. a. a, c, e   b. b
page 310

Selected Answers to Problems

3.

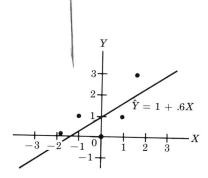

A linear regression is questionable, e.g., a quadratic regression appears better. $H_0: \beta_1 = 0$; $H_A: \beta_1 \neq 0$, Calculated $t = 2.14$. Conclusions: cannot reject H_0, $\alpha = .05$. By observation this is *not* a strong linear regression. There remains the possibility of a reasonably high β-risk. If at all possible I would use more than 5 data points for testing.   5. $1.56 < \beta_1 < 2.44$ 7. Each additional point on the test will generally relate to an increase of from $4/10$ to $6/10$ of a point in the English course. So for an increase of say 10 points (e.g., from 40 to 50) in test score, I would expect an increase of from 4 to 6 points in the English course. Also, for a test score of, say 50, the best single guess for one's English course score is $\hat{Y} = 55 + .5(50) = 80$.

1. a. r is negative, but not so close to -1.   b. Intercept 7 appears reasonable, but the slope is not $+1$; the slope is about -1.   3. The Y intercept is wrong.
5.

Review Problems page 317

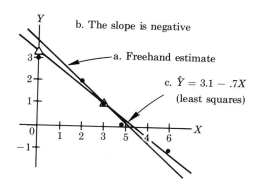

435

Selected Answers to Problems

7. Number of college graduates

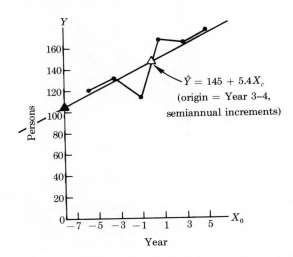

A linear trend appears moderately good, but a higher order polynomial trend could improve the fit.

9. $\hat{Y} = .55 + .63X$, the linear regression looks quite reasonable.

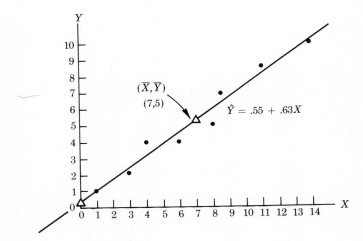

11. $r = -.97$. Both answers are the same in general; addition, subtraction, multiplication, or division operations performed on each X (or Y) value leave r unchanged.   13. a. $\hat{Y}_{72} = 70 + 1.5(72) = 178$ pounds b. $(l,u) = (1.0, 2.0 \text{ lb/in})$

Selected Answers to Problems

1. a. .99 b. .025 c. 5.02   3. a. No b. Greater than 43.2

Chapter 13 / Section 1, page 327

1. $H_0: p = .40$, $H_A: p \neq .40$, calculated $\chi^2 = 1.2$, decision: cannot reject H_0, $p = .4$ 3. a. $H_0: p = .5$, $H_A: p \neq .5$, calculated $\chi^2 = 7.2$, decision: reject H_0 b. Yes; $\alpha = .0414$ 5. $H_A: p \neq 75$, calculated $\chi^2 = 3.84$, reject H_0, $\alpha = .10$. Conclusion: the percentage of accident-free drivers in this group is not 75%.

Chapter 13 / Section 2, page 333

1. a. 24 b. 2 c. Do not reject H_0, $\alpha = .05$, tabular $\chi^2_{.95,2} = 5.99$.   3. Calculated $\chi^2 = 12.0$, cannot reject the null hypothesis of uniform admissions throughout the week.   5. Unable to reject H_0, drinking habits are independent of smoking for this group of women.

Chapter 13 / Section 3, page 338

1. a. 1.145 b. $\alpha = 1 - .75 = .25$ c. A value larger than 41.64   3. H_0: sales are independent of line, H_A: sales depend on line, Calculated $\chi^2 \doteq 7.4$, decision: reject H_0. Conclusion: encourage Ann to sell ranges and Steve to sell refrigerators.   5. The class 'New York by Other' has a short cell, $E = 4.8$. Since we have no more data, I must drop the line for New York leaving:

	Pothead	Other
Los Angeles (LA)	29	11
Miami (M)	15	9

Calculated $\chi^2 \doteq 0.7$
Conclusion: cannot reject $H_0: p_{LA} = p_M$ for p = percent of potheads among drug users.   7. H_0: frequency of accidents is independent of the hour and by days, H_A: frequency of accidents by day depends on the hour, calculated $\chi^2 = 14.3$. There is a dependence between hour and days. Since the observed frequency of accidents is substantially greater than expected for Monday 9–10, and for Friday 11–12 and 4–5, I would analyze these periods more closely.   9. $H_0: p_A = p_B$, p = percent of sales of "new" package at the indicated store, $H_A: p_A \neq p_B$, calculated $\chi^2 \doteq 19.3$, reject H_0, $\alpha = .05$. The percentages are not equal. Differences are evident by comparing observed and expected values for individual cells.   11. H_0: The advertising medium is independent of income (as concerns buying from this retail firm), calculated $\chi^2 = 30$, decision: reject H_0, $\alpha = .01$. The advertising medium that most influences their customers depends upon the customer's income.
13. $H_0: p = .40$, p = percent of arrivals from 12–2 P.M., $H_A: p \neq .40$, test: $\chi^2_{.99,1} = 6.64$, calculated $\chi^2 \doteq 0.1$, decision: cannot reject H_0. Forty percent of arrivals occur between 12–2 P.M. for $\alpha = .01$.   15. H_0: effect (or cure) is independent of treatment, calculated $\chi^2 = 3.38$, cannot reject H_0, $\alpha = .05$. This test does not support the drug as a cold remedy.

Review Problems page 341

Selected Answers to Problems

Appendix A
page 372

1. a. 0 b. 1 c. 0 d. 2 e. 0 f. 1 g. $\frac{1}{10}$ h. 0 i. undefined   2. a. .0032   b. .00001 c. 1.0201 d. .0009 e. 1,000,000 f. .1   3. a. $\frac{1}{12}$ b. $\frac{1}{2}$ c. $\frac{1}{3}$ d. $\frac{1}{24}$ e. $\frac{3}{4}$ f. 12   4. a. $27/171 = (3 \cdot 9)/(19 \cdot 9) = 3/19$ b. $15X = 6 \cdot 4$ or $X = 24/15 = 1.6$ c. $4X = 2 \cdot 7$ or $X = 7/2 = 3.5$ 5. a. 87.5% b. .005 c. $\frac{5}{8}$ d. 66.7% e. .71 f. $\frac{3}{8}$ 6. a. -1 b. 1 c. 18 d. 1/18 e. $3\sqrt{3}$ f. -1 7. a. $-1 < +1$ b. $1 > -1$ c. $-\frac{1}{2} < -\frac{1}{3}$ d. $-1.5 > -2.5$ e. $\left(\frac{1}{2}\right)^2 > \left(\frac{1}{2}\right)^3$ as $\frac{1}{4} > \frac{1}{8}$ 8. a. 2.0 b. 8.7 c. 1.110 d. 1.26 e. .25 f. 1.2   9. a. 63001 b. .0630 c. 2.9 d. .93 e. .093 f. .29   10. a. 7 b. -1 c. 5 d. .4 e. 2 f. .3   11. a. 20 b. 6 c. 0 d. 5 e. 4 f. 2   12. a. IV b. This is true (assuming that we exclude points on the axes from being "in" the quadrant). c. No, II, IV
13. a.

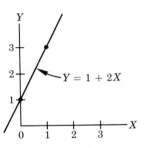

13. b.

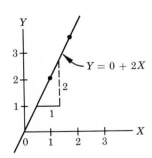

438

c. i. $Y = 1 - X$ ii. $Y = 1 + 0X = 1$ d. The sign indicates the direction of the slope; for "+" the slope (Y-values) increases with X; for "−" the slope (Y-values) decreases as X increases.   14. a. 1   b. −6   c. 8   d. $2^3 = 8$ 15. *Part 1*   a. We want the length of two pieces cut from a 9-foot board. One piece is 3 feet longer than the other.   b. Let $X =$ the length of the longer piece, $X - 3 =$ the length of the shorter piece.   c. Total length $= 9$ ft $= X + (X - 3) = 2X - 3$.   d. $9 = 2X - 3$, or $9 + 3 = 12 = 2X$, so $X = 6$ ft for the longer piece.   e. $X = 6$ ft; $X - 3 = 3$ ft gives $X + (X - 3) = 9$ ft.   *Part 2*   The longer piece is 58 inches. Use the hint given for Part 1.   16. a. We seek how many additional dollars to invest in order to accumulate \$100 interest.   b. Let $X =$ the amount (dollars) to invest.   c. $\$100 = \$1{,}000 \cdot .06 + \$X \cdot .05$ ($\$100 =$ dollars of interest).   d. $\$100 = \$60 + .05X$, $\$40 = .05X$, $\$40/.05 = X$; $X = \$800$.   17. a. We want the annual premium (cost) for a \$31,250 household insurance policy.   b. Let $X =$ the annual premium, in dollars.   c. $X/\$31{,}250 = \$2.85/\$1{,}000$.   d. $\$1000X = \$31{,}250 \cdot 2.85$, $X = \$89.06$.   e. Check: $\$89.06/\$31{,}250 = .00285 = 2.85/1000 = \$2.85$ per \$1,000 of insurance.   18. a. 6   b. 1 (by definition)   c. 120   d. $720 - 120 = 600$ e. 144   f. 12   19. $(X + Y)^4 = X^4 + 4X^3Y + 6X^2Y^2 + 4XY^3 + Y^4$. For $X = Y = \frac{1}{2}$, $\left(\frac{1}{2} + \frac{1}{2}\right)^4 = 1^4 = 1$ is a quick answer, or expanding the binomial, $\left(\frac{1}{2} + \frac{1}{2}\right)^4 = \left(\frac{1}{2}\right)^4 + 4\left(\frac{1}{2}\right)^3 \frac{1}{2} + 6\left(\frac{1}{2}\right)^2 \left(\frac{1}{2}\right)^2 + 4\left(\frac{1}{2}\right)\left(\frac{1}{2}\right)^3 + \left(\frac{1}{2}\right)^4 = \frac{16}{16} = 1$.   20. a. 10   b. 4   c. 28   d. 78   e. 455   f. 135

Index

Absolute value, 348
Addition rule of
  probabilities, 84
Aggregate index, 58
α (alpha)-risk, 205, 329
Alternative hypothesis, 203
Analysis of variance, 262
  unequal group sizes, 270
Area—as probability measure, 126, 131
Averages, 9
  mean (arithmetic), 10, 44
  median, 11
  mode, 12
Axioms, probability, 84

b_0, B_0 intercept values, 287
$b_1 B_1$, estimated & true slope, 287
Bar graph, 23
Base period, index numbers, 57
Bernoulli trials, 99
β (beta)-risk, 205, 216, 329
Binomial distribution, 102
  mean of, 116
  normal approximation to, 145
  variance of, 117
Binomial theorem, 369
Bound of error, E, 186, 193, 234
Brown, B. W., Jr., 316

Campbell, Stephen K., 34
Causation, and correlation, 296, 312
Central Limit Theorem, 168, 205
Central tendency measures
  (see averages)
Chi-square approximation, 324

Class
  boundaries, 23
  limits, 24
  mark, 26
  modal, 45
  width, 23
Cochran, William, 276
Coding in regression, 302
Coefficient of determination, 295
Combinations, 102, 370
Complement (events), 73
Compound events, 70
Conditional probability, 91
Conditioned event, 91
Confidence intervals for
  binomial p, 192
  differences in means, 252, 255
  means (large samples), 182
  means (small samples), 232
  the slope, in regression, 304
Confidence level, 182
Consumer Price Index, 57
Contingency tables, 334
  degrees of freedom, 334, 338
  expected numbers, 337
  tests for independence, 334, 337
Continuity correction, 146
Continuous random variables, 123
Correlation coefficient, 293
Countably infinite, 158
Count data inference, 323
Cox, Edwin, 152
Critical value, 206
Cumulative normal, $\phi(Z)$, 141
Cureton, Edward, 316

Data plots, 285, 363
Degrees of freedom, 228

Index

Degrees of freedom *(con't)*
    general, 18, 228
    in analysis of variance, 263, 264
    in chi-square tests, 334, 338
    in regression tests, 307
    in student's t-test, 229, 230
Dependent variable, 283
Descriptive statistics, 6, 9
Difference between means, 253
Discrete random variables, 97, 100
Distribution, definition, 17
    binomial, 102
    chi-square, 324
    frequency, 24, 38
    F-ratio, 260
    geometric, 110
    hypergeometric, 108
    normal, 130
    probability, 100
    sampling, 162
    Student's t, 225
    uniform (continuous), 125
    uniform (discrete), 111
    Z-(standard normal), 130
Dubins, Lester E., 95
Dunnett, Charles W., 219

E, bound of error, 186, 193
Edwards, Allen, 341
Engineer's rule, 358
Error
    of estimation, 186, 193
    type 1 and type 2, 204
Estimate, 158
Estimation, point
  versus interval, 181, 182
Estimator, 158
Events
    complementary, 73
    compound, 70
    independent, 98
    mutually exclusive, 73, 85
    null (empty), 86
    simple, 70
Expected cell frequency, 337
Expected value (true mean), 115

Experiment, 70
Experimental design, 196, 244, 247, 273

F-distribution and
  analysis of variance, 262
F-ratio statistic, 259
Factorials, 369
Fisher, R. A., 1, 247
Frequency
    curve, 26
    distribution, 24, 38
    polygon, 25

Geometric probability rule, 110
Gossett, W. S., 1, 225
Graphical displays, 22
Grouping data, 37

Hansen, Morris, 7
Harris, Thomas, 179
Histogram, 23
Hoel, Paul G., 177
Homogeneity of variance, 259
Hypergeometric distribution, 108
Hypothesis, definition, 203
Huff, Darryl, 34

Independent events, 97
Independent variable, 283, 334
Index numbers, 56
Inference, statistical, 6, 179
Intersection of events, 72
Interval half width, 186

Kruskal, William H., 59

Index

Law of large numbers, 166
Least squares estimation, 287
Level of significance, i.e. α, 206
Lieberman, Gerald, 121
Lindgren, B. W., 341
Linear regression, 283
Line graph, 25, 104
Lockwood, Arthur, 34

Mass points, 101, 127
Mean, arithmetic, 10
    for binomial distribution, 116
    deviations from, 11
    distribution (true), 114, 136
    for grouped data, 44
    sampling distribution of, 168
    sampling variance of, 166
    tests on, 201, 225
    weighted, 55
Mean square
    among, 263
    within, 264
Median, definition, 11
    grouped data, 49
Mid-range, 12
Mode, definition, 12
    for grouped data, 45
Models, mathematical versus
    statistical, 284
Mosteller, Frederick, 7
μ (mu), true mean, 114
Multiplication rule, 91
Mutually exclusive events, 73, 85

Nielsen, A. C., Company, 3, 180
Nominal scale, 323
Normal approximation, 145
Normal distribution,
    cumulative (ϕ), 140
    standard (Z), 130
Null (empty) event, 86
Null hypothesis, 203

Ogive, 50
Ordered pairs, 71, 364
Orlando Chamber of Commerce, 35

Paired differences test, 240
Parameter, 158
Pascal's triangle, 371
Pearson, E. S., 249
Pearson's coefficient of
    correlation, 293
Percentages, 353
Percentiles, percentile ranks, 49, 140
ϕ (phi), 141
Pictogram, 26
Pie chart, 27
Point estimate, 181
Pooled variance, 253
Population, definition, 10, 176
Probability
    addition rule, 84
    axioms, 84
    conditional, 91
    distribution, 100
    empirical, 81
    equal likelihood definition, 80
    function or rule, 101
    multiplication rule, 91
    subjective, 82
    of type 1 and type 2 errors, 205
Probability distributions
    (*see* distributions)
Proportions, estimation of, 181
    tests on p, 213

r (Pearson's correlation
    coefficient), 293
Random sampling, 159, 176
Random variable, definition, 100
    continuous, 123
    discrete, 100
Range, definition, 15
    in approximating s, 20
Regression
    assumptions for inference, 306

Regression *(con't)*
   linear model, 285
Regression coefficients
   estimators, 288
      interpretation of, 289
      tests on, 307
Relative frequency, 37
Relative frequency definition
   of probability, 81
Relatives index, 59
Reliability, 119
Research hypothesis, 203
Retrospective-prospective
   studies, 314
ρ (rho), true linear correlation, 295

s, sample standard
   deviation, 15
s^2, sample variance, 17
Sample, defined, 10
   mean, 9
   random, 159
Sample size, 185, 193
Sample space, 71
Sampling
   distributions, 162
   without replacement, 97, 163
Scientific method in testing, 202, 206
Short cell, 331
Shulte, Albert P., 316
σ (sigma), 116, 136
Σ (capital sigma),
   summations, 10, 361
Significant difference, 204
Simple events, 70
Slonim, Morris, 199
Slope, 288, 365
Snedecor, George, 2, 35
Spear, Mary, 35
Square roots, 358
Standard deviation, definition, 15, 17
   of b_1 values, 307
   of differences in means, 252, 253
   for grouped data, 45
Standard error of the mean, 166

Standard normal distribution, 130
Statistics, 10, 158
Steel, Robert G. D., 249, 276
Steger, Joseph A., 341
Student's t-distribution, 225
Subjective probability, 82
Sum of squares
   among groups, 263
   within groups, 264
Surgeon General's Report, 312
Systematic approach to
   problem solving, 88, 137, 366

t-distribution, 225
Tests of hypothesis, defined, 202
   on differences in
      means, 240, 252, 253
   for equal treatment means,
      262, 270
   for equal variances, 259
   on means, large samples, 208
   on means, small samples, 236
   one tail–two tail, 208
   paired differences, 240
   on proportions, 213
   Mon the slope, 307
   of independence, 334
Test statistic, 205
Time series, 299
Tree diagram, 75
Trend, in regression, 300
True mean, true variance, 114, 116
True regression, 287
Two sample techniques, 251
Type 1 and type 2 errors, 204

Unbiased estimator, 163
Uniform probability rule
   continuous, 125
   discrete, 111
Union, of two events, 73
Union 76 Fuel Economy Tests, 196

Variance
  of a binomial distribution, 117
  of a discrete distribution, 116
  of difference between
    means, 252, 253
  for grouped data, 45
  pooled estimator for, 253
  about regression, 306
  sample, 17, 45

Weighted mean, 55

Word problems, working, 88, 137, 366

Yamane, Taro, 59, 177
Y-intercept, 289

Z-scores, 131, 136
Z-standard normal
  distribution, 131

Contents for
Appendix A   Math Essentials

Math Symbols 348

Calculations with Zero and One 349

Setting Decimals 349

Operations with Common (Ratio) Fractions 350

Decimal and Common Fraction Conversion; Percentages ... 353

Exponents and Roots 354

Inequalities 355

Rules for Rounding 356

Use of Squares and Square Root Tables 358

Summation Notation 361

Rectangular Coordinates 363

Linear Equations 364

Word Problems 366

The Binomial Theorem 369

Problems 372